机械精度设计与检测基础

（英汉双语）

范智广　姜壹夫　田　阳　主编

U0953145

东北大学出版社
·沈　阳·

图书在版编目（CIP）数据

机械精度设计与检测基础：英、汉 / 范智广，姜壹夫，田阳主编. -- 沈阳 : 东北大学出版社，2025. 4.
ISBN 978-7-5517-3767-8

Ⅰ. TH122；TG801

中国国家版本馆CIP数据核字第2025J19H06号

出 版 者：东北大学出版社
地址：沈阳市和平区文化路三号巷11号
邮编：110819
电话：024-83683655（总编室）
024-83687331（营销部）
网址：http://press.neu.edu.cn
印 刷 者：辽宁一诺广告印务有限公司
发 行 者：东北大学出版社
幅面尺寸：210 mm × 285 mm
印　　张：16
字　　数：532千字
出版时间：2025年4月第1版
印刷时间：2025年4月第1次印刷
策划编辑：刘新宇
责任编辑：项　阳
责任校对：罗　鑫
封面设计：潘正一
责任出版：初　茗

ISBN 978-7-5517-3767-8　　定价：42.00元

前 言

机械精度设计与检测（又称互换性与测量技术基础）是高等院校本科、专科机械类及近机类专业的重要技术基础课程。在全球制造业加速智能化、标准化（如ISO、ASME等国际标准体系）的时代背景下，《机械精度设计与检测基础（英汉双语）》教材以中英文双语形式呈现核心内容，助力读者在掌握极限与配合、几何公差、表面粗糙度等理论知识的同时，精准地理解专业术语的中英文表述及应用场景，实现技术认知与语言能力的同步提升。在内容编排上，本教材充分地考虑教学需求。

首先，每章内容以英文开篇，随后辅以中文对照解析，有效地避免语言障碍对知识吸收的影响。

其次，引入大量的典型工程案例（如优先数的工程应用、轴承关键部件精度设计等），通过中英文双视角分析，将理论知识与生产实践紧密结合。

再次，本教材内容严格遵循现行国家标准，确保与工业实际高度契合。此外，特别设计双语习题模块，帮助读者在实践中巩固知识、强化专业英语应用能力。

最后，本教材适用范围广泛，既适用于多学时深度教学，也可满足少学时精简授课需求。各章节内容相对独立，可根据不同专业培养目标灵活选用，兼具教学灵活性与实用性。

本教材由沈阳工程学院范智广、姜壹夫、田阳担任主编；沈阳工程学院鲍君善、赵璐、潘萍萍、朱爽、李玉秀，上海健康医学院范智平担任副主编。范智广负责统稿。具体分工如下：第一、二、四、五、六章由范智广编写和翻译，第三、六章由姜壹夫、李玉秀编写和翻译，第七、八、九、十章由田阳编写和部分翻译。本教材图片由赵璐、潘萍萍、朱爽绘制，第七、八章英文翻译由鲍君善、范智平完成。本教材提供电子课件，主编邮箱 fzgsyz123@163.com。

本教材由沈阳工业大学孙兴伟教授、沈鼓集团齿轮公司沈玉琢高级工程师担任主审。

在本教材编写过程中，得到东北大学孙瑶、天津职业技术师范大学赵巍、沈阳工程学院张晨晨的大力帮助；沈鼓集团齿轮公司杨晖高级工程师参与本教材圆柱齿轮的精度检测的审阅工作，并结合实践经验提出了许多宝贵建议。

鉴于编译者水平有限，本教材中不妥之处在所难免，特别是英文部分可能存在中式表达，敬请读者批评指正。

编 者

2025年1月

目 录

Chapter 1　Introduction 绪论

1.1　Interchangeability　互换性

1.1.1　Definition of Interchangeability 互换性的概念

In the mechanical manufacturing, interchangeability refers to fabricating workpieces respectively according to specified tolerances of geometric, physical, and mechanical performance parameters, and then during assemblying and repairing one such workpiece can freely replace another, without further machine or custom fitting.

在机械制造中，互换性是指根据几何、物理和机械性能参数的规定公差制造工件，在装配和修理过程中，这样的工件可以自由地替换另一个工件，而不需要进一步加工或定制，配件就能装配在机器上，满足使用要求。

During daily life, there are a lot of devices with interchangeability. Such as when a light bulb fails, to buy a new one and replace the old one, then the light can continue to serve for us; a component fails in a bicycle or a watch, to buy one of the same component and to replace it, the bicycle and watch can be used again. In the mechanical engineering, there are more devices with interchangeability. For example, the rolling bearing are all manufactured and assembled according to interchangeability principle.

互换性实例在日常生活中随处可见。例如，当一个灯泡坏了，买一个新的，装上即可；自行车或手表中某个零件坏了，换上一个相同规格的新零件，便能再次使用。在机械工程领域，存在大量互换性的实例，例如，滚动轴承均按互换性原则制造和装配。

1.1.2　Classification of Interchangeability 互换性的分类

According to the interchange extent, interchangeability can be divided into two types, i.e., complete interchangeability and incomplete interchangeability.

互换性按其互换程度分为完全互换性和不完全互换性两种。

Complete interchangeability: in an assembly, any auxiliary choose and repairment are not needed.

完全互换性：要求零部件在装配时，无需挑选或辅助加工。

Incomplete interchangeability: in an assembly, selection, pre-grouping and adjustment during assembly are permitted.

不完全互换性：零部件在装配过程中，在装配前允许对零件进行挑选、预先分组或调整，然后按照对应组进行装配。

Usually the complete interchangeability is adopted for cooperation between different factories and incomplete interchangeability is adopted during workpieces are machined and assembled in the same factory.

一般来说，零部件需厂际协作时应采用完全互换性；零部件在同一厂制造和装配时，可采用不完全互换性。

1.1.3 The Role of Interchangeability 互换性的作用

The role of interchangeability in machinery manufacturing：

互换性在机械制造行业中的作用：

（1） In the designing aspect，it can reduce the workload of drawing and calculation，shortens the design cycle，and is conducive to the application of computer aided design.

从设计方面看，零部件具有互换性，可以减少绘图和计算的工作量，缩短设计周期，有利于计算机辅助设计的应用。

（2） In the manufacturing aspect，a large number of standard machine elements can be made. Then decentralized manufactures can be adopted and automatic manufacturing lines can be applied resulting in a lower costs of these standard machine elements.

从制造方面看，标准件具有互换性，可采用分散制造，并应用自动化生产线，从而降低标准机械零件成本。

（3） In the repairing aspect，a machine element can be replaced as quickly as possible when the machine element is interchangeable，to decrease the repairing time and cost，and to increase the efficiency and service life of the machine.

从维护方面看，具有互换性的零部件在磨损或损坏后可以及时更换，因而减少了机器的维修时间和费用，从而提高了机器的使用价值并延长其寿命。

1.2 Standardization and Preferred Value 标准化与优先数

1.2.1 Standardization 标准化

In order to achieve interchangeability，a means must be adopted to maintain the necessary technical unification among the various scattered and local production departments and segments，so as to form a unified whole. Standardization is defined as：the development and implementation of concepts，theories，procedures and designs to achieve and maintain the required levels of compatibility，interchangeability or commonality in the operational，procedural，material，technical and administrative fields to attain interoperability. Standardization includes all the active processes of formulating，publishing and implementation，and implementation is the key process. Moreover，standardization is an active process of continuous improvement through revision.

为了实现互换性，必须采用一种手段，使各个分散的、局部的生产部门和生产环节之间保持必要的技术统一，以形成一个统一的整体。标准化被定义为：对标准概念、理论、程序及设计的制定、贯彻执行的一系列过程，以便在操作、程序、材料、技术和管理领域获得并保持所需的兼容性、互换性或通用性水平，以实现互操作性。标准化包括制定、发布与实施的全过程，其中，实施标准是核心环节。此外，标准化也是一个不断循环提高的过程。

A standard may be developed privately or unilaterally，for example by corporations，regulatory bodies，militaries，etc. Standards can also be developed by groups such as trade unions，and trade associations. Standard organizations often have more diverse input and usually develop voluntary standards：these might become mandatory if adopted by government，business contract，etc.

一个标准可能是由私人或单方面制定的，例如，由公司、监管机构、军队等制定。标准也可以由工会和行业协会等团体制定。标准组织通常有更多样化的投入，通常制定自愿性标准，如果被政府、商业合同等采用，这些标准可能成为强制性标准。

Standardization is one of the important means of modern organization of production, the premise of professional collaboration of production, and an important part of scientific management.

标准化是组织现代化生产的重要手段之一，是实现专业化协作生产的必要前提条件，是科学管理的重要组成部分。

The standards adopted in mechanical industry mainly include international standard, national standards. The code for all international standards is ISO (International Organization for Standardization). The code for Chinese Standards is code GB (National Standards of the People's Republic of China). The code for Germany Industry Standards is DIN, the code for Japanese Industry Standards is JIS. ANSI is the code for American National Standards Institute. etc.

机械工业采用的标准主要有国际标准、国家标准。所有国际标准的代码都是ISO（国际标准化组织）。中国标准的代码为GB（中华人民共和国国家标准），德国工业标准的代码是DIN，日本工业标准的代码是JIS，ANSI是美国国家标准协会的代码，等等。

1.2.2 Series of Preferred Numbers 优先数系

It is an important part of standardization to simplify, coordinate and unify technical parameters. All technical parameters are involved in product design and technical standard formulation. After a value is selected as the parameter index of a product, the value will be transferred according to certain rules. In production, the same parameters of the same product take different values from small to large, and form product series with different specifications to meet the various needs of users. Preferred numbers and series of preferred numbers constitute a scientific standard which provides an important basis for the numerical classification of product series.

工程上各种技术参数的简化、协调和统一是标准化的一项重要内容。所有技术参数都涉及产品设计和技术标准的制定。选定一个数值作为某产品的参数指标后，这个数值就会按照一定的规律进行参数传递扩散。生产中，同一产品的相同参数从小到大取不同的值，形成不同规格的产品系列，以满足用户的各种需求。优先数和优先数系是一种科学的数值标准，它为产品系列化提供了重要基础。

R5, R10, R20, R40, R80 are priority series specified by National Standards. R5, R10, R20, R40 are basic series and R80 is additional series. Series of preferred values are the decimal geometric series and their common ratio can be calculation according to the Eq. (1-1):

国家标准规定R5，R10，R20，R40，R80是优先数系列。R5，R10，R20，R40为基本系列，R80为补充系列。优先数系是十进制几何级数，各系列的公比计算见式（1-1）：

R5：$q_5 = \sqrt[5]{10} \approx 1.5849 \approx 1.60$ (1-1a)

R10：$q_{10} = \sqrt[10]{10} \approx 1.2589 \approx 1.25$ (1-1b)

R20：$q_{20} = \sqrt[20]{10} \approx 1.1220 \approx 1.12$ (1-1c)

R40：$q_{40} = \sqrt[40]{10} \approx 1.0593 \approx 1.06$ (1-1d)

R80：$q_{80} = \sqrt[80]{10} \approx 1.0293 \approx 1.03$ (1-1e)

The most basic R5 series consists of these five rounded numbers:

最常见的R5系列5个数字为：

R5：1.00　1.60　2.50　4.00　6.30

Example 1-1： If our design constraints tell us that the two screws in our gadgets between 30 mm and 50 mm apart，we make it 40 mm，because 4.00 is in the R5 series of preferred numbers.

例1-1： 如果一个产品设计要求一个机件的两个螺丝之间的距离为30～50 mm，通常选取该尺寸数值为40 mm，因为4.00是R5系列的优先数。

The basic series of preferred values are shown in Tab. 1-1.

优先数系的基本系列见表1-1。

Tab. 1-1　The basic series of preferred values（common values）(excerpted from GB/T 321-2005）

表1-1　优先数系的基本系列常用值（摘自GB/T 321—2005)

Basic series 基本系列	The common values from 1 to 10 1~10 的常用值										
R5	1.00		1.60		2.50		4.00		6.30		10.00
R10	1.00	1.25	1.60	2.00	2.50	3.15	4.00	5.00	6.30	8.00	10.00
R20	1.00	1.12	1.25	1.40	1.60	1.80	2.00	2.24	2.50	2.80	3.15
	3.55	4.00	4.50	5.00	5.60	6.30	7.10	8.00	9.00	10.00	
R40	1.00	1.06	1.12	1.18	1.25	1.32	1.40	1.50	1.60	1.70	1.80
	1.90	2.00	2.12	2.24	2.36	2.50	2.65	2.80	3.00	3.15	3.35
	3.55	3.75	4.00	4.25	4.50	4.75	5.00	5.30	5.60	6.00	6.30
	6.70	7.10	7.50	8.00	8.50	9.00	9.50	10.00			

Derived series：if the basic series are not suitable the requirements，the derived series can be used. The derived method is as the Eq.（1-2)：

派生系列：如果基本系列的优先数不满足要求，可以使用派生系列。派生系列的推导方法见式（1-2)：

$$R\frac{i}{b} = \sqrt[b]{10^i} = 10^{i/b} \quad (1\text{-}2)$$

Where，b—the selected series value，(for example，$b = 40$ for the R40 series)；

i—the element of this series，(with the value of i between 0 and b).

式中，b——选取的基本系列，例如$b = 40$，即选取的R40优先数系；

i——间隔项数，i的取值介于0和b之间。

Example 1-2： R5/2 series are：1.00　2.50　6.30　16.00　40.00　100.00

例1-2： R5/2系列是1.00　2.50　6.30　16.00　40.00　100.00

1.3　Purpose of This Course 本课程的学习任务

"Design of Mechanical Precision and Foundation of Geometrical Capacity Survey" is the basic technical course for the students whose majors are associated with mechanics. Engineers and workers who design or fabricate machines should have the knowledge introduced in this course. For the students who study in an engineering college，this course aims to train students to master the basic theory，knowledge and skills of product precision design and quality assurance，and lay the foundation for further application of national standards and control of product quality, and the basic requirements are as follows:

“机械精度设计与检测基础”是机械类专业学生的技术基础课。设计或制造机器的工程师和工人应具备本课程介绍的知识。对于在工科院校学习的学生，本课程旨在培养其掌握产品精度设计和质量保证的基本理论、知识和技能，为进一步应用国家标准和控制产品质量奠定基础，课程的基本要求如下：

(1) To understand clearly the concepts of interchangeability, standardization and measurement technology.

清楚了解互换性、标准化和测量技术的概念。

(2) To be familiar with the basic content of every tolerance studied in this course, to know exactly the basic terms.

熟悉本课程所学习的每一个公差的基本内容，准确掌握基本术语。

(3) To analyze fit properties and fit type during reading an assembly drawing and to draw the size tolerance zone figure.

在阅读装配图和绘制装配图时，能够分析配合性质和配合类型，并绘制出尺寸公差带图。

(4) During reading a detail drawing (single part drawing), to know exactly the precision requirements about size, form, orientation, location, run-out and surface texture. To be able to calculate the limits of size, to describe the form, orientation, location requirements for every feature.

在读零件图（单一图纸）时，准确了解尺寸、形状、方位、位置、跳动、表面粗糙度等精度要求，能够计算尺寸极限，描述每个特征的形状、方向、位置等要求。

(5) During drawing an engineering drawing (assembly drawing and detail drawing), to indicate the precision requirements on the drawing correctly according to the standards.

在绘制工程图（装配图、零件图）时，能够按照标准在图纸上正确标注精度要求。

All kinds of tolerances have strict principles and regulations in the implementation of national standards, and are highly practical in application. It should be noted that it is impossible to specify the workpiece precision requirements just by finishing this course. To truly master and flexibly apply this knowledge, it is necessary to accumulate 3 to 5 years of rich practical experience in related fields, which should be documented for future reference and improvement. Therefore, this course primarily focuses on imparting fundamental theoretical knowledge of geometrical tolerance and measurement technology to students.

各类公差在国家标准的贯彻上都有严格的原则性和法规性，而在实际应用中具有很强的实践性。需要注意的是，仅仅完成这门课程是无法设计出工件的精度要求的。要真正掌握并灵活应用这些知识，还需要在相关领域积累3~5年的丰富实践经验，这些经验需以文档形式记录，以供后续参考与提升。因此，本课程的主要任务是向学生传授几何公差与测量技术的基础理论知识。

Exercise 1 习题 1

1-1 Determine the preferred numbers according to the basic series of preferred numbers:

按优先数的基本系列确定优先数：

(1) The first number is 10, to write the last five preferred numbers according to R5 series.

第一个数为10，按R5系列确定后五项优先数。

(2) The first number is 100, to write the last three preferred numbers according to R20/3.

第一个数为100，按R20/3系列确定后三项优先数。

1-2 The main parameter（drilling diameter）of the radial drilling machine is the following：25，40，63，80，100，125. Try to identify which series they belong to within series of preferred numbers ，and specify the common ratio of that series.

摇臂钻床的主参数（钻孔直径）为：25，40，63，80，100，125。试判断它们属于优先数系中哪种系列，其公比为多少。

Chapter 2 Dimensional Precision Design and Inspection 尺寸精度设计与检测

2.1 Overview 概述

The machinery industry is one of the pillar industries in China, and the fit of holes and shafts has an important impact on the performance and life of mechanical products. The basic terms and definitions of dimensions, deviations, tolerances, and fits are specified in GB/T 1800.1-2020.

机械工业是我国的支柱产业之一，孔、轴配合对机械产品的性能和寿命有重要影响。GB/T 1800.1—2020规定了有关尺寸、偏差、公差、配合的基本术语和定义。

2.2 Basic Terms and Definitions 基本术语与定义

2.2.1 The Definition of Hole and Shaft 孔和轴的概念

Hole: designates all internal features of a part, including features of size which are not cylindrical.

孔：通常指工件的圆柱形内表面，也包括非圆柱形内表面。

Shaft: designates all external features of a part, including features of size which are not cylindrical.

轴：通常指工件的圆柱形外表面，也包括非圆柱形外表面。

As shown in Fig. 2-1, the general term "hole" or "shaft" can refer to the space containing or contained by the two parallel faces (or tangent planes) of a workpiece, such as the width of a slot or the thickness of a key B.

如图2-1所示，广义的“孔”或“轴”可以看作由两平行平面（或切平面）所形成的包容或被包容面，如槽的宽度或键的厚度B。

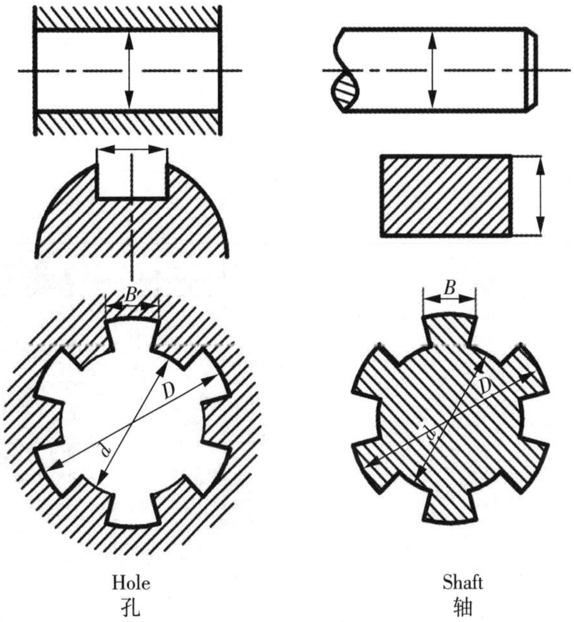

Fig. 2-1 Diagram of hole and shaft definition

图2-1 孔和轴的定义示意图

2.2.2 Terms and Definitions About Size 有关尺寸的术语及定义

2.2.2.1 Size 尺寸

Size is the numerical value of a linear dimension in a particular unit, such as diameters, radius, length, width, depth, and thickness, Usually, the unit is mm.

尺寸是以特定单位表示线性尺寸的数值，如直径、长度、宽度、高度、中心距等。通常尺寸单位是毫米（mm）。

2.2.2.2 Nominal size（D, d）公称尺寸

Nominal size refers to the size of a feature of perfect form as defined by the drawing specification. Nominal size is the theoretical dimension from which the hole and shaft limits are derived. Capital letter D is used to present hole nominal size, and lowercase letter d is used to present shaft nominal size.

公称尺寸指的是图样规范的理想形状特征的尺寸。公称尺寸是理论尺寸，孔、轴的极限值由此导出。孔的公称尺寸用大写字母D表示，轴的公称尺寸用小写字母d表示。

2.2.2.3 Actual size（D_a, d_a）实际尺寸

Actual size refers to the local size obtained by measurement. The code D_a and d_a represent the actual size of holes and shafts. Usually, the actual sizes at different places of a feature of size are different.

实际尺寸是通过测量取得的局部尺寸。孔、轴的实际尺寸分别用D_a和d_a表示。通常，同一尺寸要素的不同部位的实际尺寸不同。

2.2.2.4 Limits of size（D_{max}, D_{min}, d_{max}, d_{min}）极限尺寸

The two extreme permissible sizes of a feature, between which the actual size should be lie, are the limits of size, including upper limit of size and lower limit of size. An example is shown in Fig.2-2.

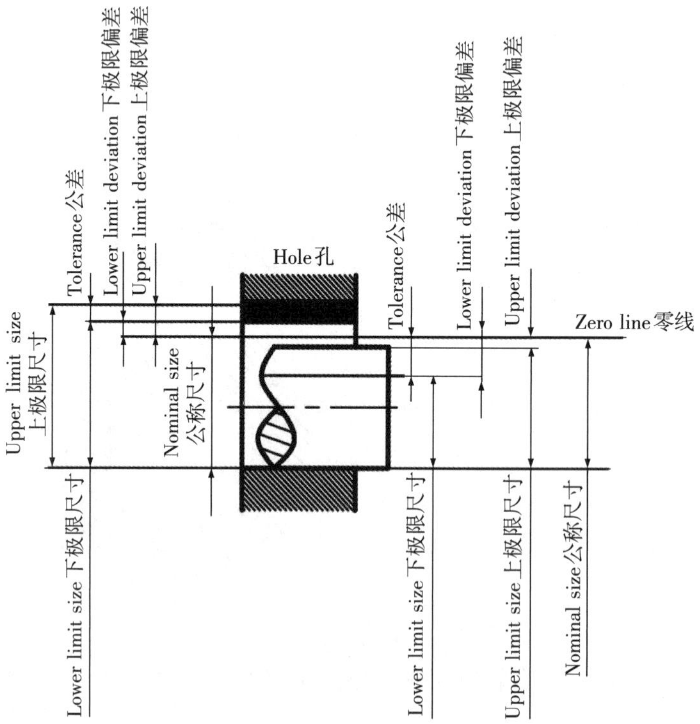

Fig. 2-2 Diagram of limits and fits

图2-2 极限与配合示意图

极限尺寸是一个孔或轴的尺寸允许的两个极端值。实际尺寸应介于极限尺寸之间，包括上极限尺寸和下极限尺寸。如图2-2所示。

（1）Maximum limit of size（D_{max}，d_{max}）： the largest permissible size of a feature of size. The code of D_{max} and d_{max} represents the maximum limit of size for holes and shafts.

最大极限尺寸（D_{max}，d_{max}）：尺寸要素允许的最大尺寸称为最大极限尺寸（或上极限尺寸），孔和轴的最大极限尺寸分别用D_{max}、d_{max}表示。

（2）Minimum limit of size（D_{min}，d_{min}）：the smallest permissible size of a feature size of size. The code of D_{min} and d_{min} represents the maximum limit of size for holes and shafts.

最小极限尺寸（D_{min}，d_{min}）：尺寸要素允许的最小尺寸称为最小极限尺寸（或下极限尺寸）。孔和轴的最大、最小极限尺寸分别用D_{min}、d_{min}表示。

Normally，any actual size after processing usually should not exceed the maximum limit size and the minimum limit size，i.e.，$D_{min}≤D_a≤D_{max}$，$d_{min}≤d_a≤d_{max}$，otherwise，the workpiece is unqualified. The limit size is used to control the actual size.

一般情况下，加工后的实际尺寸不能超过最大极限尺寸与最小极限尺寸，即满足$D_{min}≤D_a≤D_{max}$，$d_{min}≤d_a≤d_{max}$。若超出极限尺寸范围，则对应的工件尺寸是不合格的。极限尺寸被用来控制实际尺寸。

2.2.3 Terms and Definitions of Deviation，Tolerance 有关公差与偏差与公差带的术语及定义

2.2.3.1 Size deviation尺寸偏差（简称偏差）

The size deviation can be defined as the algebrac difference of a size minus the nominal size. Size deviations include actual deviation and limit deviation.

偏差是指某一尺寸减其公称尺寸所得的代数差。尺寸偏差可分为实际偏差和极限偏差。

（1）Actual deviation 实际偏差。

Actual deviation refers to the algebraic difference value between actual size and the nominal size. The code E_a and e_a respectively represent actual deviation of holes and shafts.

实际尺寸减其公称尺寸所得代数差称为实际偏差。孔的实际偏差用字母E_a表示，轴的实际偏差用字母e_a表示。

$$E_a = D_a - D \tag{2-1}$$

$$e_a = d_a - d \tag{2-2}$$

（2）Limit deviation极限偏差。

Limit deviation refers to the algebraic difference value between limits of size and nominal size，which includes upper limit deviation and lower limit deviation. Upper limit deviation is the algebraic difference value between upper limit of size and nominal size，and lower limit deviation is the algebraic difference value between lower limit of size and nominal size. The code ES and EI respectively represent upper limit deviation and lower deviation of holes，and the code es and ei respectively represent upper limit deviation and lower deviation of shafts. The limits deviations can be expressed by Eq.（2-3）to（2-6）:

极限偏差指极限尺寸与公称尺寸之间的代数差，包括上极限偏差和下极限偏差。上极限偏差是上极限尺寸与公称尺寸之间的代数差，下极限偏差是下极限尺寸与公称尺寸之间的代数差。孔的上极限偏差、下极限偏差代号分别用大写字母ES，EI表示，轴的上极限偏差、下极限偏差代号分别用小写字母es，ei表示。极限偏差可以用式（2-3）至式（2-6）表示：

$$ES = D_{max} - D \tag{2-3}$$

$$EI = D_{min} - D \tag{2-4}$$

$$es = d_{max} - d \tag{2-5}$$

$$ei = d_{min} - d \tag{2-6}$$

Any actual deviation shall not exceed the limit deviation, $EI \leqslant E_a \leqslant ES$, $ei \leqslant e_a \leqslant es$, otherwise, the component is unqualified.

合格零件的实际偏差应在规定的极限偏差范围内，即$EI \leqslant E_a \leqslant ES$，$ei \leqslant e_a \leqslant es$，否则视为不合格零件。

（3）偏差的标注。

Size deviation is the algebraic difference value, so the value of the size deviation may be positive, negative, or zero. Therefore, in front of the deviation value (except the deviation value is zero), the corresponding "＋" or "－" sign should be marked. The upper deviation is marked at the upper right corner of the nominal size, the lower deviation is marked at the lower right corner of the nominal size.

偏差是代数差，所以偏差的数值可能是正值、负值或零。故在偏差值的前面（除偏差值为零外），应标上相应的"＋"号或"－"号。上偏差标在公称尺寸右上角，下偏差标在公称尺寸右下角。

For example, $\phi 25^{0}_{-0.03}$ mm indicates a nominal size of ϕ25 mm, its upper deviation is 0, and its lower deviation is －0.03 mm.

例如，$\phi 25^{0}_{-0.03}$ mm表示公称尺寸为ϕ25 mm，上偏差为0，下偏差为-0.03 mm。

2.2.3.2 Size tolerance and size tolerance zone 尺寸公差与公差带

（1）Size tolerance尺寸公差。

Size tolerance is the difference between the upper limit of size (the maximum limit of size) and the lower limit of size (the minimum limit of size), and also is the different value between the upper limit deviation and lower limit deviation. The size tolerance of holes (T_h) and shafts (T_s) can be expressed as Eq. (2-7) to (2-8).

尺寸公差（简称公差）等于上极限尺寸（最大极限尺寸）与下极限尺寸（最小极限尺寸）之差，或上极限偏差减下极限偏差之差。孔和轴的尺寸公差分别用T_h和T_s表示，见式（2-7）、式（2-8）：

$$T_h = |D_{max} - D_{min}| = |ES - EI| \tag{2-7}$$

$$T_s = |d_{max} - d_{min}| = |es - ei| \tag{2-8}$$

Size tolerance is an absolute quantity without symbol. Tolerance value cannot be zero.

尺寸公差是一个没有符号的绝对值，且不能为零。

（2）Illustration of limits and fits (Tolerance zone) 公差带图。

For simplification reason, the body of holes and shafts is not drawn, while only the tolerance zone is drawn enlarged to analyze the relationship of the holes and the shafts, shown in Fig. 2-3. This figure is named size tolerance zone. The illustration size tolerance zone includes zero line and size tolerance zone.

为了简化，不绘制孔和轴的实际形状，而只将公差带放大绘制出来，以分析相配合的孔和轴之间的关系，如图2-3所示。此图被称为尺寸公差带。图2-3尺寸公差带图包括零线和尺寸公差带。

① Zero line: in an illustration of limits and fits, the straight line, representing the nominal size, to which the deviations and tolerance are referred. The zero line is drawn horizontally, with positive deviations shown above and negative deviations below, see Fig. 2-3. Usually, the zero line indicates the nominal size.

零线：在极限与配合的公差带图中，零线是代表公称尺寸的一条直线，以其为基准确定偏差和公差。通常，零线沿水平方向绘制，正偏差位于零线的上方，负偏差位于零线的下方，如图2-3所示。通常，零线表示公称尺寸。

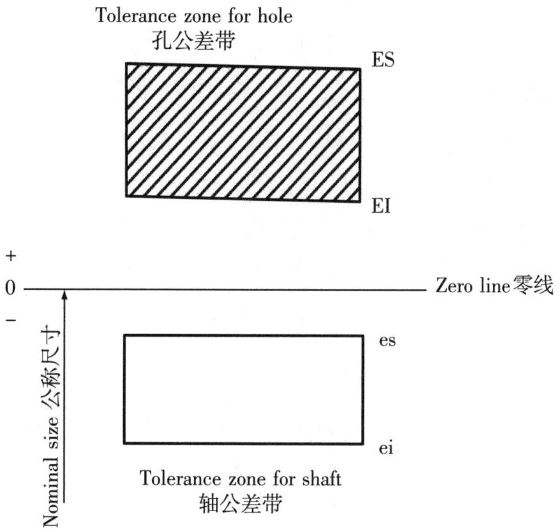

Fig. 2－3 Illustration of limits and fits

图2-3 孔与轴配合公差带图

② Size tolerance zone: in an illustration of limits and fits, the zone contained between two lines representing the upper and lower deviations, defined by the magnitude and position of the size tolerance zone. The magnitude of the tolerance zone is the width of size tolerance zone in the direction perpendicular to the zero line, which is determined by standard tolerance. The position of tolerance zone is relative to the zero line in the direction perpendicular to the zero line, which is determined by the fundamental deviation.

尺寸公差带：在公差带图中，由代表上偏差和下偏差的两条直线所限定的一个区域，其大小和位置由尺寸公差带的大小及其相对于零线的位置所定义。公差带的大小是指垂直于零线方向上尺寸公差带的宽度，由标准公差确定；公差带的位置是指垂直于零线方向上相对于零线的位置，由基本偏差确定。

Example 2-1: Given: the nominal size of hole and shaft is ϕ25 mm, the limits of size for hole is $D_{max}=\phi25.021$ mm and $D_{min}=\phi25.000$ mm, the limits of size for hole is $d_{max}=\phi24.980$ mm and $d_{min}=\phi$ 24.967 mm, to finish following questions:

① To calculate the limit deviations and size tolerance. ② To draw the illustration of size tolerance zone.

例2-1：已知孔、轴公称尺寸为ϕ25 mm，孔的极限尺寸：$D_{max}=\phi25.021$ mm，$D_{min}=\phi25.000$ mm，轴的极限尺寸：$d_{max}=\phi24.980$ mm，$d_{min}=\phi24.967$ mm。完成下列计算：

① 计算孔与轴的极限偏差和尺寸公差。② 作出公差带图。

Solution: ① According to the given condition, the nominal size $D(d)=25$ mm, then according to the Eq. (2-3) to (2-6), the deviation can be achieved:

Upper deviation of hole: $ES=D_{max}-D=25.021-25=+0.021$ mm

Lower deviation of hole: $EI=D_{min}-D=25-25=0$

Upper deviation of shaft: $es=d_{max}-d=24.980-25=-0.02$ mm

Lower deviation of shaft $ei=d_{min}-d=24.967-25=-0.033$ mm

According to the Eq. (2-7) to (2-8), the size tolerance of hole and shaft can be achieved:

The size tolerance of hole: $T_h = |D_{max} - D_{min}| = |25.021 - 25.000| = 0.021$ mm

The size tolerance of shaft: $T_s = |d_{max} - d_{min}| = |24.980 - 24.967| = 0.013$ mm

② The illustration of size tolerance zone is shown in Fig. 2-4.

解：根据已知条件公称尺寸$D(d) = 25$mm，由式（2-3）至式（2-6）得：

孔的上偏差：$ES = D_{max} - D = 25.021 - 25 = +0.021$ mm

孔的下偏差：$EI = D_{min} - D = 25 - 25 = 0$

轴的上偏差：$es = d_{max} - d = 24.980 - 25 = -0.02$ mm

轴的下偏差：$ei = d_{min} - d = 24.967 - 25 = -0.033$ mm

由式（2-7）、式（2-8）可得孔与轴的尺寸公差：

孔的尺寸公差：$T_h = |D_{max} - D_{min}| = |25.021 - 25.000| = 0.021$ mm

轴的尺寸公差：$T_s = |d_{max} - d_{min}| = |24.980 - 24.967| = 0.013$ mm

②公差带示意图如2-4所示。

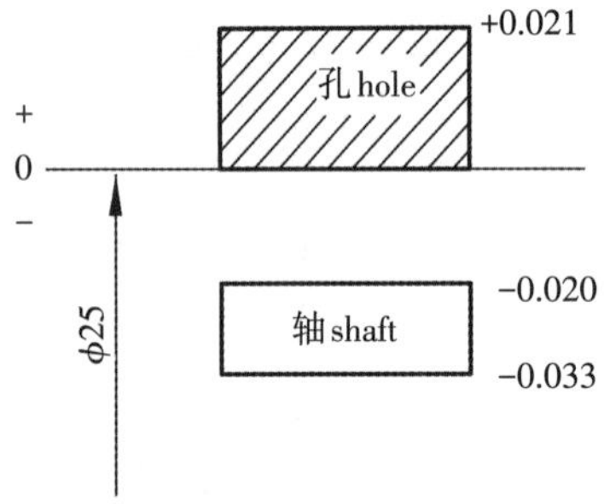

Fig. 2-4 Illustration of size tolerance zone for example 2-1

图2-4 例2-1的公差带示意图

2.2.4 Terms and Definitions About Fits 关于配合的术语与定义

2.2.4.1 Fit 配合

Fit refers to the relationship between size tolerance zones of holes and shafts assembled together whose nominal dimensions are same. According to the definition of fit, there are two fundamental conditions of fit: one is that the nominal size should be same, i.e. $D = d$; another is the holes and shafts have the relationship containing and contained, i.e. they are assembled together. At the same time, the fits are the assembly relationships of a batch of holes and shafts, other than one certain hole assembling with one certain shaft, so it is exact to describe fits using the relationship of size tolerance zone.

配合是指公称尺寸相同、相互结合的孔和轴公差带之间的关系。依据配合的定义，形成配合要有两个基本条件：一是孔和轴的公称尺寸必须相同，即$D = d$；二是具有包容和被包容的特性，即孔和轴的结合。另外，配合还指一批孔、轴的装配关系，而不是指单个孔与轴的装配。所以，用尺寸公差带的相互关系来描述配合比较确切。

2.2.4.2 Clearance and interference 间隙或过盈

Clearance is defined as the algebraic difference between the dimensions of the hole and the mating shaft; When this difference is positive, it is termed clearance. (The dimension of the shaft is smaller than the dimension of the hole),the code X is used to represent clearance value.

间隙是孔的尺寸减去相配合的轴的尺寸代数差；此值为正时，称为间隙（轴的尺寸小于孔的尺寸),用代码X来表示间隙值。

Interference is defined as the algebraic difference between the dimensions of the hole and the mating shaft; When this difference is negative, it is termed interference. (The dimension of the shaft is larger than the dimension of the hole), the code Y is used to represent interference value.

过盈是孔的尺寸减去相配合的轴的尺寸代数差；此值为负时，称为过盈（轴的尺寸小于孔的尺寸），用代码Y来表示过盈值。

2.2.4.3　Fit types 配合类型

According to the different relative positional relationships of the tolerance zones of the mating hole and shaft, there are three types of fit, i. e. clearance fit, interference fit and transition fit.

根据相互结合的孔、轴公差带不同的相对位置关系，配合可分为间隙配合、过盈配合、过渡配合三种。

（1）Clearance fit 间隙配合。

A fit that always provides a clearance between the hole and shaft when assembled, i.e. the minimum size of the hole is either greater than or, in the extreme case, equal to the maximum size of the shaft (The minimum clearance is equal to zero). The size tolerance zone of hole is totally above the size tolerance zone of shaft, shown in Fig. 2-5.

间隙配合是指在相互配合的孔和轴之间具有一定的间隙，即孔的最小直径大于轴的最大直径或在少数情况下与轴的最大直径相等（最小间隙等于零）的配合。如图2-5所示，孔的公差带在轴的公差带之上。

In a clearance fit, the limits of clearance are concerned. Limits of clearance are general term of the maximum clearance and the minimum clearance, and they are the limits of allowed clearance. The clearance varies from minimum clearance to maximum clearance.

在间隙配合中，要考虑间隙的极限。间隙极限是最大间隙和最小间隙的总称，它们是允许间隙的极限。间隙从最小间隙到最大间隙不等。

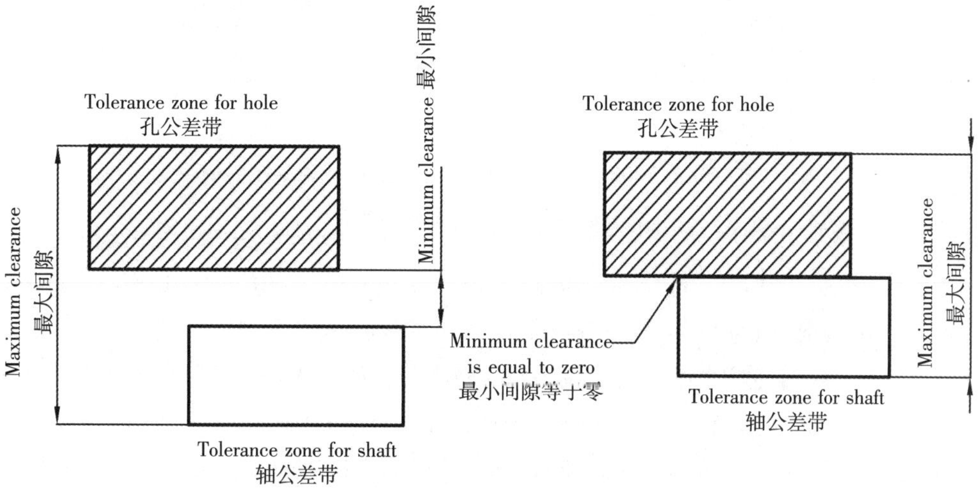

Fig. 2-5　Illustration of size tolerance zone for clearance fits

图2-5　间隙配合公差带示意图

Maximum clearance (X_{max}): in a clearance fit, the positive difference between maximum limit of size of the hole and the minimum limit of size of the shaft, is designated by X_{max}.

最大间隙（X_{max}）：在间隙配合中，孔的最大尺寸极限与轴的最小尺寸极限之间的差值为正，用X_{max}表示。

Minimum clearance (X_{min}): in a clearance fit, the positive (or zero) difference between minimum limit of size of the hole and the maximum limit of size of the shaft, is designated by X_{min}.

最小间隙（X_{min}）：间隙配合中，孔的最大尺寸极限与轴的最小尺寸极限之间的差值为正（或零），用X_{min}表示。

Maximum clearance（positive）最大间隙（为正）

$$X_{max} = D_{max} - d_{min} = ES - ei > 0 \tag{2-9}$$

Minimum clearance（positive or zero）最小间隙（为正或零）

$$X_{min} = D_{min} - d_{max} = EI - es \geqslant 0 \tag{2-10}$$

（2）Interference fit过盈配合。

A fit which provides an interference everywhere between the hole and shaft when assembled，i.e. the maximum size of the hole is either smaller than or，in the extreme case，equal to the minimum size of the shaft（The minimum interference is equal to zero）,the size tolerance zone of the hole is totally below the size tolerance zone of the shaft，the shown in Fig. 2-6.

过盈配合是指在相互配合的孔和轴具有一定的过盈，即孔的最大直径小于轴的最小直径或在少数情况下与轴的最小直径相等（最小间隙等于零）的配合。此时，孔的公差带在轴的公差带之下，如图2-6所示。

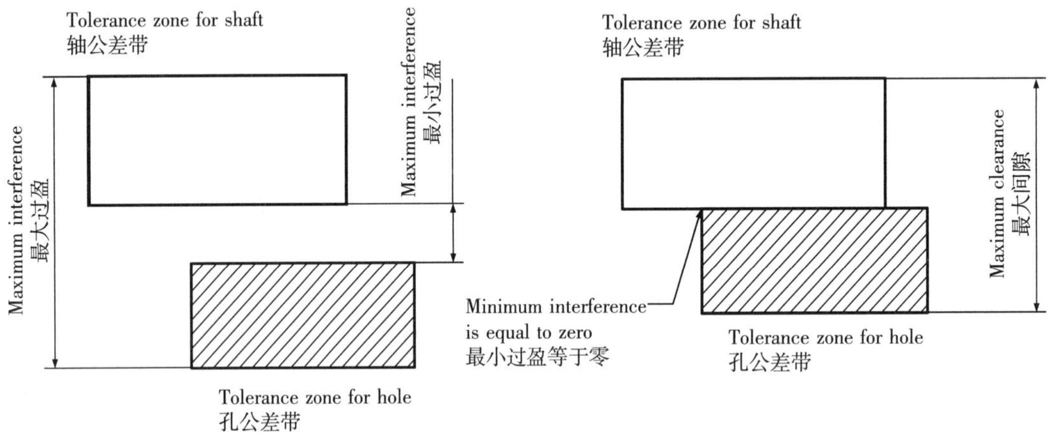

Fig. 2-6 Illustration of size tolerance zone for interference fit

图2-6 过盈配合公差带示意图

Maximum interference（Y_{max}）：in an interference fit，the negative difference between minimum limit of size of the hole and the maximum limit of size of the shaft，is designated by Y_{max}.

最大过盈（Y_{max}）：在过盈配合中，孔的最小尺寸极限与轴的最大尺寸极限之差为负，用Y_{max}表示。

Minimum interference（Y_{min}）：in an interference fit，the negative（or zero）difference between maximum limit of size of the hole and the minimum limit of size of the shaft，is designated by Y_{min}.

最小过盈（Y_{min}）：在过盈配合中，孔的最大尺寸极限与轴的最小尺寸极限之差为负差（或零），用Y_{min}表示。

Maximum interference（negative）：最大过盈（为负）：

$$Y_{max} = D_{max} - d_{min} = EI - es < 0 \tag{2-11}$$

Minimum interference（negative or zero）：最小过盈（为负或零）：

$$Y_{min} = D_{min} - d_{max} = ES - ei \leqslant 0 \tag{2-12}$$

（3）Transition fit过渡配合。

A fit that may provide either clearance or interference between the hole and the shaft when assembled， depending on the actual sizes of the hole and shaft, i.e. the tolerance zones of the hole and the shaft overlap completely or in part, shown in Fig. 2–7。

在孔和轴装配时，根据孔和轴的实际尺寸，可能会在孔和轴之间具有间隙或过盈的配合。此时，孔的公差带与轴的公差带完全或部分重叠，如图2–7所示。

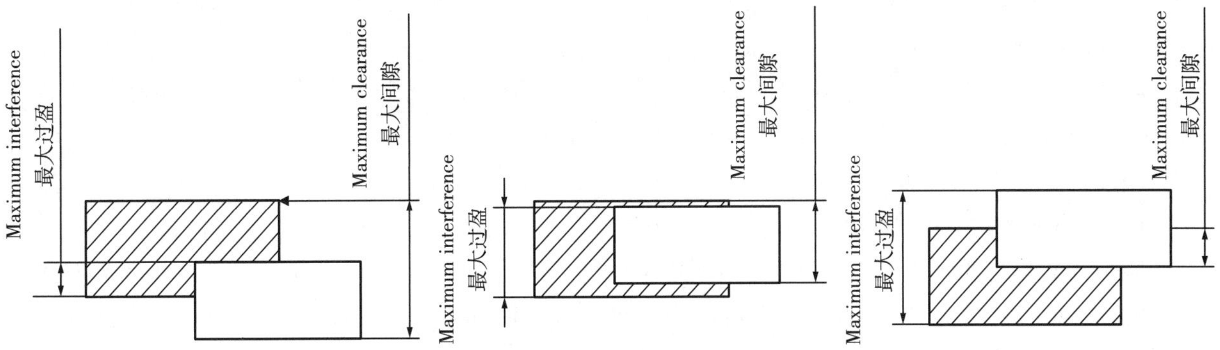

Fig. 2–7 Illustration of size tolerance zone for transition fit

图2–7 过渡配合公差带示意图

4) Fit tolerance (T_f) 配合公差

Fit tolerance is the variation of a fit of clearance (s) and/(or) interference (s), i.e. the absolute value of the difference between limit clearance (s) and/(or) limit interference (s), as Eq. (2 – 13) to (2 – 15):

配合公差指相互配合的间隙和（或）过盈的变动量，等于极限间隙和（或）极限过盈差的绝对值，见式（2 – 13）至式（2 – 15）。

Clearance fit tolerance 间隙配合公差:

$$T_f = |X_{max} - X_{min}| = T_h + T_s \tag{2–13}$$

Interference fit tolerance：过盈配合公差:

$$T_f = |Y_{min} - Y_{max}| = T_h + T_s \tag{2–14}$$

Transition fit tolerance：过渡配合公差:

$$T_f = |X_{max} - Y_{max}| = T_h + T_s \tag{2–15}$$

i.e. the equations of fit tolerance for all three fit type is same, shown as Eq. (2–14):

即三种配合类型的配合公差公式相同，见式（2 – 16）所示。

$$T_f = T_h + T_s \tag{2–16}$$

2.2.5 Fit System 配合制

The combination of the hole and the shaft is used widely in mechanical devices. According to the functional requirements, there are three fit types when a hole and a shaft are assembled together: clearance fit used for the connection that is able to move relatively, interference fit used for permanent connection, and transition fit used for the connection that is able to locate and be removed. To meet these three functional requirements, GB/T 1800.1–2020 specifies fit systems: hole-basis fit system and shaft-basis fit system.

孔与轴配合是机械装置中应用广泛的配合。根据功能要求，孔与轴配合有三种类型：间隙配合用于有相对移动的连接，过盈配合用于永久连接，过渡配合用于能够定位和拆卸的连接。为了满足这三个功能需求，GB/T 1800.1—2020规定配合制有两种基准制，即基孔制配合和基轴制配合。

2.2.5.1 Hole-basis fit system 基孔制配合

Hole-basis fit system is a fit system in which the required different fit properties are obtained by combining shafts of various fundamental deviations with the hole of a fixed fundamental deviation, shown in Fig.2–8 (a). The hole in the hole-basis fit system is a reference hole, and its code is “H”; The lower limit size of this hole is identical to the nominal size and the fundamental deviation is lower deviation, i.e. EI = 0.

基孔制配合是指基本偏差为一定的孔的公差带，与不同基本偏差的轴的公差带形成不同配合的一种制度。如图2–8（a）所示。基孔制配合中，基孔制的孔为基准孔，其代号为“H”。该孔的下极限尺寸与公称尺寸相同，且其基本偏差为下偏差，即EI = 0。

2.2.5.2 Shaft-basis fit system 基轴制配合

Shaft-basis fit system is a fit system in which the required different fit properties are obtained by combining holes of various fundamental deviations with the shaft of a fixed fundamental deviation, shown in Fig.2–8 (b). The shaft in the shaft-basis fit system is a reference shaft, and its code is “h”; The upper limit size of this shaft is identical to the nominal size and the fundamental deviation is upper deviation, i.e. es = 0.

基轴制配合是指基本偏差一定的轴的公差带，与不同基本偏差的孔的公差带形成不同配合的一种制度。如图2–8（b）所示。在基轴制配合中，轴为基准轴，其代号为“h”；该轴的上极限尺寸与公称尺寸相同，且基准轴的上偏差为零，即es = 0。

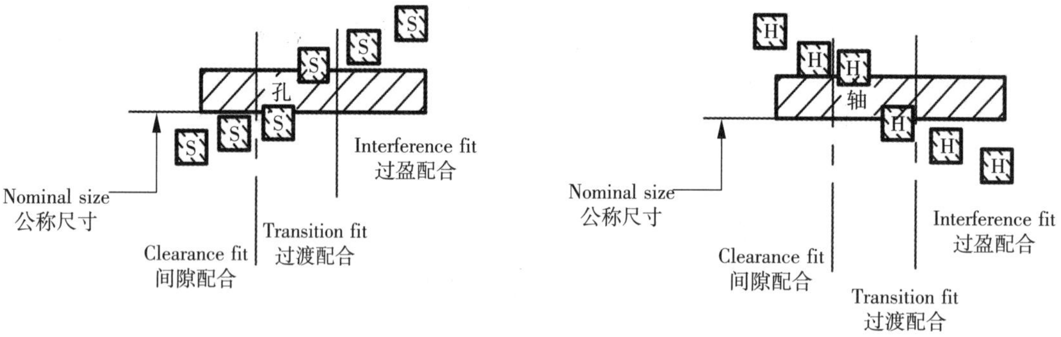

(a) Hole-basis fit system
基孔制配合

(b) Shaft-basis fit system
基轴制配合

Fig. 2–8 Schematic of fits system

图2–8 配合制示意图

Example 2–2: Given: in a hole-basis fit system, the nominal size is $D(d) = \phi 50$ mm, the minimum interference is − 45 μm, and the maximum interference is − 86 μm, the size tolerance of hole is 25 μm, to calculate the limit deviations of the hole and the shaft, then to draw the illustration of limits and fits.

例2–2：有一基孔制配合的孔、轴的公称尺寸为50 mm，最小过盈为-45 μm，最大过盈为-86 μm。孔公差为25 μm，确定孔和轴的极限偏差，并画出孔与轴公差带图。

Solution: According to the condition given, hole-basis fit system is adopted, then the fundamental deviation of hole is lower deviation EI, and EI = 0.

According to $Y_{min} = -45$ μm, $Y_{max} = -86$ μm, it is identified as an interference fit.

Because of $Y_{max} = -86\ \mu m = EI - es = 0 - es$, then es = 86 μm;

Because of $T_h = |ES - EI| = |ES - 0|$, then ES = 25 μm;

Because of $Y_{min} = -45\ \mu m = ES - ei = 25 - ei$, then ei = 70 μm.

解：根据已知条件，采用基孔制配合，孔的基本偏差为下偏差，且EI = 0.

由 $Y_{min} = -45$ μm，$Y_{max} = -86$ μm，知其为过盈配合。

由于 $Y_{max} = -86$ μm = EI − es = 0 − es，es = 86 μm；

由于 $T_h = |ES - EI| = |ES - 0|$，ES = 25 μm；

由于 $Y_{min} = -45$ μm = ES − ei = 25 − ei，ei = 70 μm。

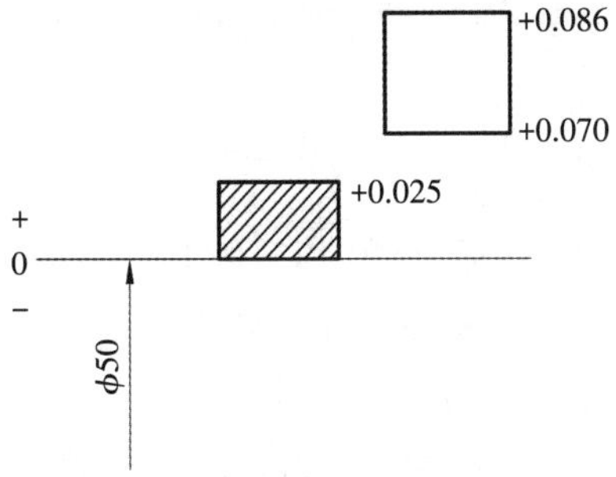

Fig. 2-9　Illustration of size tolerance zone for example 2-2

图2-9　例2-2的公差带示意图

2.3　Standard Tolerance and Fundamental Deviation 标准公差和基本偏差系列

2.3.1　Standard Tolerance Series 标准公差系列

Standard tolerance values reflect the accuracy of the workpieces, and they also reflect the machining difficulty and machining cost, so different tolerance grades are provided to meet different precision requirements. Moreover, at the same tolerance grade, the machining difficulty should be the same at different sizes. Then the tolerance values are relevant to two factors: tolerance grades and nominal sizes.

标准公差值反映了工件的精度，同时体现了加工的难易程度和加工成本，因此提供了不同的公差等级，以满足不同的精度要求。此外，公差等级相同，不同尺寸的加工难度应该是相同的。因此，公差值与两个因素有关：公差等级和公称尺寸。

The production practice and test statistical analysis demonstrate that the error in the nominal size of a batch of parts varies when different processing methods and production conditions are used. Similarly, if the processing method and production conditions remain the same but the nominal size differs, different errors will also be generated. Similarly, when machining methods and production conditions remain identical but nominal dimensions differ, errors of varying magnitudes will also arise. Since tolerance serves to control such deviations, its establishment is fundamentally based on the principles governing machining error generation and tolerances determined through experimental statistics. The corresponding equation is expressed as:

生产实践和试验统计分析证明，公称尺寸相同的一批零件，若加工方法和生产条件不同，则产生的误差也不同；同样地，若加工方法和生产条件相同，而公称尺寸不同，也会产生大小不同的误差。由于公差是控制误差的，所以制定公差的基础就是根据加工误差产生的规律和经试验统计得到的公差，计算表达式为：

$$T = ai = af(D) \tag{2-17}$$

Where, D —nominal size;

i —standard tolerance factor（or tolerance unit）, $i=f(D)$, the unit is μm;

a —coefficient of tolerance class.

式中，D——公称尺寸；

i——标准公差因子（或公差单位），$i=f(D)$；

a——公差等级系数。

When the nominal dimension is less than 500 mm, the equation for calculating the standard tolerance factor i（The unit is μm）is:

当公称尺寸小于500 mm时，标准公差因子i（单位为μm）的计算公式为：

$$i=0.45\sqrt[3]{D}+0.001D$$

Where, D —Geometric mean value of the nominal dimension range（mm）.

式中，D——公称尺寸段的几何平均值（mm）。

When the nominal dimension ranges from 500 mm to 3150 mm, the formula for calculating the standard tolerance factor I（The unit is μm）is:

当公称尺寸大于500 ~ 3150 mm时，标准公差因子I（单位为μm）的计算公式为：

$$I=0.004D+2.1$$

2.3.1.1 Standard tolerance grades 标准公差等级

According to the tolerance grade coefficients, national standards specify the following: for the size range from 0 to 500 mm（500 mm inclusive）, standard tolerances are categorized into 20 grades, designated as IT01, IT0, IT1, IT2, ..., IT18. For the size range from 500 mm to 3150 mm（inclusive）, standard tolerances are divided into 18 grades, namely IT1, IT2, ..., IT18. When "IT" is followed by Arabic numerals, "IT" denotes standard tolerance, and the numerals represent the tolerance grade. Specifically, IT01 corresponds to the highest tolerance grade, while IT18 represents the lowest. The lower the tolerance grade, the larger the tolerance value. The equation for calculating standard tolerance values under different tolerance grades is presented in Tab. 2-1.

根据公差等级系数的不同，国家标准规定：0 ~ 500 mm（包括500 mm）尺寸段标准公差分为20个等级，即IT01，IT0，IT1，IT2，…，IT18；对于500 ~ 3150 mm（包括3150 mm）尺寸段标准公差分为18个等级，即IT1，IT2，…，IT18。IT后加阿拉伯数字时，IT表示标准公差，阿拉伯数字表示标准公差等级。IT01公差等级最高，IT18公差等级最低。公差等级越低，公差越大。不同公差等级下，标准公差值的计算公式见表2-1。

Tab. 2-1 Formular for standard tolerance

表2-1 各级标准公差数值计算公式

Tolerance grades 公差等级	Nominal size 公称尺寸/mm		Tolerance grades 公差等级	Nominal size 公称尺寸/mm	
	>0 ~ 500	>500 ~ 3150		>0~500	>50 ~ 3150
IT01	$0.3+0.008D$	—	IT3	$(IT1)\left(\frac{IT5}{IT1}\right)^{1/2}$	
IT0	$0.5+0.012D$				
IT1	$0.8+0.020D$	$2I$	IT4	$(IT1)\left(\frac{IT5}{IT1}\right)^{3/4}$	
IT2	$(IT1)\left(\frac{IT5}{IT1}\right)^{1/4}$				

Tab. 2-1 (Continued) 表2-1（续）

Tolerance grades 公差等级	Nominal size 公称尺寸/mm		Tolerance grades 公差等级	Nominal size 公称尺寸/mm	
	>0 ~ 500	>500 ~ 3150		>0~500	>50 ~ 3150
IT5	$7i$	$7I$	IT12	$160i$	$160I$
IT6	$10i$	$10I$	IT13	$250i$	$250I$
IT7	$16i$	$16I$	IT14	$400i$	$400I$
IT8	$25i$	$25I$	IT15	$640i$	$640I$
IT9	$25i$	$25I$	IT16	$1000i$	$1000I$
IT10	$64i$	$64I$	IT17	$1600i$	$1600I$
IT11	$100i$	$100I$	IT18	$2500i$	$2500I$

2.3.1.2 Size Range Division 尺寸分段

According to the calculation formula of standard tolerance, for each tolerance grade, each nominal size can obtain a corresponding tolerance value. However, there are many nominal sizes in production, so the tolerance table compiled in this way will be extremely huge, which will cause difficuties to the actual production and is not conducive to the standardization and serialization of tolerance values. In order to reduce the number of tolerances, simplify the tolerance table and facilitate the application, the national standard stipulates that the nominal size is segmented as given in Tab. 2-2.

按照标准公差的计算公式，对于每个公差等级，每个公称尺寸都可计算出相应的公差值。但在实际生产中，公称尺寸种类繁多，这样编制的公差表格将会极其庞大，给生产实际带来麻烦，也不利于公差值的标准化和系列化。为了减少公差数目，简化公差表格，便于应用，国家标准对公称尺寸进行了分段，如表2-2所示。

Tab. 2-2 Segments of the nominal size

表2-2 公称尺寸分段表

Main paragraph 主段落		Middle paragraph 中间段落		Main paragraph 主段落		Middle paragraph 中间段落		Main paragraph 主段落		Middle paragraph 中间段落	
>大于	to至	>大于	to至	>大于	to至	>大于	to至	>大于	to至	>大于	to至
—	3			30	50	30 40	40 50	180	250	180 200 225	200 225 250
3	6			50	80	50 65	65 80	250	315	250 280	280 315
6	10			80	120	80 100	100 120	315	400	315 355	355 400
10	18	10 14	14 18	120	180	120 140 160	140 160 180	400	500	400 450	450 500
18	30	18 24	24 30								

The values of the standard tolerances and fundamental deviations for each nominal size segment are calculated from the geometrical mean (D) of the extreme sizes ($>D_n \sim D_{n-1}$) of that segment, as follows:

在标准公差和基本偏差的计算公式中，每个公称尺寸一律以所属分段尺寸的首尾两项的几何平均值进行计算，尺寸段内（$>D_n \sim D_{n-1}$）的几何平均值公式为：$D=\sqrt{D_n \cdot D_{n-1}}$

For the nominal size segment (≤3 mm), the geometrical mean, D, according to convention, is taken between size from 1μm to 3mm, then $D=\sqrt{1\times 3}=1.732$ mm. When the nominal size is less than or equal to 3150 mm, he values of standard tolerance calculation equation are given in Tab. 2-1. Numerical values of standard tolerances are given in Tab. 2-3.

对于不大于3 mm的尺寸段，按照惯例，几何平均值位于1 mm ~ 3 mm，所以$D=\sqrt{1\times 3}=1.732$ mm。公称尺寸不大于3150 mm时，标准公差计算公式见表2-1。标准公差数值见表2-3。

Example 2-3: Given that the nominal size is ϕ20 mm, find the tolerance values for IT7 and IT8.

例2-3：已知公称尺寸为ϕ20 mm IT6和IT7的公差值。

Solution: According to Tab. 2-2, the nominal size is 20 mm, belong to 18 ~ 30 mm, then $D=\sqrt{18\times 30}=23.24$ mm,

Tolerance unit $i=0.45\sqrt[3]{D}+0.001D=1.31$

According to Tab. 2-1, IT6 = 10i, IT7 = 16i. then

IT6 = 10 × 1.31 μm = 13.1 μm≈13 μm

IT7 = 16 × 1.31 μm = 20.96 μm≈21 μm

解：根据表2-2，公称尺寸20 mm，属于18 ~ 30 mm，则$D=\sqrt{18\times 30}=23.24$ mm，

公差单位：$i=0.45\sqrt[3]{D}+0.001D=1.31$

根据表2-1可得，IT6 = 10i，IT7 = 16i，得

IT6 = 10 × 1.31 μm = 13.1 μm≈13 μm

IT7 = 16 × 1.31 μm = 20.96 μm≈21 μm

Thought: If the tolerance values for the same tolerance grade vary with different basic sizes, is the degree of precision the same?

思考：同一公差等级对于不同的基本尺寸公差的数值不同，其精确程度相同吗？

Example 2-4: A nominal diameter of shaft is ϕ80 mm, try to determine the standard tolerance value of IT6 by Tab.2-3.

例2-4：某轴公称尺寸ϕ80 mm，查表2-3确定IT6的标准公差值。

Solution: The nominal size is 80 mm, belong to 50 ~ 80 mm.

According to Tab. 2-3, IT6 = 19 μm.

解：公称尺寸80 mm，属于50 ~ 80 mm。

查表2-3可得，IT6 = 19 μm。

2.3.2 Series of Fundamental Deviation 基本偏差系列

2.3.2.1 Symbol of fundamental deviation 基本偏差及其代号

Fundamental deviation is the upper or lower deviation used to determine the relative position between the tolerance zone and zero line. Generally, the deviation nearer to the zero line is treated as the fundamental deviation.

Tab.2-3 Numerical values of standard tolerances

表 2-3 标准公差数值

Nominal size/mm 公称尺寸/mm		Standard tolerance factor 标准公差等级																			
		IT01	IT0	IT1	IT2	IT3	IT4	IT5	IT6	IT7	IT8	IT9	IT10	IT11	IT12	IT13	IT14	IT15	IT16	IT17	IT18
>	=	Standard tolerance 标准公差/μm													Standard tolerance 标准公差/mm						
—	3	0.3	0.5	0.8	1.2	2	3	4	6	10	14	25	40	60	0.1	0.14	0.25	0.4	0.6	1	1.4
3	6	0.4	0.6	1	1.5	2.5	4	5	8	12	18	30	48	75	0.12	0.18	0.3	0.48	0.75	1.2	1.8
6	10	0.4	0.6	1	1.5	2.5	4	6	9	15	22	36	58	90	0.15	0.22	0.36	0.58	0.9	1.5	2.2
10	18	0.5	0.8	1.2	2	3	5	8	11	18	27	43	70	110	0.18	0.27	0.43	0.7	1.1	1.8	2.7
18	30	0.6	1	1.5	2.5	4	6	9	13	21	33	52	84	130	0.21	0.33	0.52	0.84	1.3	2.1	3.3
30	50	0.6	1	1.5	2.5	4	7	11	16	25	39	62	100	160	0.25	0.39	0.62	1	1.6	2.5	3.9
50	80	0.8	1.2	2	3	5	8	13	19	30	46	74	120	190	0.3	0.46	0.74	1.2	1.9	3	4.6
80	120	1	1.5	2.5	4	6	10	15	22	35	54	87	140	220	0.35	0.54	0.87	1.4	2.2	3.5	5.4
120	180	1.2	2	3.5	5	8	12	18	25	40	63	100	160	250	0.4	0.63	1	1.6	2.5	4	6.3
180	250	2	3	4.5	7	10	14	20	29	46	72	115	185	290	0.46	0.72	1.15	1.85	2.9	4.6	7.2
250	315	2.5	4	6	8	12	16	23	32	52	81	130	210	320	0.52	0.81	1.3	2.1	3.2	5.2	8.1
315	400	3	5	7	9	13	18	25	36	57	89	140	230	360	0.57	0.89	1.4	2.3	3.6	5.7	8.9
400	500	4	6	8	10	15	20	27	40	63	97	155	250	400	0.63	0.97	1.55	2.5	4	6.3	9.7
500	630	—	—	9	11	16	22	32	44	70	110	175	280	440	0.7	1.1	1.75	2.8	4.4	7	11
630	800	—	—	10	13	18	25	36	50	80	125	200	320	500	0.8	1.25	2	3.2	5	8	12.5
800	1000	—	—	11	15	21	28	40	56	90	140	230	360	560	0.9	1.4	2.3	3.6	5.6	9	14
1000	1250	—	—	13	18	24	33	47	66	105	165	260	420	660	1.05	1.65	2.6	4.2	6.6	10.5	16.5
1250	1600	—	—	15	21	29	39	55	78	125	195	310	500	780	1.25	1.95	3.1	5	7.8	12.5	19.5
1600	2000	—	—	18	25	35	46	65	92	150	230	370	600	920	1.5	2.3	3.7	6	9.2	15	23
2000	2500	—	—	22	30	41	55	78	110	175	280	440	700	1110	1.75	2.8	4.4	7	11	17.5	28
2500	3150	—	—	26	36	50	68	96	135	210	330	540	860	1350	2.1	3.3	5.4	8.6	13.5	21	33

基本偏差用来确定公差带相对于零线位置的上偏差或下偏差，一般指靠近零线的那个偏差作为基本偏差。

There are 28 fundamental deviations in total for holes and shafts in GB/T 1800.1–2020, respectively. These different basic deviations constitute a series of basic deviations. All of them are represented by Latin letters. Uppercase represents holes and lowercase represents shafts. To avoid confusion, the letter I(i)、L(l)、O(o)、Q(q) and W(w) are not used. And there are some fundamental deviation double letters CD(cd), EF(ef), FG(fg), JS(js), ZA(za), ZB(zb) and ZC(zc) to meet some requirements.The fundamental deviation for a shaft is symmetrical, in relation to the zero line, to the corresponding to the fundamental deviations of holes with the same letters. The distribution of fundamental deviations for holes and shafts is shown in Fig.2–10. At the same time, in the national standard, the holes only retain J6, J7, J8 and the shafts are only retained j5, j6, j7, j8.

GB/T 1800.1—2020规定了孔和轴各有28种基本偏差，这些不同的基本偏差构成了基本偏差系列。所有基本偏差的代号全部用拉丁字母表示，其中大写字母代表孔，小写字母代表轴。为避免混淆，不使用字母I(i)，L(l)，O(o)，Q(q)，W(w)。同时，为满足某些特殊需要的配合，部分基本偏差采用双字母组合形式：CD(cd)，EF(ef)，FG(fg)，JS(js)，ZA(za)，ZB(zb) 和ZC(zc)。轴的基本偏差与相同字母对应的孔基本偏差相对于零线呈对称分布。相对于零线，轴的基本偏差与具有相同字母的孔的基本偏差对称。孔和轴的基本偏差分布如图2-10所示。此外，在国家标准中，孔仅保留了J6，J7，J8三个公差等级；轴仅保留了j5，j6，j7，j8四个公差等级。

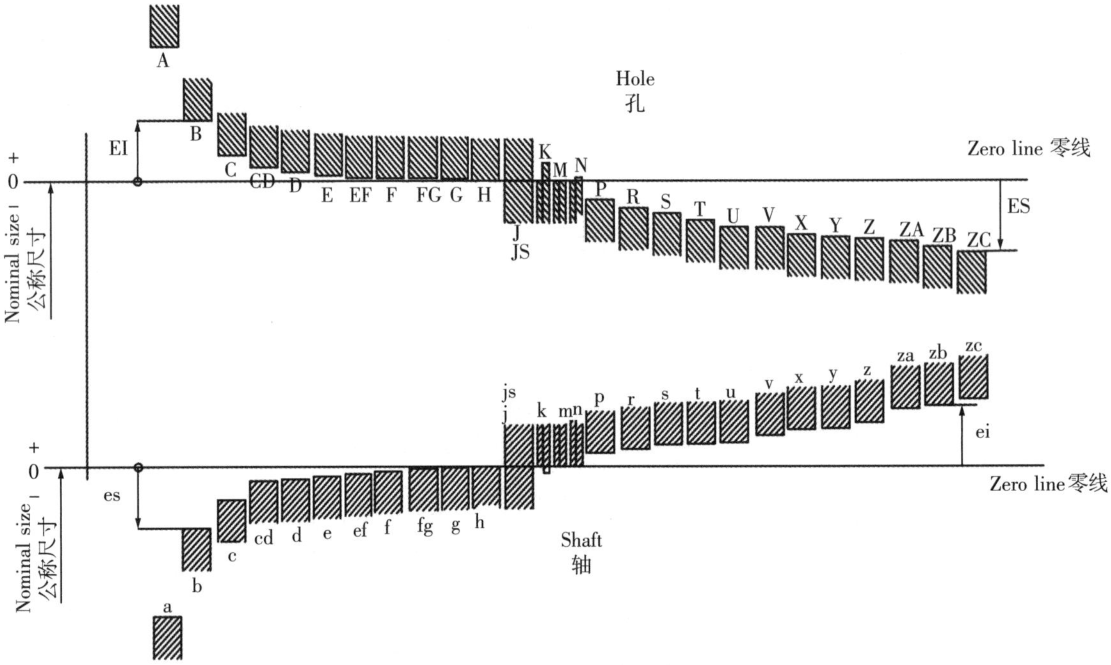

Fig. 2–10 Series of the fundamental deviations

图2–10 基本偏差系列

(1) Characteristics of fundamental deviations of holes are as follows:

孔的基本偏差有以下主要特点：

A to G: the tolerance zones are totally above the zero line; the fundamental deviations are lower deviation EI, the value is positive .

A到G：公差带在零线上方，基本偏差为下极限偏差EI，基本偏差值为正值。

H: the tolerance zone is totally above the zero line; the fundamental deviation is lower deviation, EI = 0.

At this time, the hole is the datum hole.

H：公差带在零线上方，基本偏差为下极限偏差EI，EI = 0。此时，孔为基准孔。

JS: the tolerance zone is totally symmetrical about the zero line; the fundamental deviation is the lower deviation or upper deviation. The absolute value of the limit deviation is equal to half of the tolerance.

JS：公差带完全对称于零线，基本偏差可以是上极限偏差，也可以是下极限偏差。基本偏差值的绝对值为公差值的一半。

J to ZC: the fundamental deviations are upper deviation ES, most values are negative.

J到ZC：基本偏差为上极限偏差ES，基本偏差值多为负值。

（2）Characteristics of fundamental deviations of shafts are as follows:

轴的基本偏差有以下主要特点：

a to g: the tolerance zones are totally below the zero line; the fundamental deviations are upper deviation es, the value is negative.

a到g：公差带在零线下方，基本偏差为上极限偏差es，基本偏差值为负值。

h: the tolerance zone is below the zero line; the fundamental deviation is upper deviation, es = 0. At this time, the shaft is the datum shaft.

h：公差带在零线下方，基本偏差为上极限偏差es，es = 0。此时，轴为基准轴。

js: the tolerance zone is totallical symmetrical about the zero line, the fundamental deviation is the lower deviation or upper deviation. The absolute value of the limits deviation equals to half of the tolerance value.

js：公差带完全对称于零线，基本偏差可以是上极限偏差，也可以是下极限偏差。基本偏差值的绝对值为公差值的一半。

j to zc: the tolerance zone is above the zero line , the fundamental deviations are lower deviation ei and the fundamental deviation is mostly positive.

j到zc：公差带在零线上方，基本偏差为下极限偏差ei，基本偏差值多为正值。

2.3.2.2 Fundamental deviation values of shafts and holes 孔和轴的基本偏差值

The values of fundamental deviations for shaft are calculated based on the hole-basis fit system. According to different fit requirements, a series of equations are achieved through statistical analysis based on production practices and a large number of experiments. One need not calculate the values when designing; the engineers just directly look up the table. The basic deviation of the shaft is obtained by looking up the table, and the other deviation of the shaft is calculated according to the fundamental deviation and the standard tolerance of the shaft. As shown in Eq.（2-18）. The values of fundamental deviations for shafts are given in Tab.2-4.

轴的基本偏差值是基于基孔制配合体系计算得出的。根据不同的配合需求，结合生产实践和大量实验数据，通过统计分析建立了一系列公式。在设计过程中无需手动计算这些数值，工程师可直接通过查表确定轴的基本偏差。轴的另一个极限偏差则根据已查表得到的基本偏差和轴的标准公差进行计算，具体方法如式（2-18）所示。轴的基本偏差值见表2-4。

$$ei = es - IT_n \quad 或 \quad es = ei + IT_n \tag{2-18}$$

Tab. 2-4 The values of fundamental

表2-4 轴的基本偏差数值

Nominal size 公称尺寸 / mm		The values of fundamental deviations 基本偏差数值																
		Upper limit deviation 上极限偏差（es）												Lower limit deviation 下极限偏差（ei）				
大于	至	All standard tolerance grades 所有公差等级												IT5，IT6	IT7	IT8	IT7~IT4	≤IT3，>IT7
		a[①]	b[①]	c	cd	d	e	ef	f	fg	g	h	js[②]	j			k	
—	3	−270	−140	−60	−34	−20	−14	−10	−6	−4	−2	0	Deviation 偏差=±IT$_n$/2	−2	−4	−6	0	0
3	6	−270	−140	−70	−46	−30	−20	−14	−10	−6	−4	0		−2	−4		+1	0
6	10	−280	−150	−80	−56	−40	−25	−18	−13	−8	−5	0		−2	−5		+1	0
10	14	−290	−150	−95	−70	−50	−32	−23	−16	−10	−6	0		−3	−6		+1	0
14	18																	
18	24	−300	−160	−110	−85	−65	−40	−25	−20	−12	−7	0		−4	−8		+2	0
24	30																	
30	40	−310	−170	−120	−100	−80	−50	−35	−25	−15	−9	0		−5	−10		+2	0
40	50	−320	−180	−130														
50	65	−340	−190	−140		−100	−60		−30		−10	0		−7	−12		+2	0
65	80	−360	−200	−150														
80	100	−380	−220	−170		−120	−72		−36		−12	0		−9	−15		+3	0
100	120	−410	−240	−180														
120	140	−460	−260	−200		−145	−85		−43		−14	0		−11	−18		+3	0
140	160	−520	−280	−210														
160	180	−580	−310	−230														
180	200	−660	−340	−240		−170	−100		−50		−15	0		−13	−21		+4	0
200	225	−740	−380	−260														
225	250	−820	−420	−280														
250	280	−920	−480	−300		−190	−110		−56		−17	0		−16	−26		+4	0
280	315	−1050	−540	−330														
315	355	−1200	−600	−360		−210	−125		−62		−18	0		−18	−28		+4	0
355	400	−1350	−680	−400														
400	450	−1500	−760	−440		−230	−135		−68		−20	0		−20	−32		+5	0
450	500	−1650	−840	−480														

Notes: ① Fundamental deviations a and b shall not be used for nominal sizes less than or equal to 1mm;

② For tolerance grades js7 to js11, if the IT value number, n, is an odd number, then the value = ±(IT$_n$−1)/2.

注：① 公称尺寸小于或等于1 mm时，不使用基本偏差a和b；

② 对于公差等级j7至js11，如果IT值n为奇数，则该值 = ±(IT$_n$−1)2。

deviations for shafts（GB/T 1800.1-2020）

（摘自GB/T 1800.1—2020） 单位：μm

The values of fundamental deviations基本偏差数值													
Lower limit deviation 下极限偏差(ei)													
All standard tolerance grades 所有标准公差等级													
m	n	p	r	s	t	u	v	x	y	z	za	zb	zc
+2	+4	+6	+10	+14		+18		+20		+26	+32	+40	+60
+4	+8	+12	+15	+19		+23		+28		+35	+42	+50	+80
+6	+10	+15	+19	+23		+28		+34		+42	+52	+67	+97
+7	+12	+18	+23	+28		+33		+40		+50	+64	+90	+130
							+39	+45		+60	+77	+108	+150
+8	+15	+22	+28	+35		+41	+47	+54	+63	+73	+98	+136	+188
					+41	+48	+55	+64	+75	+88	+118	+160	+218
+9	+17	+26	+34	+43	+48	+60	+68	+80	+94	+112	+148	+200	+274
					+54	+70	+81	+97	+114	+136	+180	+242	+325
+11	+20	+32	+41	+53	+66	+87	+102	+122	+144	+172	+226	+300	+405
			+43	+59	+75	+102	+120	+146	+174	+210	+274	+360	+480
+13	+23	+37	+51	+71	+91	+124	+146	+178	+214	+258	+335	+445	+585
			+54	+79	+104	+144	+172	+210	+254	+310	+400	+525	+690
+15	+27	+43	+63	+92	+122	+170	+202	+248	+300	+365	+470	+620	+800
			+65	+100	+134	+190	+228	+280	+340	+415	+535	+700	+900
			+68	+108	+146	+210	+252	+310	+380	+465	+600	+780	+1000
+17	+31	+50	+77	+122	+166	+236	+284	+350	+425	+520	+670	+880	+1150
			+80	+130	+180	+258	+310	+385	+470	+575	+740	+960	+1250
			+84	+140	+196	+284	+340	+425	+520	+640	+820	+1050	+1350
+20	+34	+56	+94	+158	+218	+315	+385	+475	+580	+710	+920	+1200	+1550
			+98	+170	+240	+350	+425	+525	+650	+790	+1000	+1300	+1700
+21	+37	+62	+108	+190	+268	+390	+475	+590	+730	+900	+1150	+1500	+1900
			+114	+208	+294	+435	+530	+660	+820	+1000	+1300	+1650	+2100
+23	+40	+68	+126	+232	+330	+490	+595	+740	+920	+1100	+1450	+1850	+2400
			+132	+252	+360	+540	+660	+820	+1000	+1250	+1600	+2100	+2600

The fundamental deviation of the hole is obtained by the conversion of the fundamental deviation of the shaft. The principle of conversion is based on two principles of national standards: process equivalence and homonymic fit. Process equivalence: in the hole-basis fit system and shaft-basis system, the process equivalence of hole and the shaft should be ensured, that is, the machining difficulty of the hole and the shaft is equal. Same Designation Fit: using the same letter to represent the tolerance zone formed by the fundamental deviation of the hole and the shaft. The fit formed according to the hole-basis system and the shaft-basis system is called same designation fit (i.g. H9/9 and F9/h9、H7/p6 and PT/h6). The same designation fit that satisfies the process equivalence has the same fit properties, that is, the type of fit is the same and the limit clearance amount or the limit interference amount is the same. According to the principles above, the fundamental deviation of the hole is converted according to the following two rules .

孔的基本偏差是由轴的基本偏差换算得到的。换算的原则是基于国家标准的两条原则：工艺等价和同名配合。工艺等价：在标准的基孔制和基轴制配合中，应保证孔和轴的工艺等价，即孔和轴的加工难易程度相同。同名配合：用同一字母表示孔和轴的基本偏差所组成的公差带。按照基孔制形成的配合和按照基轴制形成的配合称为同名配合（如H9/9与F9/h9、H7/p6与PT/h6）。满足工艺等价的同名配合，其配合性质相同，即配合种类相同且极限间隙量或极限过盈量相等。根据上述原则，孔的基本偏差按以下两种规则换算。

（1）General rule 通用规则。

The absolute values of the fundamental deviations of the hole and the shaft represented by the same letter are equal, and the signs are opposite. The general rules apply in the following cases.

同一字母的孔的基本偏差与轴的基本偏差绝对值相等，符号相反。通用规则在以下情况下适用：

① For A to H, since the absolute values of the fundamental deviation EI of the hole and the fundamental deviation es of the corresponding shaft (a to h) are equal to the minimum clearance, the conversion principle is EI = −es.

对于A ~ H，因其基本偏差EI与对应轴（a ~ h）的基本偏差es的绝对值都等于最小间隙，故其换算原则为：EI = −es。

② For K to ZC, the cases are applicable for fundamental deviations K, M and N in standard tolerance grades up to IT8, and deviation from P to ZC in standard tolerance grades up to IT7. The conversion principle is ES = −ei.

对于K ~ ZC：基本偏差代号K，M，N：适用于标准公差等级大于IT8；基本偏差代号P ~ ZC：适用于标准公差等级大于IT7，其转换原则为：孔的极限偏差ES = −ei

With the exception of the nominal size over 3 mm for fundamental deviation N in standard tolerance grades up to *IT*8, the fundamental deviation is ES = 0.

标准公差大于IT8、公称尺寸大于3 mm的N除外，其基本偏差ES = 0。

（2）Special rule 特殊规则。

When using the same letter to represent the fundamental deviation of the hole and the shaft, the fundamental deviations ES of the hole and the fundamental deviation ei of the shaft are opposite, and the absolute value differ by a Δ value. The basic deviation of the hole is obtained:

用同一字母表示孔、轴基本偏差时，孔的基本偏差ES和轴的基本偏差ei符号相反，且绝对值相差一个Δ值。由此得出孔的基本偏差为：

$$ES = -ei + \Delta, \text{ Here } \Delta = T_h - T_s = IT_n - IT_{n-1}$$

Where，IT_n—the standard tolerance of a hole of a certain grade;

IT_{n-1}—the standard tolerance of a shaft one grade higher than a hole of a certain grade.

式中，IT_n——某一级孔的标准公差；

IT_{n-1}——比某一级孔高一级的轴的标准公差。

Because it is more difficult to process the hole than the shaft of the same tolerance grade in the higher tolerance grades，the hole fit of one grade lower than the shaft is often used，and the fit properties formed by the two fit systems（or homonymic fit）are required to be the same.

因为在较高的公差等级中，同一公差等级的孔比轴加工困难，所以配合中常采用比轴低一级的孔，并要求两种配合制度（或称同名配合）所形成的配合性质相同。

The special cases are only applicable for the nominal size from 3mm（including）up to 500 mm（including）for fundamental deviations K，M，and N in standard tolerance grades up to and including IT8，and deviations P to ZC in standard tolerance grades up to and including IT7. The values of fundamental deviations for shafts are given in Tab. 2-5.

特殊规则的适用范围：公称尺寸大于或等于3 mm，小于或等于500 mm，标准公差等级不大于IT8的代号K，M，N和标准公差等级不大于IT7的代号为P～ZC的孔的基本偏差计算。孔的基本偏差见表2-5。

The other deviation of the hole is calculated according to the basic deviation and standard deviation of the hole，and the following relationship.

根据孔的基本偏差和标准偏差，孔的另一个偏差按以下关系式计算，即

$$EI = ES - IT \tag{2-19}$$

$$ES = EI + IT \tag{2-20}$$

Example 2-5：Find the basic deviation value for the size designation ϕ25e mm.

例2-5：查尺寸代号ϕ25e mm的基本偏差值。

Solution：According to Fig. 2-10，ϕ25e mm is a shaft.

The basic deviation of ϕ25e mm is the upper deviation. Referring to Tab. 2-4，for ϕ25 within the range of 18～30 mm，the basic deviation is found in Tab. 2-4：

$$es = 40\ \mu m$$

解：根据图2-10可知：ϕ25e是轴。

ϕ25e的基本偏差为上偏差，即$\phi 25 \in 18 \sim 30$ mm，查表2-4，得基本偏差：

$$es = 40\ \mu m$$

Example 2-6：According to the fundamental deviations table for hole or shaft，to determine the limit deviations about ϕ25H7/r6，ϕ25R7/h6，try to calculate limit clearances or interferences，and then draw the tolerance zone diagram.

例2-6：根据孔或轴的基本偏差表，确定ϕ25H7/r6、ϕ25R7/h6极限偏差，计算两个配合的极限间隙或过盈，然后绘制公差带图。

Solution：（1）ϕ25H7/r6.

① Look up the table to determine the standard tolerances values for hole and the shaft.

Look up Tab. 2-3，$IT6 = 13\ \mu m$，$IT7 = 21\ \mu m$.

② Determine limit deviations for hole or shaft.

ϕ25r6：from Tab. 2-4，the fundamental deviation of ϕ25r6 is lower deviation $ei = +28\ \mu m$; then $es = ei + IT6 = (+28\ \mu m) + (+13\ \mu m) = +41\ \mu m$.

Tab. 2–5 The values of fundamental

表2–5　孔

Nominal size 公称尺寸/mm		The values of fundamental deviations 基本偏差数值																				
		Lower limit deviation 下极限偏差(EI)												Upper limit deviation 上极限偏差(ES)								
大于	至	All standard tolerance grades 所有公差等级												IT6	IT7	IT8	≤IT8	>IT8	≤IT8	>IT8	≤IT8	>IT8
		A①	B①	C	CD	D	E	EF	F	FG	G	H	JS②	J			K③④		M②③④		N①②	
—	3	+270	+140	+60	+34	+20	+14	+10	+6	+4	+2	0	Deviation 偏差 $=\pm IT_n/2$	+2	+4	+6	0	0	−2	−2	−4	−4
3	6	+270	+140	+70	+46	+30	+20	+14	+10	+6	+4	0		+5	+6	+10	−1 + Δ		−4 + Δ	−4	−8 + Δ	0
6	10	+280	+150	+80	+56	+40	+25	+18	+13	+8	+5	0		+5	+8	+12	−1 + Δ		−6 + Δ	−6	−10 + Δ	0
10	14	+290	+150	+95	+70	+50	+32	+23	+16	+10	+6	0		+6	+10	+15	−1 + Δ		−7 + Δ	−7	−12 + Δ	0
14	18																					
18	24	+300	+160	+110	+85	+65	+40	+28	+20	+12	+7	0		+8	+12	+20	−2 + Δ		−8 + Δ	−8	−15 + Δ	0
24	30																					
30	40	+310	+170	+120	+100	+80	+50	+35	+25	+15	+9	0		+10	+14	+24	−2 + Δ		−9 + Δ	−9	−17 + Δ	0
40	50	+320	+180	+130																		
50	65	+340	+190	+140		+100	+60		+30		+10	0		+13	+18	+28	−2 + Δ		−11 + Δ	−11	−20 + Δ	0
65	80	+360	+200	+150																		
80	100	+380	+220	+170		+120	+72		+36		+12	0		+16	+22	+34	−3 + Δ		−13 + Δ	−13	−23 + Δ	0
100	120	+410	+240	+180																		
120	140	+460	+260	+200		+145	+85		+43		+14	0		+18	+26	+41	−3 + Δ		−15 + Δ	−15	−27 + Δ	0
140	160	+520	+280	+210																		
160	180	+580	+310	+230																		
180	200	+660	+340	+240		+170	+100		+50		+15	0		+22	+30	+47	−4 + Δ		−17 + Δ	−17	−31 + Δ	0
200	225	+740	+380	+260																		
225	250	+820	+420	+280																		
250	280	+920	+480	+300		+190	+110		+56		+17	0		+25	+36	+55	−4 + Δ		−20 + Δ	−20	−34 + Δ	0
280	315	+1050	+540	+330																		
315	355	+1200	+600	+360		+210	+125		+62		+18	0		+29	+39	+60	−4 + Δ		−21 + Δ	−21	−37 + Δ	0
355	400	+1350	+680	+400																		
400	450	+1500	+760	+440		+230	+135		+68		+20	0		+33	+43	+66	−5 + Δ		−23 + Δ	−23	−40 + Δ	0
450	500	+1650	+840	+480																		

Notes: ① Fundamental deviations A and B for all grades and N for grades above IT8 shall not be used for nominal sizes less than or

② The Δ values for fundamental deviations with designations K, M, N（for standard tolerance grades≤IT8）and P to ZC（for

−22 + 8 = −14 μm.

③ Special case: for tolerance grade M6 in the range from 250 mm to 315 mm, ES = −9 μm instead of −11μm.

④ For tolerance grades JS7 to JS11, if the IT value number, n, is an odd number, then the value =±(IT_n−1)2.

注：① 所有等级的基本偏差A和B，以及IT8以上等级的基本偏差N，不得用于小于或等于1mm的公称尺寸。

② 标准公差等级不大于IT8的代号K，M，N和标准公差等级不大于IT7的代号为P~ZC的基本偏差中的Δ值从表内右

③ 特殊情况：当公称尺寸大于250~315mm时，M6的ES = 9 μm 代替 −11 μm

④ 对于公差等级JS7至JS11，如果IT值n为奇数，则该值 = ±(IT_n−1)2.

deviations for holes
的基本偏差

The values of fundamental deviations 基本偏差数值													Δ Value Δ值/μm Standard tolerance grades 标准公差等级					
Upper limit deviation 上极限偏差(ES)																		
≤IT7	>IT7 的标准公差等级																	
P~ZC①	P	R	S	T	U	V	X	Y	Z	ZA	ZB	ZC	IT3	IT4	IT5	IT6	IT7	IT8
Values as for standard greades above IT7 incrcaseΔ 在大于IT7的标准公差等级的基本偏差数值上增加一个Δ值	−6	−10	−14		−18		−20		−26	−32	−40	−60	0	0	0	0	0	0
	−12	−15	−19		−23		−28		−35	−42	−50	−80	1	1.5	1	3	4	6
	−15	−19	−23		−28		−34		−42	−52	−67	−97	1	1.5	2	3	6	7
	−18	−23	−28		−33		−40		−50	−64	−90	−130	1	2	3	3	7	9
						−39	−45		−60	−77	−108	−150						
	−22	−28	−35		−41	−47	−54	−63	−73	−98	−136	−188	1.5	2	3	4	8	12
				−41	−48	−55	−64	−75	−88	−118	−160	−218						
	−26	−34	−43	−48	−60	−68	−80	−94	−112	−148	−200	−274	1.5	3	4	5	9	14
				−54	−70	−81	−97	−114	−136	−180	−242	−325						
	−32	−41	−53	−66	−87	−102	−122	−144	−172	−226	−300	−405	2	3	5	6	11	16
		−43	−59	−75	−102	−120	−146	−174	−210	−274	−360	−480						
	−37	−51	−71	−91	−124	−146	−178	−214	−258	−335	−445	−585	2	4	5	7	13	19
		−54	−79	−104	−144	−172	−210	−254	−310	−400	−525	−690						
	−43	−63	−92	−122	−170	−202	−248	−300	−365	−470	−620	−800	3	4	6	7	15	23
		−65	−100	−134	−190	−228	−280	−340	−415	−535	−700	−900						
		−68	−108	−146	−210	−252	−310	−380	−465	−600	−780	−1000						
	−50	−77	−122	−166	−236	−284	−350	−425	−520	−670	−880	−1150	3	4	6	9	17	26
		−80	−130	−180	−258	−310	−385	−470	−575	−740	−960	−1250						
		−84	−140	−196	−284	−340	−425	−520	−640	−820	−1050	−1350						
	−56	−94	−158	−218	−315	−385	−475	−580	−710	−920	−1200	−1550	4	4	7	9	20	29
		−98	−170	−240	−350	−425	−525	−650	−790	−1000	−1300	−1700						
	−62	−108	−190	−268	−390	−475	−590	−730	−900	−1150	−1500	−1900	4	5	7	11	21	32
		−114	−208	−294	−435	−530	−660	−820	−1000	−1300	−1650	−2100						
	−68	−126	−232	−330	−490	−595	−740	−920	−1100	−1450	−1850	−2400	5	5	7	13	23	34
		−132	−252	−360	−540	−660	−820	−1000	−1250	−1600	−2100	−2600						

equal to 1 mm.

standard tolerance grades≤IT7) shall be selected from the right-hand side of the table. Such as, ϕ30P7, ES = −22 + Δ =

侧选取。

ϕ25H7: the fundamental deviation of ϕ 25H7 is lower deviation; EI = 0,

then ES = EI + IT7 = 0 + 21 μm = +21 μm.

(2) ϕ25R7/h6.

① Look up the table to determine the standard tolerances values for hole and the shaft.

Look up Tab. 2-3, IT6 = 13 μm, IT7 = 21 μm.

② Determinate limit deviations for hole or shaft.

ϕ25h6: from Tab. 2-4, the fundamental deviation of ϕ25h6 is upper deviation: es = 0; then ei = es − IT6 = 0 − (+13 μm) = −13 μm.

ϕ25R7: the fundamental deviation of ϕ25R7 is upper deviation ES.

According to deviations P to ZC in standard tolerance grades up to and including IT7.

From Table 2-5, the fundamental deviation of ϕ25R7 is:

ES = −28μm + Δ = −28 μm + 8 μm = −20 μm,

then EI = ES − IT7 = −20 μm − 21 μm = −41 μm.

(3) Limit clearances or interferences.

For ϕ25H7/r6:

Y_{max} = ES − ei = + 21 μm − (+28μm) = −7μm,

Y_{min} = EI − es = 0 − (+41μm) = −41μm,

For ϕ25R7/h6: Y_{max} = ES − ei = −20 μm − (−13 μm) = −7 μm,

Y_{min} = EI − es = 0 − (+41 μm) = −41 μm.

As demonstrated by the above mentioned calculations and illustrated in Fig. 2-11, both the ϕ25H7/r6 and ϕ25R7/h6 fits exhibit identical maximum and minimum clearances. This equivalence indicates that the ϕ25H7/r6 and ϕ25R7/h6 fits share equivalent fitting characteristics.

解：（1）ϕ25H7/r6。

① 查表确定孔和轴的标准公差。

查表2-3得：ϕ25 μm的IT6 = 13 μm，IT7 = 21 μm。

②确定孔和轴的极限偏差。

ϕ25r6：查表2-4，基本偏差为下极限偏差ei = +28 μm，

则es = ei + IT6 = +28 μm + 13 μm = +41 μm。

ϕ25H7：基本偏差为下偏差EI = 0，则ES = EI + IT7 = 0 + 0.021 μm = +21 μm。

（2）ϕ25R7/h6。

① 查表确定孔和轴的标准公差。

查表2-3得：ϕ25μm的IT6 = 13 μm，IT7 = 21 μm。

②确定孔和轴的极限偏差。

ϕ25h6：基本偏差为上偏差es = 0；则ei = es − IT6 = 0 − (+13 μm) = −13μm。

ϕ25R7：查表2-5，基本偏差为上极限偏差ES，准公差等级不大于IT7的代号为P ~ ZC的孔的基本偏差计算。

ES = −28 μm + Δ = −28 μm + 8 μm = −20 μm，

则EI = ES − IT7 = −20 μm − 21 μm = −41 μm。

（3）极限间隙或过盈计算。

对于ϕ25H7/r6：Y_{max} = ES − ei = +21 μm − (+28 μm) = −7 μm，

$Y_{min}=EI-es=0-(+41\ \mu m)=-41\ \mu m$,

对于ϕ25R7/h6：$Y_{max}=ES-ei=-20\ \mu m-(-13\ \mu m)=-7\ \mu m$,

$Y_{min}=EI-es=0-(+41\ \mu m)=-41\ \mu m$.

由上述计算和从图2-11中可见，ϕ25H7/r6和ϕ25R7/h6两对配合的最大间隙和最小间隙均相等，即ϕ25H7/r6和ϕ25R7/h6配合性质相同。

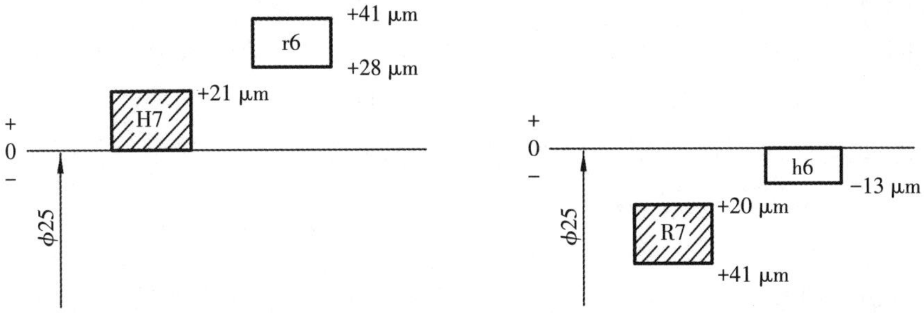

Fig. 2-11 Illustration of size tolerance zone for example 2-6

图2-11　例2-6的公差带示意图

例2-7：利用标准公差数值表和图的基本偏差数值表，确定ϕ30F8/h7和ϕ30H8/f7配合中孔和轴的极限偏差，并比较它们的配合性质是否相同。

Solution：(1) ϕ30F8/h7。

① Look up the table to determine the standard tolerances values for hole and the shaft.

from Table 2-3，IT7 = 21μm = 0.021mm，IT8 = 33 μm = 0.033mm.

② Determinate limit deviations for hole or shaft:

ϕ30h7： the fundamental deviation of ϕ30h7 is upper deviation es = 0; then

$ei=es-IT7=0-(+0.021)=-0.021$mm.

ϕ30F8：from Table 2-5，the fundamental deviation of ϕ30F8 is lower deviation:

$EI=+20\mu m=+0.020$mm

Then $ES=EI+IT8=+0.020+0.033=+0.053$ mm,

Then，$\phi 30F8=\phi 30^{+0.053}_{-0.020}$，$\phi 30h7=\phi 30^{\ 0}_{-0.021}$.

(2) ϕ30H8/f7。

Determinate limit deviations for hole or shaft:

ϕ30H8：the fundamental deviation of ϕ30H8 is lower deviation EI = 0; then $ES=EI+IT8=+0.033$ mm.

ϕ 30f7：from Tab. 2-4 and Fig. 2-10，The fundamental deviation of ϕ 30f7 is upper deviation es = −0.020 mm:

es = −0.020mm，another limit deviation is $ei=es-IT7=-0.020-0.021=-0.041$ mm.

Then，$\phi 30H8=\phi 30^{+0.033}_{\ 0}$，$\phi 30f7=\phi 30^{-0.020}_{-0.041}$

(3) limit clearance (s) and (or) interference (s)。

For ϕ30F8/h7：$X_{max}=ES-ei=(+53)-(-21)=+0.074$ mm

$X_{min}=EI-es=(+20)-0=+0.020$ mm

For ϕ30H8/f7：$X_{max}=ES-ei=(+33)-(-41)=+0.074$ mm,

$X_{min}=EI-es=0-(-20)=+0.020$ mm

As demonstrated by the above mentioned calculations and illustrated in Fig.2-12，both the ϕ30F8/h7 and ϕ30H8/f7 fits exhibit identical maximum and minimum clearances. This equivalence indicates that the ϕ30F8/h7 and ϕ30H8/f7 fits share equivalent fitting characteristics.

解：（1）ϕ30F8/h7。

① 查表确定孔和轴的标准公差。

由表2-3查得：公称尺寸为ϕ30 mm的标准公差数值：

IT7 = 21μm = 0.021mm，IT8 = 33 μm = 0.033 mm。

② 确定孔和轴的极限偏差。

ϕ30h7：基本偏差为上偏差es = 0；则 ei = es − IT7 = 0 − (+21) = −21 μm = −0.021 mm。

ϕ30F8：查表2-5，基本偏差为下极限偏差：EI = +20 μm = +0.020 mm

另一极限偏差为ES = EI + IT8 = +20 + 33 = +53 μm = +0.053 mm，

$\phi 30F8=\phi 30^{+0.053}_{-0.020}$，$\phi 30h7=\phi 30^{\ 0}_{-0.021}$。

（2）ϕ30H8/f7。

确定孔和轴的极限偏差。

ϕ30H8：基准孔的基本偏差为下偏差EI = 0，另一极限偏差为

ES = EI + IT8 = +33 μm = +0.033 mm。

ϕ30f7：由表2-4查得ϕ30f7轴的基本偏差为上偏差es = −20 μm，另一极限偏差为

ei = es − IT7 = −20−21 = −41 μm = −0.041 mm。

所以，$\phi 30H8=\phi 30^{+0.033}_{\ 0}$，$\phi 30f7=\phi 30^{-0.020}_{-0.041}$

（3）极限间隙计算：

对于ϕ30F8/h7：X_{max} = ES − ei = (+53) − (−21) = +74 μm = +0.074 mm

X_{min} = EI − es = (+20) − 0 = +20 μm = +0.020 mm

对于ϕ30H8/f7：X_{max} = ES − ei = (+33) − (−41) = +74 μm = +0.074 mm

X_{min} = EI − es = 0 − (−20) = +20 μm = +0.020 mm

由上述计算和从图2-12中可见，ϕ30F8/h7和ϕ30H8/f7两对配合的最大间隙和最小间隙均相等，即ϕ30F8/h7和ϕ30H8/f7配合性质相同。

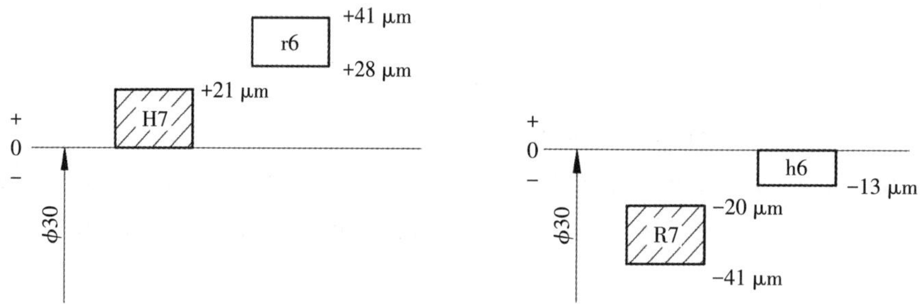

Fig. 2-12 Illustration of size talerance zone for example 2-7

图2-12 例2-7的公差带示意图

2.3.2.3 Marking method for hole and shaft or fits 孔和轴及其配合的标注方法

Marking of limits and fits（GB/T 4458.5-2020）has been learned in *Engineering Drawing*.

极限与配合（GB/T 4458.5—2020）的标注方法在《工程制图》中已经学过。

2.4 The General, Common and Prefer Tolerance Zones 一般、常用和优先的孔、轴公差带

In the standards of limits and fits, for the nominal size up to 500 mm (including), the fundamental deviation of J is used with only three tolerance grades (J6, J7, J8) and the fundamental deviation of j is used with only four tolerance grades (j5, j6, j7, j8).

在极限和配合标准中，对于公称尺寸不大于500 mm的范围，孔的基本偏差J仅与三个公差等级（J6，J7，J8）配合使用，而轴的基本偏差j则与四个公差等级（j5，j6，j7，j8）配合使用。

In the standards of limits and fits, the national standard specifies 28 fundamental deviations for both holes and shafts. For nominal sizes up to and including 500 mm, there are 20 standard tolerance grades; for nominal sizes greater than 500 mm and up to and including 3150 mm, the national standard specifies 18 standard tolerance grades. They can form numerous deviation zones (543 for holes and 544 for shafts). The tolerance zones of holes and shafts can further combine to create a vast number of fits. However, in production practice, the extensive use of a large number of tolerance zones would inevitably complicate standards and hinder production. To minimize the variety and specifications of cutting tools, measuring tools, and process equipment while meeting the actual needs of our country and considering future production development, the national standard imposes necessary restrictions on the selected tolerance zones and fits. The national standard has formulated three standards for selection (GB/T 1801-1999, GB/T 1804-2000 and GB/T 1803-2003). In these standards, tolerance zones for holes and shafts are recommended respectively. For sizes less than 500mm, the national standard specifies a total of 116 tolerance zones for shafts, including 59 commonly used tolerance zones enclosed in boxes and 13 preferred tolerance zones marked with circles, as shown in Fig. 2-13 and Fig. 2-14. The national standard also specifies 105 tolerance zones for holes, including 44 commonly used tolerance zones enclosed in boxes and 13 preferred tolerance zones marked with circles.

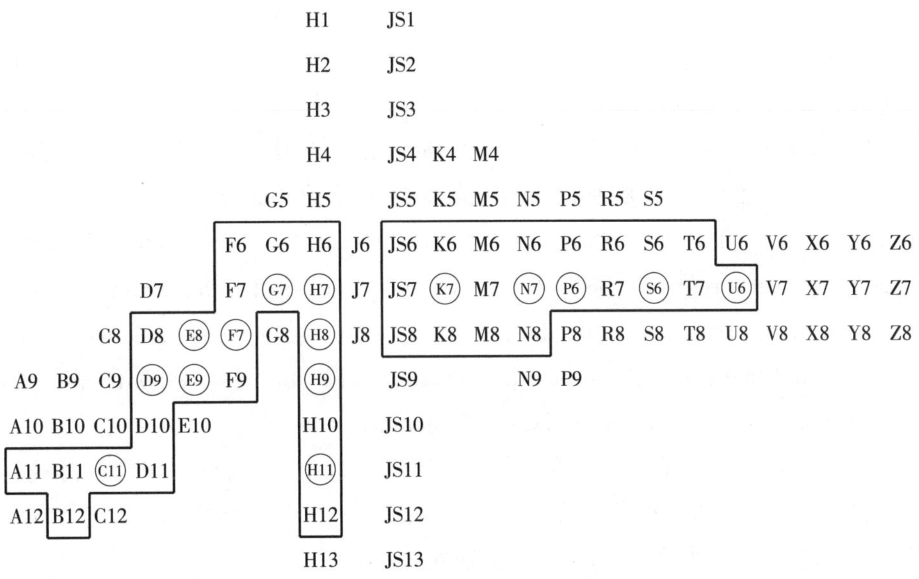

Fig. 2-13 Geneanal, common and prefer tolerance zones for hole (≤500 mm)

图2-13 一般、常用和优先的孔公差带（≤500 mm）

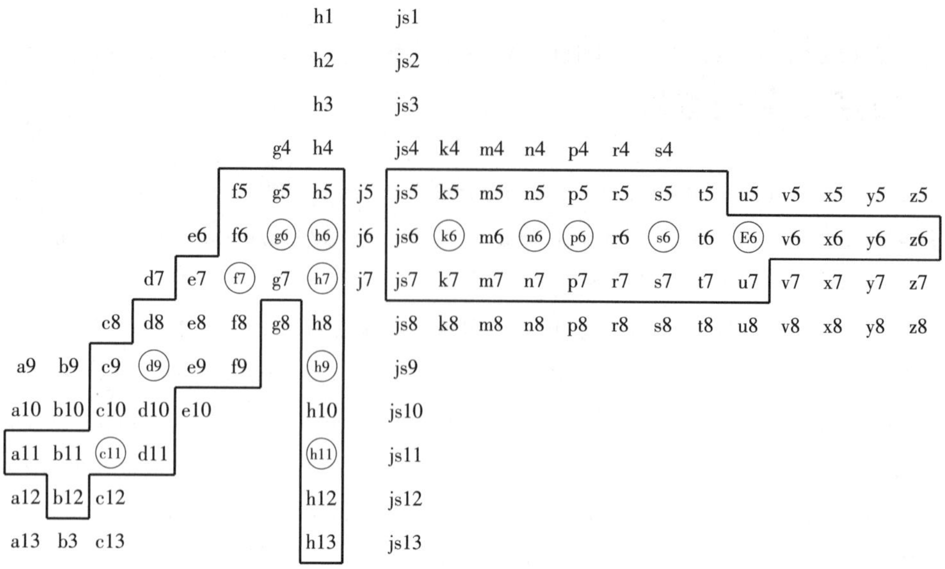

Fig. 2-14 Geneanal, common and prefer tolerance zones for shaft（≤500 mm）

图2-14 一般、常用和优先的轴公差带（≤500 mm）

在极限和配合标准中，国家标准规定孔、轴各28种基本偏差，对于公称尺寸不大于500 mm，有20个标准公差等级；对于公称尺寸大于500 mm且不大于3150 mm国家标准规定有18个标准公差等级。它们可以组成很多种差带（孔有543种，轴有544种）。由孔、轴公差带又能组成大量的配合。但是，在生产实践中，公差带的数量使用很多，势必使标准繁杂，不利于生产。国家标准在满足我国实际需要和考虑生产发展需要的前提下，为了尽可能减少刀具、量具和工艺装备的品种和规格，对所选用的公差带与配合做了必要的限制，国家标准制订了（GB/T 1801—1999，GB/T 1804—2000及GB/T 1803—2003）三个以供选用的标准，在这些标准中，分别推荐了孔、轴公差带。对于尺寸不大于500 mm，国家标准规定了一般、常用和优先的轴公差带共116种，其中方框内的59种为常用公差带，带圆圈的13种为优先的公差带，如图2-13和图2-14所示。国家标准还规定了一般、常用和优先的孔公差带105种，其中方框内的44种为常用公差带，带圆圈的13种为优先公差带。

Tab. 2-6 and Tab. 2-7 show the common and preferred fits for hole-basis fits and shaft-basis fits respectively whose basic size is up to 500 mm（ including 500 mm). The fits marked with circles are preferred fits，13 fits for hole-basis fits and 13 fits for shaft-basis fits.

表2-6和表2-7分别列出了公称尺寸小于等于500mm的基孔制配合和基轴制配合的常用和优先配合。标有圆圈的配合为优先配合，分别为13种基孔制配合和13种基轴制配合。

It must be noted that in Tab. 2-6，when the standard tolerance of the shaft is less than or equal to IT7，it is compatible with the reference hole of the lower level; When it is higher than or equal to IT8，it matches the reference hole of the same grade. In Tab. 2-7，when the standard tolerance of the hole is less than IT8 or a few standard loterances of the hole are equal to IT8，it is matched with the reference shaft of the higher level，and the rest is matched with the same tolerance grade.

必须注意到，在表2-6中，当轴的标准公差不大于IT7级时，它与低一级的基准孔相配；当轴的标准公差不小于IT8级时，它与同级基准孔相配。表2-7中，当孔的标准公差小于IT8级或少数等于IT8级时，它与高一级的基准轴相配，其余则孔、轴同级相配。

Tab. 2-6 Common and preferred fits for hole-basis system

表2-6 基孔制常用的优先配合

Reference hole 基准孔	Shaft轴																				
	a	b	c	d	e	f	g	h	js	k	m	n	p	r	s	t	u	v	x	y	z
	Clearance fit间隙配合								Transition fit 过渡配合			Interference fit过盈配合									
H6						H6/f5	H6/g5	H6/h5	H6/js5	H6/k5	H6/m5	H6/n5	H6/p5	H6/r5	H6/s5	H6/t5					
H7						H7/f6	(H7/g6)	(H7/h6)	H7/js6	(H7/k6)	H7/m6	(H7/n6)	(H7/p6)	H7/r6	(H7/s6)	H7/t6	(H7/u6)	H7/v6	H7/x6	H7/y6	H7/z6
H8					H8/e7	(H8/f7)	H8/g7	(H8/h7)	H8/js7	H8/k7	H8/m7	H8/n7	H8/p7	H8/r7	H8/s7	H8/t7	H8/u7				
				H8/d8	H8/e8	H8/f8		H8/h8													
H9			H9/c9	(H9/d9)	H9/e9	H9/f9		(H9/h9)													
H10			H10/c10	H10/d10				H10/h10													
H11	H11/a11	H11/b11	(H11/c11)	H11/d11				(H11/h11)													
H12		H12/b12						H12/h12													

Notes: ① The fits in '( )' are preferred, totally 13 preferred fits.

②H6/n5 and H7/p6 which nominal size up to and including 3 mm and H8/r7 which nominal size up to and including 100 mm are transition fits.

注：① 带括号的配合为优先配合，共有13种优先配合。

② 公称尺寸不大于3 mm的H6/n5，H7/p6和公称尺寸不大于100 mm的H8/r7为过渡配合。

Tab. 2-7 Common and preferred fits for shaft-basis system

表2-7 基轴制常用的优先配合

Reference shaft 基准轴	Hole孔																				
	A	B	C	D	E	F	G	H	JS	K	M	N	P	R	S	T	U	V	X	Y	Z
	Clearance fit间隙配合								Transition fit 过渡配合			Interference fit过盈配合									
h5						F6/h5	G6/h5	H6/h5	JS6/h5	K6/h5	M6/h5	N6/h5	P6/h5	R6/h5	S6/h5	T6/h5					
h6						F7/h6	(G7/h6)	(H7/h6)	JS7/h6	(K7/h6)	M7/h6	(N7/h6)	(P7/h6)	R7/h6	(S7/h6)	T7/h6	(U7/h6)				
h7					E8/h7	(F8/h7)		(H8/h7)	JS8/h7	K8/h7	M8/h7	N8/h7									
h8				D8/h8	E8/h8	F8/h8		H8/h8													
h9				(D9/h9)	E9/h9	F9/h9		(H9/h9)													
h10				D10/h10				H10/h10													
h11	A11/h11	B11/h11	(C11/h11)	D11/h11				(H11/h11)													
h12		B12/h12						H12/h12													

Note: the fit in '( )' are preferred totally 13 preferred fits.

注：带括号的配合为优先配合，共有13种优先配合。

The fit of hole-basis system or shaft-basis system can also get a variety of different fit types.

基孔制或基轴制的配合也可以得到多种不同的配合类型。

For example:

例如：

(1) When A ~ H or a ~ h matches with the datum part (shaft or hole), then the fit is a formed clearance fit;

当A ~ H或a ~ h与基准件（轴或孔）配合时，所形成的配合为间隙配合；

(2) J ~ N or j ~ n matches with the datum part (shaft or hole), then the fit is basically a formed transition fit;

J ~ N或j ~ n与基准件配合时，所形成的配合基本上为过渡配合；

(3) P ~ ZC or p ~ zc matches with the datum part (shaft or hole), then the fit is basically a formed interference fit.

P ~ ZC或p ~ zc与基准件配合时，所形成的配合基本上为过盈配合。

Based on the above rules, the type of fit can be directly determined. For example, we can directly judge the fit ϕ40H8/f7 is a clearance fit, ϕ80K7/h6 is a transition fit.

根据上述规律可以直接判断配合类型。例如，ϕ40H8/f7是间隙配合；ϕ80K7/h6是过渡配合。

2.5 Selection of Tolerances and Fits 公差与配合的选用

It is a very important work in mechanical designing to select limits and fits. Selection of basic system of fit, standard tolerance grades and fits is reasonable or not, not only influence the product performance and quality, but also affects the economic benefits. Therefore, the principle of selection is to obtain the best technical and economic benefits on the premise of satisfying the use requirements.

尺寸公差与配合的选择是机械设计与制造中的一个重要环节。基准制、公差等级和配合种类是否合理，不但会影响零件的性能和质量，还会影响经济效益。因此，选择的原则是在满足使用要求的前提下能够获得最佳的技术经济效益。

2.5.1 Selection of Basic System of Fit 基准制的选用

When selecting the basic fit system, factors such as performance, mechanical structure, processing and assembly processes, and economic considerations should be fully taken into account.

选择基本配合制度时，应充分考虑性能、机械结构、加工装配工艺及经济性等因素。

2.5.1.1 The hole-basis fit system is preferred 优先选用基孔制配合

When machining small-sized and high-precision holes, adopting the hole-basis fit system can reduce the number and specifications of fixed cutting tools and measuring instruments.

加工小尺寸高精度的孔，采用基孔制配合可以减少定值刀具和测量工具的数目和规格。

2.5.1.2 In the following cases, the shaft-basis fit system should be selected 下列情况应选用基轴制

(1) The shafts are manufactured with cold-drawn steel with a certain tolerance level (IT9 ~ IT11) and are not subjected to further machining (This type of steel is produced with the base shaft system). To achieve various fits, different hole tolerance band positions can be selected. This situation is mainly applied in agricultural machinery and textile machinery.

轴采用具有一定公差等级（IT9～IT11）而不再进行机械加工的冷拔钢材（这种钢材按基准轴的公差带制造）来制作。若需要各种不同的配合时，可选择不同的孔公差带位置来实现。这种情况主要应用在农业机械和纺织机械中。

（2）Precision shafts with machining dimensions smaller than 1 mm are more challenging than holes of the same grade. Therefore，in the manufacturing of instruments，watch production，and radio engineering，steel wire that has been formed through light rolling is often directly used as shafts. In these cases，it is more economical that shaft-basis system fit.

加工尺寸小于1 mm的精度要求较高的轴比同级加工孔要困难，因此在仪器制造、钟表生产、无线电工程中，常使用经过轻微滚压的钢丝直接做轴。这时，采用基轴制较经济。

（3）According to structural requirements，when assembling several hole components with different fit requirements on shafts of the same nominal size，the shaft-basis fit system should be used. For example，the fit between the piston pin shaft of an engine and the connecting rod bushing hole and piston hole is as shown in Fig. 2-15（a). Based on operational needs and ease of assembly，a transition fit is used between the piston pin shaft and the piston hole，while a clearance fit is used with the connecting rod bushing hole. If the hole-basis system is used for the fit，as shown in Fig. 2-15（b），the shaft will be made in a stepped shape. In contrast，using a shaft-basis system，as shown in Fig. 2-15（c），the shaft can be made as a straight shaft. This choice is advantageous for both shaft processing and assembly convenience.

根据结构需要，在同一公称尺寸的轴上装配有不同配合要求的几个孔件时应采用基轴制配合。例如，发动机的活塞销轴与连杆衬套孔和活塞孔之间的配合，如图2-15（a）所示。根据工作需要及装配性，活塞销轴与活塞孔采用过渡配合，而与连杆衬套孔采用间隙配合。若采用基孔制配合，轴将做成阶梯状，如图2-15（b）所示。相比之下，采用基轴制配合，销轴可做成光轴，如图2-15（c）所示。这种选择既有利于轴的加工，又便于装配。

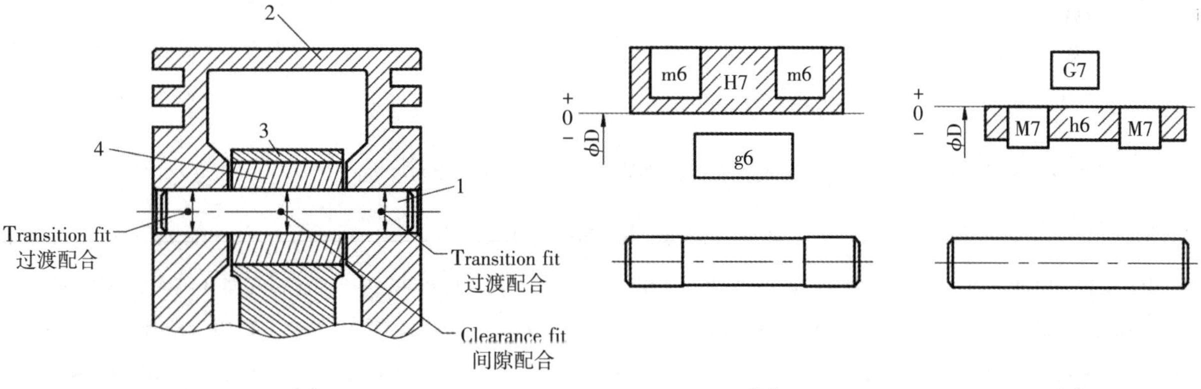

(a) (b) (c)

1—piston pin 活塞销；2—piston 活塞；3，4—connecting rod连杆

Fig. 2-15 Example for selection of basis fit system

图2-15 基准制选择示例

（4）Selection of the reference system for fitting with standard workpieces. When fitting with standard workpieces（parts or assemblies），the decision to use the hole-basis system or a shaft-basis system should be based on the standard workpieces. For example，in the fit between the shaft and the bearing，as shown in Fig. 2-16，the tolerance zone for the housing hole（ϕ30mm）is chosen as K7，and the tolerance zone for the shaft journal（ϕ10mm）is chosen as r6.

与标准件配合的基准制选择。若与标准件（零件或部件）配合，应以标准件为基准件来确定采用基孔制还是基轴制。如图2-16所示轴与轴承的配合，选择箱体孔（ϕ30mm）的公差带为K7，选择轴颈（ϕ10mm）的公差带为r6。

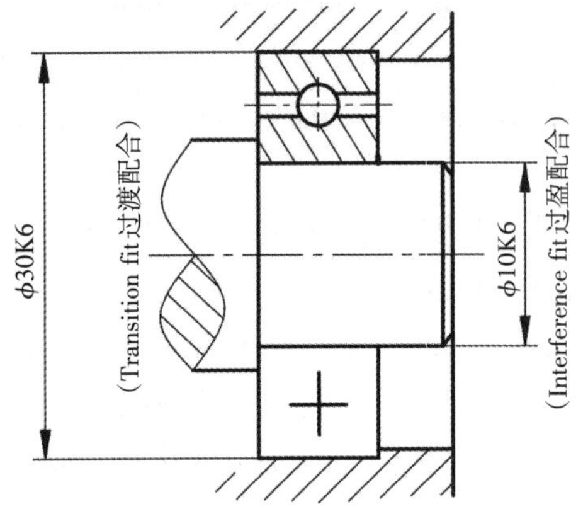

Fig. 2-16 Selecting the fit system based on the datum part

图2-16 根据基准件选择配合制

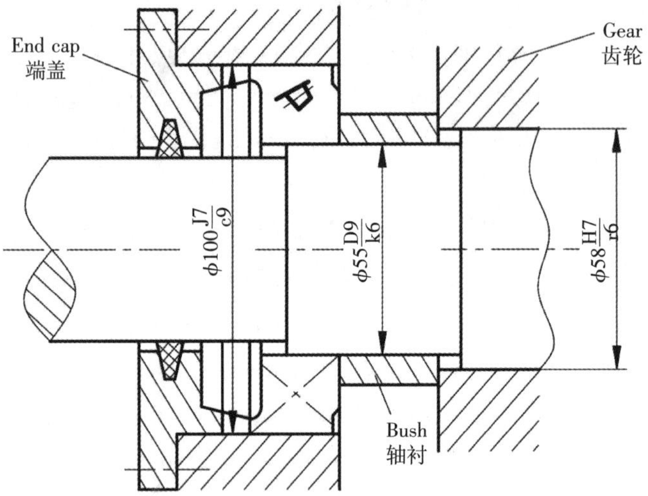

Fig. 2-17 Selecting non-basis system fit

图2-17 选择非基准制配合

2.5.1.3 Selection of non-basis fit system 非基准制配合的选用

In order to meet the requirements of special fits, the use of non-basis system fits is permissible. When a hole is to be fitted with several shafts or a shaft is to be fitted with several holes, additionally, the fit requirements for each are different, then some of the fits will involve non-basis system fits. Take, for example, the fit between holes and shafts in a typical housing , as shown in Fig. 2-17.

为满足特殊配合的要求，允许选用非基准制配合。当一个孔与几个轴相配合或一个轴与几个孔相配合，此外，其配合要求各不相同时，则有的配合要出现非基准制的配合，以典型箱体中孔轴配合为例，如图2-17所示。

2.5.2 Selection of Size Tolerance Grades公差等级的选用

To select size tolerance grades, it is to handle the contradiction between the functional requirements and the machining cost, complexity of manufacturing process. Because the dimensional accuracy is related to the difficulty of processing, the cost of processing and the quality of the work of the parts, the basic principle of choosing the tolerance grade is to choose the lower tolerance grade as far as possible under the premise of meeting the requirements of use. The selection methods of tolerance grades include analogy method and calculation method. In practical applications, the analogy method is often used. The following aspects should be considered when selecting size tolerance grades.

尺寸公差等级的选用就是解决机械零件功能要求、加工成本及制造工艺复杂性之间的矛盾。由于尺寸精度与加工的难易程度、加工的成本以及零件的工作质量有关，因此选择公差等级的基本原则是，在满足使用要求的前提下，尽量选取较低的公差等级。公差等级的选用方法有类比法和计算法。在实际应用中，常采用类比法。选择尺寸公差等级时，应考虑以下几方面。

2.5.2.1 Processing equivalence for a fit between shaft and hole 轴与孔配合的工艺等价原则

Processing equivalence refers to the manufacture difficulties should be the same for a shaft and a hole fit together, i.e. for the hole and the shaft which nominal size no more than 500 mm and the tolerance grades higher than IT8 (≥IT8), the shaft should be finer by one grade than the hole grade. But lower grades (< IT8) or nominal size greater than 500 mm, the same tolerance grade of the and hole is selected. For non-basis system of fit, if the accuracy requirement of one parts is not high, then the tolerance grade can lower 2 ~ 3 grade to reduce the machining cost.

工艺等价原则是指相互配合的孔和轴，它们的加工难易程度相同。例如，对于公称尺寸不超过500 mm且公差等级不小于IT8的孔与轴配合，国家标准推荐孔比轴低一级相配合。但对公差等级或公称尺寸大于500 mm 的配合，推荐采用孔、轴同级配合。对于非基准制配合，如果对某一零件的精度要求不高，则可将公差等级降低2 ~ 3个等级，以降低加工成本。

2.5.2.2 Application of tolerance grades 公差等级的应用

In mechanical design and manufacturing, the selection of tolerance grades requires consideration of the functional requiements of parts, manufacturing economy, and fit characteristics. The selection principles and typical applications of common tolerance grades (IT5 ~ IT12) are shown in Tab. 2-8.

在机械设计与制造中，公差等级的选用需结合零件的功能要求，加工经济性及配合性质，常用公差等级（IT5 ~ IT13）的选用原则及典型应用如表2-8所示。

Tab. 2-8 Application common dimensional tolerances grades

表2-8 常用尺寸公差等级的应用

Tolerance grades 公差等级	Application 应用
IT5	Used for very small size tolerance and geometrical tolerance, fit properties are stable. Usually IT5 used in important fits in machine tools, engine, instruments etc, Such as housing hole fit with rolling bearing which grade is 5, spindle, tailstock and sleeve in machine tool, precision machinery and high speed shaft, precision screw shaft 用于非常小的尺寸公差和几何公差，其配合性能稳定。通常用于机床、发动机、仪器等中的重要配合，例如与5级滚动轴承配合的箱体孔、机床中的主轴、尾座和套筒、精密机械和高速轴、精密丝杠等
IT6	Fit properties are unify. Used for holes or shafts fitted with class 6 rolling bearings; the axle fitted with gears, worms, couplings, belt pulleys, cams etc. The column of radial drilling machine; guide workpieces of machine jigs; basic hole of gears which grades are 6; basic shaft of gears which are grade 7 or 8 配合性质统一。用于与6级滚动轴承配合的孔或轴；与齿轮、蜗杆、联轴器、皮带轮、凸轮等配合的轴；钻床的立柱；机床夹具的导向件；6级齿轮的基本孔；7级或8级齿轮的基准轴
IT7	The accuracy of IT7 is lower than IT6, and it used general mechanical elements, such as holes of IT7 coupling, belt pulley, cams; holes of chuck seat in machine tools, fixed grill bush and changeable drill bush; basic holes of gears which grade are 7 or 8; basic shafts of gear which grade are 9 or 10 IT7的精度低于IT6，用于一般机械零件，如联轴器、皮带轮、凸轮的孔；机床卡盘座孔、固定钻套和可换钻套；7级或8级齿轮的基准孔；9级或10级齿轮的基准轴
IT8	It is a middle precision, used for the width size of bearing seats, basic holes of gears which grade are 9~12; basic shaft of gears which grade are 11~ 12 在机械制造中属于中等精度，用于轴承座的宽度尺寸、9~12级齿轮的基本孔；11~12级齿轮的基准轴
IT9 ~ IT10	Mainly used for external surface and hole of shaft bush; operation elements and shaft; belt pulley and shaft which axle is hollow; key and springs 主要用于轴套的外表面和孔；操作元件和轴：空心轴的皮带轮和轴；键和弹簧
IT11 ~ IT12	The accuracy is very low, and the clearance is very large after assembly, so it is suitable used for the condition where no fit requirements. Such as flange, slide block and slide gears; the in-process size during machining; fit workpieces by puching 精度很低，装配后间隙很大。适用于无配合要求的场合，如法兰、滑块和滑移齿轮；加工过程中的工序尺寸；冲压配合的工件

The tolerance grades that can be achieved by different machining methods are shown in Tab. 2-9. These methods are available for reference.

各种加工方法能够达到的公差等级见表2-9。这些方法可供选择时参考。

Tab. 2-9 The tolerance grades that can be achieved by different machining methods

表2-9 各种加工方法能够达到的公差等级

Machining methods 加工方法	公差等级（IT）																	
	01	0	1	2	3	4	5	6	7	8	9	10	11	12	13	14	15	16
Lapping 研磨	+	+	+	+	+	+	+											
Honing 珩磨						+	+	+	+									
Cylindrical grinding 圆磨							+	+	+	+								
Surface grinding 平磨							+	+	+	+								
Diamond turning 金刚石车							+	+	+									
Diamond boring 金刚石镗							+	+	+									
Broaching 拉销							+	+	+	+								
Reaming 铰孔								+	+	+	+	+						
Precision turning and precision boring 精车精镗									+	+	+							
Rough turning 粗车												+	+	+				
Rough boring 粗镗												+	+	+				
Milling 铣										+	+	+	+					
Planing and shaping 刨，插												+	+					
Drilling 钻削												+	+	+	+			
Stamping 冲压												+	+	+	+	+		
Rolling and extruding 滚压，挤压												+	+					
Forging permanent 锻造																	+	
Sand casting and gas cutting 砂型铸造，气割																		+
Mold casting 金属型铸造																+	+	

2.5.3　Methods of Selecting Symbol of Fundamental Deviation 基本偏差代号的选择方法

There are three methods to select the symbol of fundamental deviation: calculation, experiment and analogy.

基本偏差代号的选择方法有三种：计算法、试验法和类比法。

The calculation method is to calculate the required clearance or interference size according to the theoretical formulas to select the fit symbol. For example, according to the theory of liquid lubrication, the minimum clearance required to guarantee the state of liquid friction is calculated, Because there are many factors affecting the clearance and interference, theoretical calculation gets only approximate results, so the actual selection needs to be determined through experiments. In general, it is rare to use the calculation method.

计算法就是根据理论公式，计算出满足使用要求的间隙或过盈的大小，来选定配合的方法。例如，根据液体润滑理论，可计算保证液体摩擦状态所需要的最小间隙。由于影响间隙和过盈的因素很多，理论计算得到的只是近似结果，所以实际使用的时候还需要通过试验来确定。一般情况下，很少用计算法。

The experimental method is to determine the clearance or interference range satisfying the working performance of the product through experiments. This method is mainly used in the situation where these values have great influence on the product performance and people lack experience. The experimenal method is more reliable, but the cycle is long and the cost is high, so it is only used for very important occasions.

试验法是用试验的方法确定满足产品工作性能的间隙或过盈范围。该方法主要用于这些参数对产品性能影响大而人们又缺乏经验的场合。试验法比较可靠，但是周期长、成本高，所以只应用于非常重要的场合。

The analogy method involves using proven and similar machinery, mechanisms, and components as templates to select the fit type. It is the most commonly used method for determining the fit type in machinery and instruments.

类比法就是以经过生产验证的且类似的机械、机构和零部件为样板，来选用配合种类。类比法是确定机械和仪器配合种类最常用的方法。

2.5.4　Selection of Fits 配合的选择

The fit types should be to select after the basic system of fit and tolerance grades have been selected. Actually, selection of fits is to determine the fundamental deviation symbol of the non-basis parts, i.e., for the hole-basis system of fit, it is to determine the fundamental deviation symbol of the shaft, and for the shaft-basis system of fit, it is to determine the fundamental deviation symbol of the hole. The following aspects should be considered to select fits:

在确定了配合制和公差等级后，再进行配合种类的选择。实际上，配合的选择是确定非基件的基本偏差符号的选择，即对于基孔制配合，是确定轴的基本偏差符号；对于基轴制配合，是确定孔的基本偏差符号。选择配合时应考虑以下方面：

(1) When the basic system is established, the preferred fit should be selected as far as possible, followed by the common fit. If preferred fits and commonly used fits cannot meet the requirements, the necessary fitting

relationships should be configured according to application demands. For this purpose, standard-recommended general-purpose hole and shaft tolerance zones are selected. Tab. 2-10 lists the properties of fundamental deviations and applications of common fits in common size segments.

在确定了基准制的基础上，应尽可能地选用优先配合，其次是常用配合。如果优先配合和常用配合不能满足要求时，可按使用要求组成需要的配合，选标准推荐的一般用途的孔、轴公差带。表2-10列出了常见尺寸段中常用配合的各种基本偏差的特性和应用。

Tab. 2-10 Properties of fundamental deviations and applications

表2-10 各种基本偏差的特征和应用

Fits type 配合类型	Fundamental deviations 基本偏差	Properties and applications of fits 特征和应用
Clearance fit 间隙配合	a（A） b（B）	Can be particularly large gap, rarely used. Mainly used when working with high temperature, thermal deformation of large parts of the fit, such as combustion engine piston and pin is H9/a9 可得到特大的间隙，应用很少。主要用于高温工况下、热变形较大的零件配合，例如内燃机活塞与活塞销的配合采用 H9/a9
	c（C）	Can be a large gap. Generally used for slow and relaxed dynamic connections and poor working conditions (such as agricultural machinery), work stress and large deformation and the assembly process of bad parts fitting, also apply to high temperature working clearance. Such as an combustion engine exhaust valve and tube H8/c7 间隙可能较大。一般用于缓慢且宽松的动态连接和工作条件（如农业机械），工作应力大、受力变形大以及装配过程中配合不良的零件，也适用于高温工作间隙。例如，内燃机排气阀与导管，配合采用H8/c7
	d（D）	With tolerance grade IT7 ~ IT11, applicable to a loose clearance (such as a pulley, the pulley and the looper shaft), as well as the large size of the sliding bearing and the shaft neck tie (such as turbines, ball mill bearing) . Piston ring and piston ring groove fit can be H9/d9 一般用于公差等级为IT7 ~ IT11，适用于较松的间隙配合（如皮带轮、滑轮与导轮轴），以及大尺寸滑动轴承与轴颈的配合（如涡轮机、球磨机轴承）。活塞环与活塞环槽的配合可采用H9/d9
	e（E）	With tolerance grade IT6 ~ IT9, have obvious gap, for large span and multi pivot shaft neck and bearing with high speed, heavy load, and the large size shaft and bearings, such as the main bearing of large motor or combution engine is H8/e7 适用于公差等级为IT6 ~ IT9，具有明显的间隙，大跨度、多支点轴颈与高速、重载且尺寸较大的轴和轴承的配合，例如大型电机或内燃机的主轴承采用H8/e7配合
	f（F）	With tolerance grade IT6 ~ IT8, used for general rotation fit, a fit between shaft and sliding bearings which affected little by temperature, using ordinary lube, such as H7/f6 for rotor shaft and the sliding bearing in a gear box, a small motor, pump 用于公差等级为IT6 ~ IT8，一般转动配合，适用于受温度影响较小的轴与用普通润滑剂润滑的滑动轴承之间的配合，例如齿轮箱、小型电机、泵中的转子轴与滑动轴承采用H7/f6配合
	g（G）	With tolerance grade IT5 ~ IT7, smaller clearance, used for rotation fit with light load, precision device, and slide valve, a link pin, such as G6 used for a drill guide bush 用于公差等级为IT5 ~ IT7，间隙较小、轻载、精密装置以及滑阀、连杆销等的转动配合，例如钻套所用的G6。
	h(H)	With tolerance grade IT4 ~ IT11. Widely used for a fit without relative rotation or general location fit without temperature and deformation affected; may also use in the accuracy sliding bearing, such as H6/h5 for lathe tailstock guide hole and sliding sleeve 广泛用于公差等级为IT4 ~ IT11，无相对转动或一般定位配合，且不受温度和变形影响的情况；也可用于精密滑动轴承，例如车床尾座导孔与滑动套采用配合为H6/h5

Tab. 2-10（Continued）表2-10（续）

Fits type 配合类型	Fundamental deviations 基本偏差	Properties and applications of fits 特征和应用
Transition fit 过渡配合	js (JS)	With tolerance grade IT4 ~ IT7. A transition fit with an average gap; slight interference positioning fit, such as JS7 for couplings, gear and wheel hub, rolling bearing outer ring and the housing. Usually assembly by hand or mallet 用于公差等级为IT4 ~ IT7。这是一种带有平均间隙的过渡配合；也用于有轻微过盈的定位配合，例如联轴器、齿轮与轮毂、滚动轴承外圈与外壳采用JS7配合。通常通过手工或木槌进行装配
	k(K)	With tolerance grade IT4 ~ IT7. A transition fit with average gap close to 0. Used for concentrical fit, such as rolling bearing inner and outer rings respectively in shaft neck, housing. Assembly with a mallet 适用于公差等级为IT4 ~ IT7,平均间隙接近0的过渡配合。用于同心配合,例如滚动轴承的内圈和外圈分别与轴颈、外壳的配合。装配时使用木槌
	m(M)	With tolerance grade IT4 ~ IT7. Transition fit with small average interference, accuracy consentrical fit, such as H7/m6 used worm wheel rim and bronze wheel 适用于公差等级为IT4 ~ IT7.,平均过盈量较小的过渡配合,属于精密同心配合,例如蜗轮轮缘与青铜轮毂采用H7/m6配合
	n(N)	With tolerance grade IT4 ~ IT7. Transition fit with large average interference, rarely gap. To transmit greater torque with keys, such as a gear in punch with shaft. Assembly with the fading or press machining 适用于公差等级为IT4 ~ IT7,平均过盈量较大的过渡配合,极少出现间隙。用于通过键传递较大扭矩的情况,例如冲床上的齿轮与轴的配合。装配时采用热胀或压装工艺
Interference fit 过盈配合	p(P)	Interference fit with a H6 or H7 hole, but a transition fit with H8 hole. Standard press-fit with parts made of carbon steel and cast iron, such as H7/P6 for rope pulley and ring gear in winch can used for small interference parts made of alloy steel 与H6或H7孔的配合时是过盈配合，而与H8孔配合时为过渡配合。对于碳素钢和铸铁零件使用标准压入配合。例如绞车上绳轮和齿圈采用的H7/P6配合，也可用于合金钢制成的小过盈量零件
	r(R)	For large torque transfer or subjected to shock loading and the need key connection, such as H7/r6 used for worm hole and shaft. Must pay attention to，H8/r7 is transition fit when basic size less than 100mm 适用于需要传递大扭矩或承受冲击载荷且需要键连接的情况，例如蜗杆孔与轴采用的H7/r6配合。必须注意的是，当基本尺寸小于100mm时，H8/r7为过渡配合
	s(S)	Used for steel and cast iron parts of the permanent and semipermanent binding, can produce considerable binding force, such as H7/s6 used for a ring bush pressed into the a valve seat, or onto a shaft 用于钢和铸铁零件的永久或半永久结合，可产生相当大的结合力，例如H7/s6配合用于将环形衬套压入阀座或压到轴上
	t(T)	Used for steel and cast iron parts combination of permanent, can transfer torque without keys, with thermal sleeve and cooling axial assembly, such as H7/t6 used for a coupling with a shaft 用于钢和铸铁零件的永久性组合，可在无键的情况下传递扭矩，采用热套和冷却轴向装配方法，例如联轴器与轴采用的H7/t6配合
	u(U)	For large interference fit, maximum interference should be checked, assembly with thermal expansion, such as H6/u5 used for a train wheel hub and a shaft 对于大过盈配合，应验算最大过盈量，并采用热膨胀法进行装配，例如火车轮毂与轴采用的H6/u5配合
	v(V), x(X) y (Y), z(Z)	For very large interference fit, so for rarely example used, with little experience and information, subject to trial before the application. Generally not recommended 对于极大过盈配合的情况，由于应用实例极少，相关经验和信息有限，因此在使用前需进行试验。通常不建议采用

(2) The national standard indicated the common and preferred fits for shaft-basis fit system and hole-basis fit system, which are shown in Tab. 2-7 and Tab. 2-8 respectively. First the preferred fits should be chosen. If the preferred fits cannot meet the function, then the common fits are selected. And if the preferred and common fits cannot meet the function, then the other fits can be selected. Usually, common fits application refers to Tab. 2-11.

国家标准规定了基孔制配合和基轴制配合的常用配合和优先配合，分别见表2-7和表2-8。首先，应该选择优先配合。如果优先配合不能满足功能要求，则选择常用配合。如果优先配合和常用配合都不能满足功能要求，则可以选择其他配合。一般情况下，常用配合的应用见表2-11。

Tab. 2-11 Properties and applications of common fits (≤500 mm)

表2-11 常用配合的特征和应用（≤500 mm）

Fit types 配合类别	Fit properties 配合特征	Symbols for fit 配合代号	Applications 应用场合
Clearance fit 间隙配合	Greatest clearance 特大间隙	H11/a11, H11/b11, H12/b12	Used for high temperature or great clearance fit needed 用于高温环境或需要大间隙配合的情况
	Greater clearance 很大间隙	H11/c11, H11/d11	Used for very poor work condition, deformed under force, or great clearance need for assembly and high temperature work condition 用于工作条件非常恶劣，受力易变形，或需要大间隙进行装配以及高温工作条件的情况
	Great clearance 较大间隙	H9/c9, H10/c10, H8/d8, H9/d9, H10/d10, H7/e7, H8/e8, H9/e9	Used for sliding bearing under high speed and heavy load, for sliding bearing of big diameter, and for large span or multiple points supporting 用于高速重载下的滑动轴承，大直径滑动轴承，以及大跨度或多支点支承的配合
	Common clearance 一般间隙	H6/f5, H7/f6, H8/f7, H8/f8, H9/f9	Used for running fit under common velocity, and widely used for sliding bearing under common lubrication, when temperature has little affection 用于普通转速的间隙配合，在普通润滑条件下广泛用于滑动轴承，且受温度影响较小
	Smaller clearance 较小间隙	H6/g5, H7/g6, H8/g7	Used for precision sliding fit, or intermittence rotating fit worked at lowly speed 用于精密滑动配合，或低速运转下的间歇旋转的配合
	Minimum clearance 零间隙	H6/h5, H7/h6, H8/h7, H8/h8, H9/h9, H10/h10, H11/h11, H12/h12	Used for common alignment fit, low moving fit or swing fit 用于普通定位配合，低运动配合或摆动配合
Transition fit 过渡配合	Slightly clearance entirely 绝大部分有微小间隙	H6/js5, H7/js6, H8/js7	Used for concentrical fit with which easily mount and dismount or fit which transmits static load after fastening pieces added 用于易于装拆的同轴配合，或在加紧固件后用于传递静载荷的配合
	Slightly clearance mostly 大部分有微小间隙	H6/k5, H7/k6, H8/k7	Used for concentrical fit with a lite vibration. Load can be transmitted after fastening pieces added, and easily mount and dismount; usually mounted by mallet 用于有轻微振动的同轴配合。在加紧固件后可传递载荷，且易于装拆；通常使用木槌进行安装
	Slightly interference mostly 大部分有微小过盈	H6/m5, H7/m6, H8/m7	Used for high precision concentrical and shake proof fits. Large load can be transmitted after fastening pieces mostly added. Usually mounted by copper hammers 用于高定位精度较高且能抗振的同轴配合。大多数加紧固件后可传递大载荷。通常使用铜锤进行安装

Tab. 2-11 (Continued) 表2-11（续）

Fit types 配合类别	Fit properties 配合特征	Symbols for fit 配合代号	Applications 应用场合
Transition fit 过渡配合	Slightly interference entirely 绝大部分有微小过盈	H7/n6，H8/n7	Used for extra precision concentrical fit or tight fit; heavy load or impacted load can be transmitted after fastening piece added such as key; dismounted only at overhaul 用于超精密同轴配合或紧配合；加键等紧固件后可传递重载或冲击载荷；仅在检修时拆卸
	Smaller interference entirely 绝大部分有较小过盈	H8/p7	Heavy load can be transmitted after key added, and vibration and impact load can be sustained. Never dismounted 加键后能传递重载荷,用于承受振动和冲击载荷。永不拆卸
Interference fit 过盈配合	Slightly interference 轻型	H6/n5，H6/p5，H7/p6，H6/r5，H7/r6，H8/r7	Used for exact concentrical fit. Usually can bot transmit large load, if load need to transmit, fastening piece is still needed 用于精确的同心配合,一般不能靠传递大载荷。要传递力矩需要加紧固件
	Common interference 中型	H6/s5，H7/s6，H8/s7，H6/t5，H7/t6，H8/t7	Small load can be transmitted without fastening piece. Large load can be transmitted after fastening piece added 无需紧固件即可传递小载荷。加紧固件后可传递大载荷
	Great interference 重型	H7/u6，H8/u7，H7/v6	Large load can be transmitted without fastening piece and the material strength should be very high 无需紧固件即可传递大载荷，要求材料有非常高强度
	Greatest interference 特重型	H7/x6，H7/y6，H7/z6	Extra load can be transmitted, however, experiments needed before these fits are used 能传递和承受额外载荷，须经试验后方可应用

Example 2-8: *A* fit $D=d=\phi100$ mm, $X_{max}=+0.041$ mm, and $Y_{max}=-0.048$ mm. To determine the fit type, the symbol fundamental deviations of hole and shaft, and the grade of tolerance.

例2-8： 有一孔、轴配合，公称尺寸为ϕ100mm，要求配合的过盈或间隙为 $-0.048\sim+0.041$ mm。试确定配合类型、孔和轴基本偏差代号、配合代号和公差等级。

Solution: (1) Grade of tolerance.

Given the specified conditions, the combination of this hole and shaft is a transition fit, with an allowable fit tolerance:

$T_f=X_{max}-Y_{max}=+0.041\text{ mm}-(-0.048)\text{ mm}=0.087\text{ mm}$

$T_f=T_h+T_s$,

If the tolerance grade is greater than 8, the tolerances grade of shaft is same with the tolerance grade of hole, $T_h=T_s\approx T_f/2$, else if the tolerance grade is lower than or equal to grade 8, the tolerance grade of shaft is higher one grade than the tolerance grade of hole, the $T_h>T_s\approx T_f/2$. Anyway, we can select the tolerance grades of hole and shaft according to the $T_f/2$. In this example, $T_f/2=0.0445$ mm.

From Tab. 2-3, IT8 = 0.054 mm, IT7 = 0.035 mm, then

$T_h=\text{IT8}=0.054\text{ mm}$, $T_s=\text{IT7}=0.035\text{ mm}$

From Tab. 2-3, it can be seen that 44.5 μm falls between IT7 = 35 μm and IT8 = 54 μm. Within this range of tolerance grades, the tolerance grade for the hole is chosen as IT8, and the tolerance grade for the shaft is chosen as IT7.

(2) Determine the tolerance zone symbols for the hole and the shaft.

No special requirements needed, the basis-hole fit system is preferentially selected, the fundamental deviation symbol of hole is H, Then the tolerance zone of hole is H8. The limit deviations are EI = 0 and ES = +0.054 mm.

For a transition fit of hole basis fit, the fundamental deviation of shaft is lower deviation ei, from the equation $ES - ei \leqslant X_{max} = 0.041$mm, then

$ei = ES - X_{max} = (0.054 - 0.041)$ mm = +0.013 mm,

From Tab. 2-4, the fundamental deviation value of shaft for nominal size 100 mm neartly equal to ei = +0.013 mm, the tolerance zone symbol is m, then the tolerance zone of shat is m7; then es = ei + IT7 = (+0.013 + +0.035) mm = +0.048 mm.

(3) Then the fit is ϕ100H8/m7.

(4) Checking calculation.

$X'_{max} = ES - ei = (0.054 - 0.013)$ mm = 0.041mm

$Y'_{max} = EI - es = (0 - 0.048)$ mm = −0.048 mm

So the fit meets the requirements.

In practical application, the calculated tolerance value and limit deviation value may not be exactly consistent with the data in the table. We should select appropriate fit in according to the actual accuracy requirements.

解：（1）确定孔、轴公差等级。

由给定条件可知，此孔、轴结合为过渡配合，其允许的配合公差为

$T = |X_{max} - Y_{max}| = |0.041 - (-0.048)|$ mm = 0.089 mm

$T_f = T_h + T_s$

假设孔与轴为同级配合，则 $T_h = T_s \approx T_f/2 = 0.089$ mm/2 = 0.0445 mm = 44.5 μm

如果公差等级大于8级，则轴的公差等级与孔的公差等级相同，即 $T_h = T_s \approx T_f/2$；否则，如果公差等级小于或等于8级，则轴的公差等级比孔的公差等级高一级，此时 $T_h > T_s \approx T_f/2$。无论如何，我们都可以根据 $T_f/2$ 来选择孔和轴的公差等级。在本例中，$T_f/2 = 0.0445$mm。

从表2-3可知，44.5 μm介于IT7 = 35 μm和IT8 = 54 μm之间，而在这个公差等级范围内，于是取孔公差等级为IT8，轴公差等级为IT7。

（2）确定孔和轴的公差带代号。

无特殊要求时，选用基孔制，孔的基本偏差为H，则孔的公差带为H8。孔的极限偏差为下偏差EI = 0和上偏差ES = +0.054 mm。

对于基孔制的过渡配合，轴的基本偏差为下偏差ei，

由公式 $ES - ei \leqslant X_{max} = 0.041$mm 最大间隙 $X_{max} = 0.041$mm，可得

$ei = ES - X_{max} = (0.054-0.041)$ mm = +0.013 mm，

查表2-4，公称尺寸为100 mm时，轴的基本偏差值接近ei = +0.013 mm，公差带代号为m，则轴的公差带为m7；那么上偏差es = ei + IT7 = (+0.013 + +0.035) mm = +0.048 mm。

（3）选择的配合为ϕ100H8/m7。

（4）验算。

$X'_{max} = ES - ei = (0.054 - 0.013)mm = 0.041\ mm$

$Y'_{max} = EI - es = (0 - 0.048)mm = -0.048\ mm$

因此，满足要求。

实际应用时，计算出的公差数值和极限偏差数值不一定与表中的数据正好一致。应按照实际的精度要求，适当选择。

2.6 General Tolerance of Linear Sizes 一般线性尺寸公差

General tolerance of linear sizes (also known as general tolerance) refers to the tolerance that can be guaranteed by the machine tool equipment under the general process condition of the workshop, and is usually not inspected. According to the national standard GB/T 1804-2020, there are four tolerance grades: precision, medium, rough and coarse, respectively, expressed by the letters f, m, c and v.

线性尺寸未注公差（也称一般公差）指在车间普通工艺条件下，机床设备所能保证的公差，通常不进行检验。根据国家标准GB/T 1804—2020对线性尺寸的一般公差规定了4个公差等级——精密级、中等级、粗糙级和最粗级，分别用字母f，m，c和v表示。

Limit deviation values of linear dimensions are shown in Tab. 2-12. Limit deviation values for fillet radii and chamfer height are shown in Tab. 2-13.

线性尺寸（倒角边除外）的允许偏差数值见表2-12。国家标准同时也对倒角边（对倒圆半径与倒角高度尺寸）的允许偏差数值做了规定，见表2-13。

The general tolerances are specified on the drawing by a general tolerance note, usually in the title block or near technical requirements, e.g. referring to a standard specifying the general tolerances, as shown in Fig. 2-18, indication of general tolerances. When the intermediate level m is selected, it is expressed as GB/T 1804-m.

一般公差在图纸上的标注是通过一般公差说明进行标注，通常位于标题栏或技术要求附近。例如，当选用中等级m时，则表示为GB/T 1804-m，如图2-18所示。

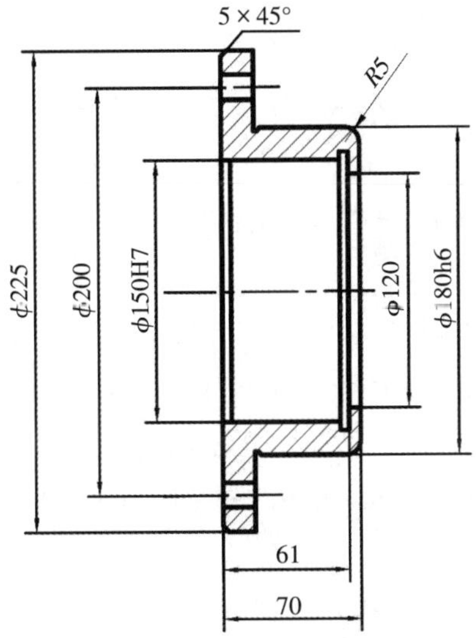

Technical requirement: general tolerance GB/T 1804-m

技术要求：未注尺寸公差按GB/T 1804—m

Fig. 2-18 General tolerance GB/T 1084-m

图2-18 一般公差GB/T 1084—m

Tab. 2-12 Limit deviation values of linear dimensions（GB/T 1804-2020）

表2-12 线性尺寸的极限偏差数值（GB/T 1804—2020）

单位：μm

Tolerance grades 公差等级		Permissible deviations for basic size range 公称尺寸分段							
Designation 代号	Description 说明	0.5~3	>3~6	>6~30	>30~120	>120~400	>400~1000	>1000~2000	>2000~4000
f	Fine 精密级	±0.05	±0.05	±0.1	±0.15	±0.2	±0.3	±0.5	—
m	Medium 中等级	±0.1	±0.1	±0.2	±0.3	±0.5	±0.8	±1.2	±2
c	Coarse 粗糙级	±0.2	±0.3	±0.5	±0.8	±1.2	±2	±3	±4
v	Very coarse 最粗级	—	±0.5	±1	±1.5	±2.5	±4	±6	±8

Tab. 2-13 Limit deviation values for fillet radii and chamfer heights（GB/T 1804-2020）

表2-13 倒圆半径与倒角高度尺寸的极限偏差数值（GB/T 1804—2020）

Tolerance grades 公差等级		Permissible deviations for basic size range 公称尺寸分段			
Designation 代号	Description 说明	0.5~3	>3~6	>6~30	>30
f	Fine 精密级	±0.2	±0.5	±1	±2
m	Medium 中等级				
c	Coarse 粗糙级	±0.4	±1	±2	±4
v	Very coarse 最粗级				

Exercise 2 习题2

2-1 How many types of the fit divided to? What is the characteristic of every size tolerance zone position for shaft and hole?

配合分为多少种类型？轴和孔的每种尺寸公差带位置的特点是什么？

2-2 Why can not judge the fit type according to the actual size of shaft and hole which fitted together?

为什么不能根据相互配合的轴和孔的实际尺寸来判断配合类型？

2-3 How many size tolerance grades are there in ISO and GB standard, how to represent the grade level?

ISO和GB标准中有多少个尺寸公差等级，如何表示等级？

2-4 How many size tolerance fundamental deviations are there in the ISO and GB standard? Write out their symbol. How to determine the fundamental deviation value for shafts? How to determine the fundamental deviation value for holes?

ISO和GB标准中有多少个尺寸公差基本偏差？写出它们的符号。如何确定轴的基本偏差值？如何确定孔的基本偏差值?

2-5 Fill the following table，unit：mm.

填充下表，单位：mm.

Nominal size 公称尺寸	Maximum limits of size 最大极限尺寸		Minimum limits of size 最小极限尺寸		Upper deviation 上偏差		Lower deviations 下偏差		Size tolerance 尺寸公差	
	Symbol 符号	Value 值	Symbol 符号	Value 值	Symbol 符号	Value 值	Symbol 符号	Value 值	Symbol 符号	Value 值
hole $\phi10$		10.040		10.025						
shaft $\phi40$						−0.060				0.046
hole $\phi50$				50.025						0.039
shaft $\phi60$						+0.051		+0.026		

2-6 Given the nominal size is $\phi40$ mm for a hole and a shaft，the maximum limit of size is $\phi40.045$ mm and the minimum limit of size is $\phi40.020$ mm for the hole，the maximum limit of size is $\phi40$ mm and the minimum limit of size is $\phi39$，084 mm for the shaft. To calculate the limits of deviations，fundamental deviations，and the size tolerances for the hole and the shaft，then to draw the size tolerance zone diagram.

已知孔和轴的基本尺寸为$\phi40$ mm，孔的最大极限尺寸为$\phi40.045$ mm，最小极限尺寸为$\phi40.020$ mm，轴的最大极限尺寸为$\phi40$ mm，最小极限尺寸为$\phi39.084$ mm。计算孔和轴的极限偏差、基本偏差以及尺寸公差，并绘制尺寸公差带图。

2-7 Given the nominal size is $\phi20$ mm for a fit，the maximum limit of size is

$\phi20.021$ mm and the minimum limit of size is $\phi20$ mm for the hole，the maximum limit of size is $\phi20.002$ mm and the minimum limit of size is $\phi19.989$ mm for the shaft. To calculate the limits of deviations，fundamental deviations，the size tolerances for the hole and the shaft，to calculate the limit clearance （s）/ interference （s）,fit tolerance for the fit，then to draw the size tolerance zone diagram.

已知配合的公称尺寸为$\phi20$ m，孔的最大极限尺寸为$\phi20.021$ mm，最小极限尺寸为$\phi20$ mm，轴的最大极限尺寸为$\phi20.002$ mm，最小极限尺寸为$\phi19.989$ mm。计算孔和轴的极限偏差、基本偏差、尺寸公差，计算极限间隙/过盈、配合公差，并绘制尺寸公差带图。

2-8 Fill the following table，unit：mm.

填充下表，单位：mm.

Nominal size 公称尺寸	Hole孔			Shaft轴			X_{max}/Y_{min}	X_{min}/Y_{max}
	ES	EI	T_h	es	ei	T_s		
$\phi15$			0.018	0				−0.008
$\phi25$		0				0.013	+0.074	
$\phi45$			0.025	0				−0.050

2-9 Given the nominal size is $\phi30$ mm，according to the following conditions to design the fit symbols respectively.

给定公称尺寸为$\phi30$ mm，根据以下条件分别确定配合代号。

（1）The limit clearances are between + 0.021 ~ + 0.056 mm.

极限间隙在 + 0.021 ~ + 0.056 mm之间。

（2）The limit clearance is + 0.033 mm，and the limit interference is − 0.023 mm.

极限间隙为 + 0.033 mm，极限过盈为−0.023 mm。

（3）The limit interferences are between $-0.010 \sim -0.070$ mm.

极限过盈在 $-0.010 \sim -0.070$ mm之间。

2-10 Fill the following table.

填下表：

Fit symbol 配合代号	Tolerance symbol 公差代号	Fundamental deviation 基本偏差		Standard tolerance value 标准公差值	Another limit deviation 另一个极限偏差	
		Symbol 符号	Value 值		Symbol 符号	Value 值
$\phi 40\frac{G7}{h6}$	ϕ40G7	EI			ES	
	ϕ40h6	es			ei	
$\phi 50\frac{H7}{p6}$	ϕ50H7					
	ϕ50p6					
$\phi 37\frac{M7}{h6}$	ϕ37M7					
	ϕ37h6					
$\phi 25\frac{P7}{h6}$	ϕ25P7					
	ϕ25h6					

Chapter 3 Fundamentals of Geometric Quantity Measurement 几何量测量基础

3.1 Overview 概述

The geometric quantity measurement includes the concepts of measurement and inspection, measurement units and length benchmarks, length measurement value transmission systems, gauge blocks used as standard measuring tools in production practice and their applications, as well as the classification of measuring instruments and measurement methods.

几何量测量技术方面的基础知识，包括测量与检验的概念、计量单位与长度基准、长度量值传递系统、在生产实践中作为标准量具的量块及其应用，还包括计量器具和测量方法的分类。

3.1.1 Concept of Measurement and Inspection 测量与检验的概念

Measurement is the process of comparing the measured quantity with a standard quantity expressed in units of measurement to determine the measured value. If L represents the measured value, E represents the standard quantity, and the ratio of the two is $X = L/E$, then there is

测量是将被测的量与用计量单位表示的标准量进行比较，从而确定被测量值的过程。以L表示被测量的量值，E表示标准量，二者的比值为$X = L/E$，则有

$$L = XE \tag{3-1}$$

The measured value is the numerical value of the measured quantity expressed in units of measurement.

即测量所得量值为用计量单位表示的被测的量的数值。

Inspection refers to the experimental process of determining whether a measurement is qualified.

检验是指判断测量是否合格的实验过程。

Any complete measurement process must have the object being measured and the unit of measurement used, and a measurement method that is suitable for the object being measured must be adopted, and the measurement results must achieve the required measurement accuracy. Therefore, the measurement process should include four elements: measurand, measurement unit, measurement method, and measurement accuracy.

任何一个完整的测量过程必须有被测对象和所采用的计量单位，同时要采用与被测对象相适应的测量方法，并使测量结果达到所要求的测量精度。因此，测量过程应包括被测对象、计量单位、测量精度和测量方法4个要素。

3.1.1.1 Measurand 被测对象

In geometric measurement, the measured object mainly refers to the geometric parameters of the part, such as size, geometric error, and surface roughness. Due to the wide variety and complexity of the measured object, it is very important to be familiar with and master the definition of the measured object, analyze and study its characteristics

在几何量测量中，被测对象主要是指零件的尺寸、几何误差及表面粗糙度等几何参数。由于被测对象种类繁多，复杂程度各异，因此熟悉和掌握被测对象的定义，分析和研究被测对象的特点十分重要。

3.1.1.2 Measurement unit 计量单位

The length unit in China's legal measurement system is the meter (m). In the mechanical manufacturing industry, the commonly used unit of length is millimeters (mm); When conducting precision measurements, micrometers (μm) are often used; When conducting ultra precision measurements, nanometers (nm) are often used.

我国的法定计量单位中长度单位为米（m）。在机械制造业中，常用的长度单位为毫米（mm）；精密测量时，多采用微米（μm）；超精密测量时，多采用纳米（nm）。

3.1.1.3 Measurement accuracy 测量精度

Measurement accuracy refers to the degree of consistency between the measured value and its true value. Measurement errors are inevitable during the measurement process. Small measurement error implies high measurement accuracy; Large measurement error implies low measurement accuracy. Only when the measurement error is small enough can it indicate that the measurement results are reliable. Therefore, measurement results without knowing their measurement accuracy are meaningless. The measurement accuracy is usually represented by the limit error or uncertainty of the measurement.

测量精度即测得值与其真值的一致程度。测量过程中不可避免地存在测量误差。测量误差小，测量精度高；测量误差大，测量精度低。只有测量误差足够小，才表明测量结果是可靠的。因此，不知道测量精度的测量结果是没有意义的。通常用测量的极限误差或测量的不确定度来表示测量精度。

3.1.1.4 Measurement method 测量方法

Measurement method refers to the sum of measurement principles, measurement tools and measurement conditions used during measurement.

测量方法是指在进行测量时所采用的测量原理、计量器具及测量条件的总和。

3.1.2 Measurement Units and Length Standards 计量单位与长度基准

3.1.2.1 Measurement unit 计量单位

To ensure the accuracy of measurement, it is first necessary to establish a unified and reliable measurement unit. In 1984, the State Council issued *the Legal Measurement Units of the People's Republic of China*, which stipulated that "meters" (m) were the basic units of length. The commonly used unit of length in mechanical manufacturing is millimeter (mm).

为了保证计量的准确度，首先需要建立统一、可靠的计量单位。1984年国务院发布了《中华人民共和国法定计量单位》，其中规定"米"（m）为长度的基本单位，机械制造中常用的长度单位为毫米（mm）。

3.1.2.2 Length reference 长度基准

The current definition of "meter", adopted at the 17th International Conference on Metrology in October

1983, is the length of the distance traveled by light in a vacuum within a time interval of 1/299792458 seconds. This is the theoretical definition of "meter", and when using it, it is necessary to reproduce the definition of "meter" in order to obtain the length standards of each country.

1983年10月，在第十七届国际计量大会上通过的“米”的现行定义为：米是光在真空中1/299792458秒的时间间隔内所经过的路程的长度。这是“米”在理论上的定义，使用时需要对“米”的定义进行复现，才能获得各自国家的长度标准。

3.1.3　Length Measurement Value Transmission Systems 长度量值传递系统

In engineering, it is generally not possible to directly measure the geometric parameters of parts by using light waves according to the definition of "meter", but various measuring instruments are used. In order to ensure the accuracy and uniformity of measurement values, it is necessary to establish a measurement value transmission system from the length datum to the measured part. The main standard devices for length measurement value transmission in China are gauge blocks (end face measuring tools) and line rulers (marking measuring tools), and their transmission systems are shown in Fig. 3-1.

在工程上，一般不能直接按照“米”的定义用光波来测量零件的几何参数，而是采用各种计量器具。为了保证量值的准确和统一，必须建立从长度基准到被测零件的量值传递系统。我国长度量值传递的主要标准器是量块（端面量具）和线纹尺（刻线量具），其传递系统如图3-1所示。

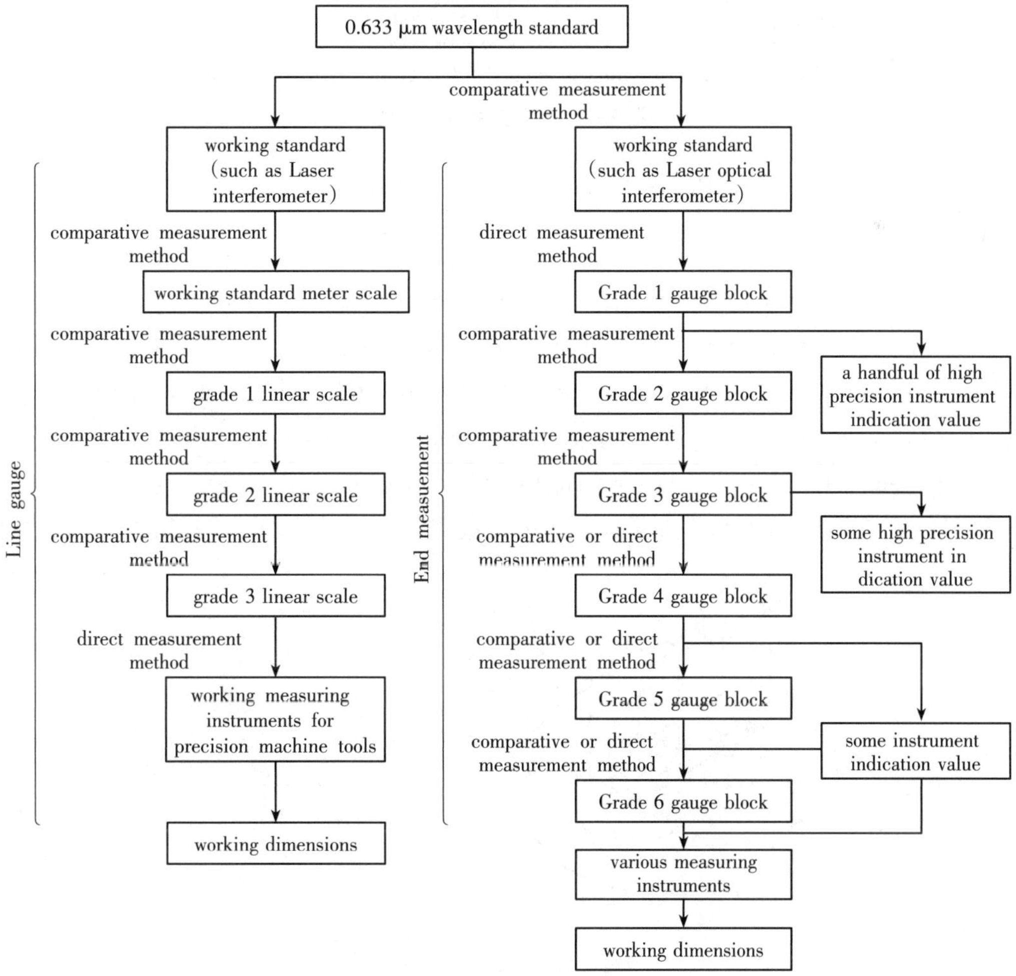

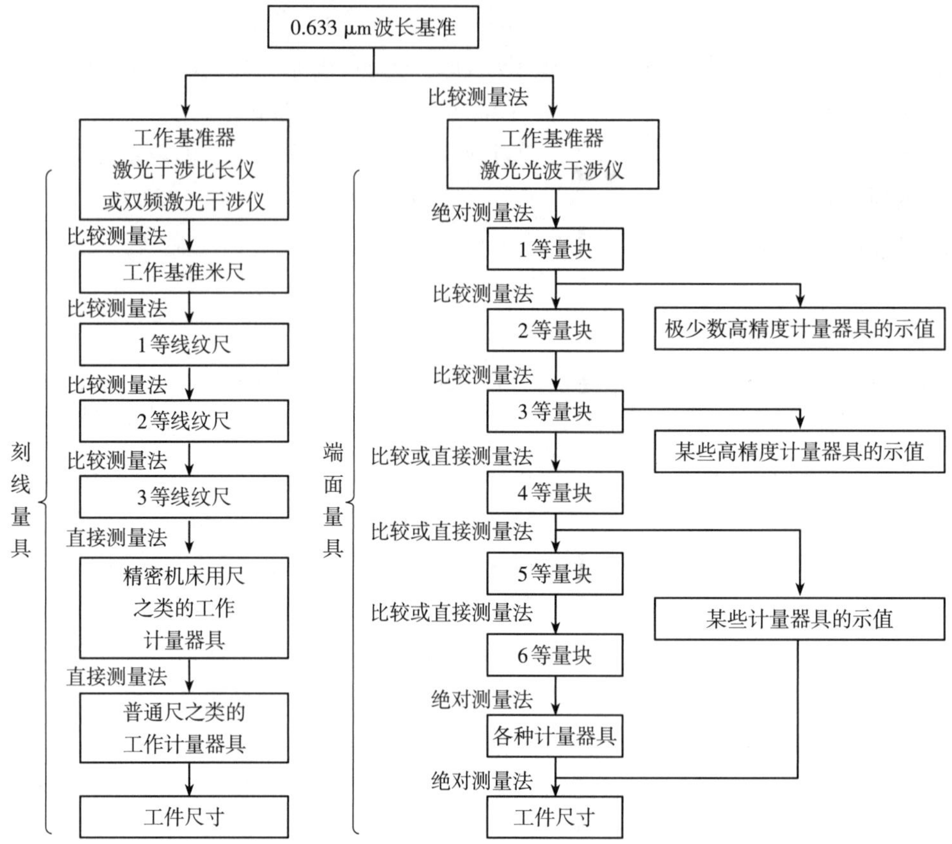

Fig. 3-1　Length measurement value transmission systems

图3-1　长度量值传递系统

3.1.4　Gauge Blocks and Their Applications 量块及其应用

A gauge block, also known as a block gauge, is a flat parallel end face measuring tool made of chromium manganese steel. It has the characteristics of stable size, less deformation, and good wear resistance. gauge blocks have a wide range of uses. In addition to serving as standard instruments for transmitting length values, they can also be used to adjust instruments, machine tools, and other equipment. They can also be used to directly measure parts. Gauge blocks are usually made into rectangular shapes, as shown in Fig. 3-2 (a). The size marked on the gauge block is called the nominal size of the gauge block, as shown in Fig. 3-2 (b).

量块又称块规，它是平面平行端面量具，多由铬锰钢制成。其具有尺寸稳定、不易变形和耐磨性

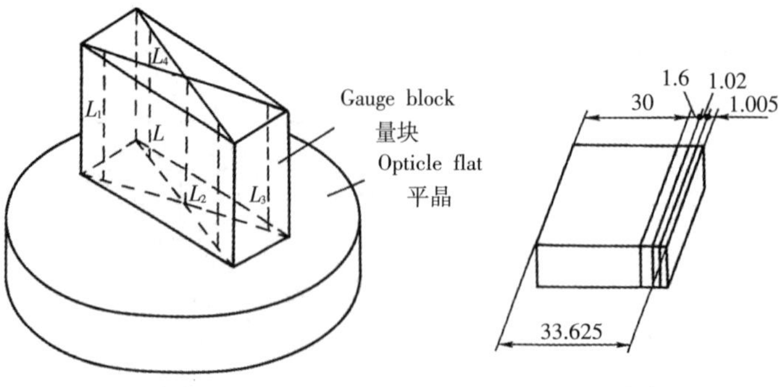

(a)　　(b)

Fig 3-2　Gauge blocks

图3-2　量块

好等特点。量块的用途广泛。除作为标准器具进行长度量值的传递外，还可用来调整仪器、机床和其他设备。此外，还可以用来直接测量零件。量块通常制成长方体，如图3-2（a）所示。量块上标出的尺寸称为量块的标称尺寸，如图3-2（b）。

Two parallel planes with smooth surfaces （R_x≤0.08 μm） and minimal flatness error are called measurement surfaces or working surfaces. The accuracy of the gauge block is extremely high，but the two working surfaces are not absolutely parallel. Therefore，the size of the gauge block is specified as the vertical distance from the center of one working surface to the parallel working plane when one working surface of the gauge block is fitted onto a parallel working plane，as shown in Fig 3-2（b）, The size marked on the gauge block is called the nominal size of the gauge block. For gauge blocks with marked sizes（ie. nominal sizes）less than 6 mm，the side with numbers is the upper measurement surface; Gauge blocks with a nominal size greater than or equal to 6 mm have a digital surface on the right side as the upper measuring surface. In order to meet different production requirements，gauge blocks are classified into grades 00，0，K，1，2，and 3 based on their manufacturing accuracy. Among them，grade 00 has the highest accuracy，grade 3 has the lowest accuracy，and grade K is the calibration level. The maximum allowable values for the nominal length deviation（limit deviation）and length variation of each grade of gauge block when used in stages are shown in Tab. 3-1.

其中两个不粗糙（R_x≤0.08 μm）且平面度误差很小的平行平面，称为测量面或工作面。尽管量块的精度极高，但是两个工作面也不能保证绝对平行。因此，量块的尺寸被规定为把量块的一个工作面研合在平行的工作平面上时，从该工作面中心到平行工作平面的垂直距离，如图3-2（b）所示。量块上标出的尺寸称为量块的标称尺寸。标称尺寸（即名义尺寸）小于6 mm的量块，有数字的一面为上测量面；标称尺寸大于等于6 mm的量块，有数字面的右侧面为上测量面。为了满足不同的生产要求，量块按其制造精度分为00，0，K，1，2，3级。其中，00级精度最高，3级精度最低，K级为校准级。按级使用时，各级量块的标称长度偏差（极限偏差）和长度变动量的最大允许值见表3-1。

Tab. 3-1　The maximum allowable values for the nominal length limit deviation（t_e）and length variation（t_v）of each level of gauge block

表3-1　各级量块测量面上任意点的长度极限偏差t_e和长度变动量的最大允许值t_v

Nominal length 标称长度(l_n)/mm	Grade K K级		Grade 0 0级		Grade 1 1级		Grade 2 2级		Grade 3 3级	
	t_e	t_v	t_e	t_v	t_e	t_v	t_e	t_v	t_e	t_v
l_n≤10	±0.20	0.05	±0.12	0.1	±0.20	0.16	±0.45	0.3	±1.0	0.5
10 < l_n≤25	±0.30	0.05	±0.14	0.1	±0.30	0.16	±0.60	0.3	±1.2	0.5
25 < l_n≤50	±0.40	0.06	±0.20	0.1	±0.40	0.18	±0.80	0.3	±1.6	0.55
50 < l_n≤75	±0.50	0.06	±0.25	0.12	±0.50	0.18	±1.00	0.35	±2.0	0.55
75 < l_n≤100	±0.60	0.07	±0.30	0.12	±0.60	0.2	±1.20	0.35	±2.5	0.6
100 < l_n≤150	±0.80	0.08	±0.40	0.14	±0.80	0.2	±1.6	0.4	±3.0	0.65
150 < l_n≤200	±1.00	0.09	±0.50	0.16	±1.00	0.25	±2.0	0.4	±4.0	0.7
200 < l_n≤250	±1.20	0.1	±0.60	0.16	±1.20	0.25	±2.4	0.45	±5.0	0.75
250 < l_n≤300	±1.40	0.1	±0.70	0.18	±1.40	0.25	±2.8	0.5	±6.0	0.8
300 < l_n≤400	±1.80	0.12	±0.90	0.2	±1.80	0.3	±3.6	0.5	±7.0	0.9
400 < l_n≤500	±2.20	0.14	±1.10	0.25	±2.20	0.35	±4.4	0.6	±9.0	1
500 < l_n≤600	±2.60	0.16	±1.30	0.25	±2.6	0.4	±5.0	0.7	±11.0	1.1

Tab. 3-1 (Continued) 表3-1（续）

Nominal length 标称长度(l_n)/mm	Grade K K级		Grade 0 0级		Grade 1 1级		Grade 2 2级		Grade 3 3级	
	t_e	t_v	t_e	t_v	t_e	t_v	t_e	t_v	t_e	t_v
$600 < l_n \leq 700$	±3.00	0.18	±1.50	0.3	±3.0	0.45	±6.0	0.7	±12.0	1.2
$700 < l_n \leq 800$	±3.40	0.2	±1.70	0.3	±3.4	0.5	±6.5	0.8	±14.0	1.3
$800 < l_n \leq 900$	±3.80	0.2	±1.90	0.35	±3.8	0.5	±7.5	0.9	±15.0	1.4
$900 < l_n \leq 1000$	±4.20	0.25	±2.00	0.4	±4.2	0.6	±8.0	1	±17.0	1.5

Measuring blocks are classified according to their calibration accuracy into level 1, 2, 3, 4, 5, 6, etc. Among them, level 1 has the highest accuracy and level 6 has the lowest accuracy. The maximum allowable values of measurement uncertainty and length variation for each of equal block length are shown in Tab. 3-2.

量块按照检定精度分为1，2，3，4，5，6等。其中，1等精度最高，6等精度最低。各等量块长度测量不确定度和长度变动量最大允许值见表3-2。

Tab. 3-2 The maximum allowable values of measurement uncertainty and length variation for each equal block length

表3-2 各等量块长度测量不确定度和长度变动量最大允许值

Nominal length 标称长度(l_n)/mm	Level 1 1等		Level 2 2等		Level 3 3等		Level 4 4等		Level 5 5等	
	Measurement uncertainty 测量不确定度	Length variation 长度变动量	Measurement uncertainty 测量不确定度	Length variation 长度变动量	Measurement uncertainty 测量不确定度	Length variation 长度变动量	Measurement uncertainty 测量不确定度	Length variation 长度变动量	Measureme uncertainty 测量不确定度	Length variation 长度变动量
$l_n \leq 10$	0.022	0.05	0.06	0.10	0.11	0.16	0.22	0.30	0.60	0.50
$10 < l_n \leq 25$	0.025	0.05	0.07	0.10	0.12	0.16	0.25	0.30	0.60	0.50
$25 < l_n \leq 50$	0.030	0.06	0.08	0.10	0.15	0.18	0.30	0.30	0.80	0.55
$50 < l_n \leq 75$	0.035	0.06	0.09	0.12	0.18	0.18	0.35	0.35	0.90	0.55
$75 < l_n \leq 100$	0.040	0.07	0.10	0.12	0.20	0.20	0.40	0.35	1.00	0.60
$100 < l_n \leq 150$	0.05	0.08	0.12	0.14	0.25	0.20	0.50	0.40	1.20	0.65
$150 < l_n \leq 200$	0.06	0.09	0.15	0.16	0.30	0.25	0.60	0.40	1.50	0.70
$200 < l_n \leq 250$	0.07	0.10	0.18	0.16	0.35	0.25	0.70	0.45	1.80	0.75
$250 < l_n \leq 300$	0.08	0.10	0.20	0.18	0.40	0.25	0.80	0.50	2.00	0.80
$300 < l_n \leq 400$	0.10	0.12	0.25	0.20	0.50	0.30	1.00	0.50	2.50	0.90
$400 < l_n \leq 500$	0.12	0.14	0.30	0.25	0.60	0.35	1.20	0.60	3.00	1.00
$500 < l_n \leq 600$	0.14	0.16	0.35	0.25	0.70	0.40	1.40	0.70	3.50	1.10
$600 < l_n \leq 700$	0.16	0.18	0.40	0.30	0.80	0.45	1.60	0.70	4.00	1.20
$700 < l_n \leq 800$	0.18	0.20	0.45	0.30	0.90	0.50	1.80	0.80	4.50	1.30
$800 < l_n \leq 900$	0.20	0.20	0.50	0.35	1.00	0.50	2.00	0.90	5.00	1.40
$900 < l_n \leq 1000$	0.22	0.25	0.55	0.40	1.10	0.60	2.20	1.00	5.50	1.50

Notes: ① The area within 0.8mm from the edge of the measurement surface is excluded.
② A confidence probability of the measurement uncertainty is 0.99 in the Tab. 3-2.

注：①距离测量面边缘0.8 mm范围内不计。
② 表3-2内测量不确定度置信概率为0.99。

When using gauge blocks in grades, the nominal length of the gauge block should be taken as the working dimension, which includes the dimensional error during manufacturing.

量块按级使用时，应以量块的标称长度为工作尺寸，该尺寸包含了制造时的尺寸误差。

When using gauge blocks in levels , the actual size of the center length of the calibrated block should be taken as the working size, which is not affected by manufacturing errors and only includes smaller measurement errors during calibration. Therefore, the accuracy of using gauge blocks in levels is higher than when using them in grades.

量块按等使用时，应以经检定所得到的量块中心长度的实际尺寸为工作尺寸，该尺寸不受制造误差的影响，只包含检定时较小的测量误差。因此，量块按等使用比按级使用时的精度高。

For example, when using gauge blocks in grades, using a grade 1, 30 mm gauge block, the nominal length limit deviation is (30 ± 0.0004) mm; When using gauge blocks according to equality, use level 3 gauge blocks with a calibrated size of 30.0002 mm. The maximum deviation of the measurement uncertainty for the center length of the gauge block is (30.0002±0.00015) mm.

例如，按级使用量块时，使用1级、30 mm的量块，标称长度极限偏差为（30±0.0004）mm，按等使用量块时，使用3等量块，该量块检定尺寸为30.0002 mm，其中心长度的测量不确定度的极限偏差为（30.0002±0.00015）mm。

In order to synthesize the required size using a smaller block array, the gauge blocks are produced in sets based on specifics. There are 17 sets of measuring block series specified in GB/T 6093-2001, including 91 blocks, 83 blocks, 46 blocks, 38 blocks, and 10 blocks. Two sets of measuring block size series are listed in Tab. 3-3.

为了能用较小的块数组合成所需要的尺寸，量块按一定的尺寸系列成套生产。GB/T 6093—2001中规定的量块系列有91块、83块、46块、38块、10块等17套。表3-3列出了两种成套量块尺寸系列。

Tab. 3-3 Size series of two sets of gauge blocks

表3-3 两种成套量块尺寸系列

Set type 套别	Total number of blocks 总块数	Grade 级别	Size series 尺寸系列/mm	Spacing 间隔/mm	Number of blocks 块数
1	83	0, 1, 2	0.5	—	1
			1	—	1
			1.005	—	1
			1.01, 1.02, ···, 1.49	0.01	49
			1.5, 1.6, ···, 1.9	0.1	5
			2.0, 2.5, ···, 9.5	0.5	16
			10, 20, ···, 100	10	10
2	46	0, 1, 2	1	1	9
			1.001, 1.002, ···, 1.009	0.001	9
			1.01, 1.02, ···, 1.09	0.01	9
			1.1, 1.2, ···, 1.9	0.1	9
			10, 20, ···, 100	10	10

Due to the small flatness error and surface roughness values of the measuring surface of the gauge block, when there is an extremely thin oil film on the measuring surface, the measuring surfaces of the two gauge blocks come into contact with each other and slide tangentially relative to each other under not too much pressure, which can make the two gauge blocks stick together. So, different sized measuring blocks can be combined within a certain size range to form the desired size. In order to reduce the combination error of gauge blocks and ensure measurement accuracy, the number of gauge blocks should be minimized as much as possible, generally not exceeding 4, and the center length of each gauge block should be on the same straight line. In actual combination, the minimum tail of the required size should be eliminated, and for each selected measuring block, at least one decimal place of the required size should be reduced.

由于量块测量面的平面度误差和表面粗糙度数值均很小，所以当测量面上有一层极薄的油膜时，两个量块的测量面相互接触，在不大的压力下做切向相对滑动，就能使两个量块黏附在一起。因此，可用不同尺寸的量块在一定尺寸范围内组合成所需要的尺寸。为了减少量块的组合误差，保证测量精度，应尽量减少量块的数目，一般不应超过4块，并使各量块的中心长度在同一直线上。实际组合时，应从消去所需尺寸的最小尾数开始，每选一块量块，应至少减少所需尺寸的一位尾数。

Example 3-1: Using a set of 83 measuring blocks to form a size of 28.785mm, the combination method is as follows:

例3-1：用83块一套的量块，组成尺寸28.785 mm，其组合方法如下：

（1）The size of the first measuring block is 1.005 mm, remaining size 27.78 mm（size 1.005 mm）

第一块量块的尺寸为1.005 mm，剩余尺寸27.78 mm（尺寸1.005 mm ）

（2）The size of the second measuring block is 1.28 mm, remaining size 26.50 mm（with an interval of 0.01 mm）

第二块量块的尺寸1.28 mm，剩余尺寸26.50 mm（间隔0.01 mm）

（3）The size of the third measuring block is 6.5 mm（spaced 0.5 mm apart）

第三块量块的尺寸6.5 mm（间隔0.5 mm）

（4）Remaining size（i.e. the size of the fourth measuring block）20 mm（spaced 10 mm apart）

剩余尺寸（即第四块量块的尺寸）20 mm（间隔10 mm）

3.1.5 Classification of Measuring Instruments and Methods 计量器具和测量方法的分类

3.1.5.1 Classification of measuring instruments 计量器具的分类

Measuring instruments can be classified into the following categories based on their measurement principles, structural characteristics, and uses:

计量器具按其测量原理、结构特点和用途可分为以下几类:

（1）Benchmark measuring tools 基准量具。

Reference measuring tools are tools used to adjust and calibrate some measuring instruments or to serve as standard dimensions for comparative measurement. They are also divided into fixed value reference measuring tools（such as measuring blocks, angle blocks, etc.）and variable value reference measuring tools（such as linear rulers, etc.).

基准量具是用来调整和校准一些计量器具或作为标准尺寸进行比较测量的器具，又分为定值基准量具（如量块、角度块等）、变值基准量具（如线纹尺等）。

（2）Limit gauge极限量规。

A limit gauge is a specialized measuring instrument without scales used to check the size and geometric errors of parts，such as smooth limit gauges，position and thread gauges，etc. It can only be used to determine whether the measured geometric quantity is qualified，and cannot be used to obtain the specific value of the measured geometric quantity.

极限量规是一种没有刻度的、用于检验零件的尺寸和几何误差的专用计量器具，如光滑极限量规、位置和螺纹量规等。它只能用来判断被测几何量是否合格，而不能得到被测几何量的具体数值。

（3）Inspection fixture检验夹具。

Inspection fixtures are also specialized measuring instruments that，when used in conjunction with relevant measuring instruments，can conveniently and quickly measure multiple geometric parameters of the parts. The specialized inspection fixture for inspecting rolling bearings can simultaneously measure the dimensions of the inner and outer rings，as well as the radial and end face circular run out errors.

检验夹具也是一种专用计量器具，它与相关计量器具配合使用，可方便、快速地测得零件的多个几何参数。如检验滚动轴承的专用检验夹具可同时测得内、外圈尺寸和径向与端面圆跳动误差等。

（4）Universal measuring instruments通用计量器具。

Universal measuring instruments refer to instruments that can convert the measured geometric quantities into directly observable indicator values or equivalent information. According to their different working principles，they can be divided into the following types:

通用计量器具是指能将被测几何量的量值转换成可直接观测的指示值或等效信息的器具。按其工作原理不同，可分为以下几种：

① Vernier measuring tools，such as vernier calipers，vernier depth gauges，and vernier protractors.

游标量具，如游标卡尺、游标深度尺和游标量角器等。

② Micro helix measuring tools，such as outer diameter micrometers and inner diameter micrometers.

微动螺旋量具，如外径千分尺和内径千分尺等。

③ A mechanical comparator refers to a measuring instrument that uses mechanical transmission methods to achieve information conversion，such as lever gear comparator，torsion spring comparator.

机械比较仪即用机械传动方法实现信息转换的量仪，如杠杆齿轮比较仪、扭簧比较仪等。

④ Optical measuring instruments refer to measuring instruments that use optical methods to convert information，such as optical comparators，tool microscopes，measurement projectors，and light wave interferometers.

光学量仪是指用光学方法实现信息转换的量仪，如光学比较仪、工具显微镜、测量投影仪和光波干涉仪等。

⑤ Electric measuring instruments refers to devices that converts original information into circuit parameters.，such as inductance micrometers，capacitance micrometers，and profilometers，convert raw information into circuit parameters.

电动量仪是指将原始信息转换成电路参数的量仪，如电感测微仪、电容测微仪和轮廓仪等。

⑥ Pneumatic instruments，such as vernier type，thin film type，and corrugated tube type，are used to convert raw information through changes in flow or pressure in pneumatic systems.

气动量仪用来指通过气动系统的流量或压力的变化来实现原始信息转换的量仪，如游标式气动量仪、薄膜式气动量仪和波纹管式气动量仪等。

（5）Microcomputerized measuring instrument 微机化量仪。

Microcomputerized measuring instrument refers to an electromechanical integrated measuring instrument that can automatically collect, process, display, and print measurement results under the control of a microcomputer system, such as a computer roundness meter, a computer shape and position error measuring instrument, and a computer surface roughness measuring instrument.

微机化量仪是指在微机系统控制下，可实现数据的自动采集、自动处理、自动显示和打印测量结果的机电一体化量仪，如电脑圆度仪、电脑形位误差测量仪和电脑表面粗糙度测量仪等。

3.1.5.2 Technical performance indicators of measuring instruments 计量器具的技术性能指标

（1）Scale spacing refers to the distance between the centerlines of two adjacent lines on a scale or dial. For the convenience of visually estimating the decimal part of a scale value, the scale spacing is generally set to 1 ~ 3.5 mm.

刻度间距是指刻度尺或刻度盘上相邻两刻线中心线间的距离。为便于目测估读一个分度值的小数部分，一般将刻度间距取为1 ~ 3.5 mm。

（2）Graduation value, also known as scale value, refers to the quantity represented by the spacing between each scale on the scale or dial. The commonly used division values for geometric measuring instruments include 0.1, 0.05, 0.02, 0.01, 0.002 mm. Generally speaking, the smaller the division value, the higher the accuracy of the measuring instrument.

分度值，又称刻度值，它是指刻度尺或刻度盘上每一刻度间距所代表的量值。几何量计量器具的常用分度值有0.1，0.05，0.02，0.01，0.002 mm等。一般来说，分度值越小，计量器具的精度越高。

（3）The range of indicated values is the range from the lowest value to the highest value displayed or indicated by a measuring instrument. For example, the reading range of a mechanical comparator is ± 0.1 mm.

示值范围是指由计量器具所显示或指示的最低值到最高值的范围。例如，某一机械比较仪的示值范围是±0.1 mm。

（4）Measurement range refers to the range of minimum and maximum measured values that a measuring instrument can measure within the allowable error limit. For example, the measurement range of a certain micrometer is 50 ~ 75 mm.

测量范围是指在允许误差极限内计量器具所能测量的最小和最大被测量值的范围。例如，某一千分尺的测量范围是50 ~ 75 mm。

（5）Sensitivity and amplification ratio are the ability of measuring instruments to respond to measured changes. For measuring instruments of general length, sensitivity is also known as amplification ratio; For measuring instruments with graduated scales or dials that have equal divisions, the amplification ratio K is equal to the ratio of the scale spacing a to the division value i, ie. $K = a/i$.

灵敏度和放大比表示计量器具对被测量变化的反应能力。对于一般长度的计量器具，灵敏度又称放大比；对于具有等分刻度的刻度尺或刻度盘的量仪，放大比K等于刻度间距a与分度值i之比，即$K = a/i$。

（6）Sensitivity limit（Sensitivity range）is the minimum change in the measured value that causes a noticeable change in the indication of a measuring instrument. It represents the ability of the measuring instrument to reflect small changes being measured.

灵敏限（灵敏阈）是指引起计量器具示值可察觉变化的被测量的最小变化值。它表示量仪反映被测量微小变化的能力。

(7) Measurement force refers to the contact force between the measuring instrument and the measured surface during the measurement process. In contact measurement, the measurement force can ensure reliable contact, but excessive measurement force can cause deformation and wear of the measuring instrument and the measured parts, and changes in measurement force can make the reading unstable, affecting measurement accuracy.

测量力是指在测量过程中，计量器具与被测表面之间的接触力。在接触测量时，测量力可保证接触可靠，但过大的测量力会使量仪和被测零件变形和磨损，而测量力的变化会使示值不稳定，影响测量精度。

(8) Indication error, is the difference between the indication of a measuring instrument and the measured true value.

示值误差是指计量器具的示值与被测量真值之差。

(9) The change in indication refers to the maximum difference between the measured values when multiple repeated measurements (usually 5 ~ 10 times) are taken on the same subject under unchanged measurement conditions.

示值变动是指在测量条件不变的情况下，对同一被测量进行多次重复测量（一般5 ~ 10次）时，各测得值的最大差值。

(10) Return error refers to the range of variation in the measurement indication when measuring the same object in both directions under the same conditions.

回程误差是指在相同条件下，对同一被测量进行往、返两个方向测量时，测量示值的变化范围。

(11) Correction value, a numerical value added algebraically to an uncorrected measurement result in order to eliminate or reduce system errors. The correction value is equal to the negative value of the indication error. For example, if the indication error is +0.003 mm, the correction value is −0.003 mm.

修正值是指为了消除或减少系统误差，用代数法加到未修正测量结果上的数值。修正值等于示值误差的负值。例如，若示值误差为+0.003 mm，则修正值为−0.003 mm。

(12) Measurement uncertainty, is the degree to which measurement results cannot be determined due to the influence of measurement errors. The uncertainty is represented by the error limit.

测量不确定度是指由于测量误差的影响而使测量结果不能肯定的程度。不确定度用误差极限表示。

3.1.5.3 Types and characteristics of measurement methods 测量方法的种类及其特点

Measurement method refers to the sum of measurement principles, measuring instruments, and measurement conditions. In practical work, the types of measurement methods are often divided based on the way in which measurement results are obtained.

测量方法是指测量原理、测量器具、测量条件的总和。在实际工作中，往往从获得测量结果的方式来对测量方法的种类进行划分。

(1) According to whether the indicated value of the measuring instrument is the total value being measured, it can be divided into absolute measurement and relative measurement.

按计量器具的示值是否为被测量的全值，可分为绝对测量和相对测量。

① Absolute measurement, is a measurement where the indication of a measuring instrument is the total value being measured. For example, measuring the diameter of the shaft and hole with a vernier caliper or micrometer is considered an absolute measurement.

绝对测量是指计量器具的示值就是被测量的全值。例如，用游标卡尺、千分尺测量轴和孔的直径就属于绝对测量。

② Relative measurement, also known as comparative measurement, refers to the indication of a measuring instrument that only represents the deviation value of the measured relative to a known standard quantity, and the measured is the algebraic sum of the known standard quantity and the deviation value. For example, when measuring the diameter size of a shaft using a comparator, first set the comparator to zero using a gauge block that is the same as the nominal size of the measured shaft diameter, then replace the measured shaft and measure the deviation of the measured shaft diameter relative to the gauge block. The algebraic sum of this deviation value and the gauge block size is the actual size of the measured shaft diameter.

相对测量，又称比较测量，是指计量器具的示值只表示被测量相对于已知标准量的偏差值，而被测量为已知标准量与该偏差值的代数和。例如，用比较仪测量轴的直径尺寸，首先用与被测轴径的公称尺寸相同的量块将比较仪调零，然后换上被测轴，测得被测轴直径相对于量块的偏差。该偏差值与量块尺寸的代数和就是被测轴直径的实际尺寸。

（2）According to whether the actual measurement is being measured, the measurement can be divided into direct measurement and indirect measurement.

按实测量是否被测量，可分为直接测量和间接测量。

① Direct measurement is a measurement method that directly measures the value without the need for auxiliary calculations of functional relationships between the measured value and other actual measurements. For example, measuring the diameter of a shaft with an outside micrometer is considered a direct measurement.

直接测量是一种无须对被测量与其他实测量进行函数关系的辅助计算，而直接测得被测量值的测量方法。例如，用外径千分尺测量轴的直径就属于直接测量。

② Indirect measurement refers to the method of calculating the measured value by using other quantities that have a known functional relationship between the actual measurement and the measured value. For example, using the "bow height chord length method" to indirectly measure the radius R of the circular arc template, as long as the values of the bow height h and chord length L are measured, and then substituting into relevant formulas, the radius R of the template can be obtained. This method is called the square method.

间接测量是指实测量与被测量之间有已知函数关系的其他量，经过计算求得被测量值的方法。例如，采用“弓高弦长法”间接测量圆弧样板的半径R，只要测得弓高h和弦长L的量值，然后按照有关公式进行计算，就可获得样板的半径R的量值。

（3）According to whether multiple measurements are taken simultaneously on the part, the measurement can be divided into single measurement and comprehensive measurement.

按是否同时测量零件的多个被测量，可分为单项测量和综合测量。

① Single measurement refers to the measurement of each individual quantity being measured. For example, using a tool microscope to measure the actual values of pitch diameter、thread pitch, and tooth flank angle.

单项测量指对被测量分别进行的测量。例如，用工具显微镜分别测量中径、螺距和牙侧角的实际值。

② Comprehensive measurement is the comprehensive result of measuring the errors of some related geometric parameters on the part. For example, the measurement of comprehensive deviation of gears.

综合测量是对零件上一些相关联的几何参数误差的综合结果进行测量。例如，齿轮的综合偏差的测量。

(4) According to whether there is contact between the surface of the measured workpiece and the measuring head of the measuring instrument, it can be divided into contact measurement and non-contact measurement.

根据被测工件的表面与测量仪器的测量头之间是否存在接触，可分为接触式测量和非接触式测量。

① Contact measurement refers to the measurement of measuring force in which the measuring head of a measuring instrument comes into contact with the surface being measured and has a mechanical effect. For example, measuring the shaft diameter with a comparator.

接触测量是指计量器具的测量头与被测表面相接触并有机械作用的测量力的测量。例如，用比较仪测量轴径。

② Non-contact measurement refers to the measurement of measuring force without mechanical action, as the measuring head of the measuring instrument is not in contact with the measured surface. For example, measuring surface roughness using a light cutting microscope.

非接触测量指计量器具的测量头与被测表面不接触，因而不存在机械作用的测量力的测量。例如，用光切显微镜测量表面粗糙度。

(5) According to the effect of measurement results on the process, it can be divided into passive measurement and active measurement.

按测量结果对工艺过程所起的作用，可分为被动测量和主动测量。

① Passive measurement refers to the measurement of completed parts. The measurement results are limited to discovering and removing non-conforming products.

被动测量指对完工零件进行的测量。测量结果仅限于发现并剔除不合格品。

② Active measurement refers to the measurement conducted during the machining process of a part. At this point, the measurement results can be directly used to control the processing process and prevent the generation of waste. For example, during the process of grinding the outer and inner raceways of the inner and outer rings of rolling bearings, the measuring head measures the grinding diameter size. When the size meets the qualified range, the grinding is stopped.

主动测量指在零件加工过程中所进行的测量。此时，测量结果可直接用来控制加工过程，以防止废品的产生。例如，在磨削滚动轴承内圈和外圈的外滚道与内滚道过程中，测量头测量磨削直径尺寸。当尺寸达到合格范围时，则停止磨削。

(6) According to the motion state of the measured part, it can be divided into static measurement and dynamic measurement.

按被测零件的运动状态，可分为静态测量和动态测量。

① Static measurement refers to the measurement where the measured surface is relatively stationary with respect to the measuring head during measurement. For example, measuring the diameter of a part with a micrometer.

静态测量指在测量时被测表面与测头相对静止的测量。例如，用千分尺测量零件的直径。

② Dynamic measurement refers to the measurement of relative motion between the measured surface and the measuring head during measurement. It can reflect the process of changes in the measured parameters. For example, when using an electric profilemeter to measure surface roughness.

动态测量指在测量时被测表面与测头之间有相对运动的测量。它能反映被测参数的变化过程。例如，用电动轮廓仪测量表面粗糙度。

Active measurement and dynamic measurement are the main development directions of measurement technology. The former can closely integrate processing and measurement, fundamentally changing the passive situation of measurement technology, while the latter can greatly improve measurement efficiency and ensure the quality of parts.

主动测量和动态测量是测量技术的主要发展方向。前者能将加工和测量紧密结合起来，从根本上改变测量技术的被动局面；后者能较大地提高测量效率和保证零件的质量。

3.2 Measurement Errors 测量误差

3.2.1 Measurement Error and Its Representation 测量误差及其表示

3.2.1.1 Measurement error 测量误差

Measurement error refers to the difference between the measurement result (measured value) and the measured true value. δ represents measurement error (also referred to as error in this chapter); x represents measured value, x_0 represents the measured true value, in actual measurement work, since the measured true value cannot be determined, the measurement error is approximated using the agreed true value (a quantity that is closer to the true value than the measured value, such as the arithmetic mean of multiple measurement results, the value of high-precision gauge blocks, etc.). In addition, the measurement error is an algebraic difference, which may be positive, negative, or zero.

测量误差是指测量结果（测得值）与被测量真值的差值。δ表示测量误差（本章中也简称误差），x表示测得值，x_0表示被测量真值，则在实际测量工作中，由于被测量真值无法确定，因此实际上是用约定真值（比测得值更接近真值的量值，如多次测量结果的算术平均值、高精度量块的量值等）近似确定测量误差。此外，测量误差是代数差，可能为正、负或零。

3.2.1.2 Representation of measurement errors 测量误差的表示

There are two ways to represent measurement errors: absolute error and relative error.

测量误差有绝对误差和相对误差两种表示方法。

Absolute error (Δ) refers to the defined measurement error, and its magnitude can reflect the level of measurement accuracy, to some extent. However, when comparing the measurement accuracy of different sizes, it is not reasonable to use the absolute error of the measurement. In this case, the factor of the measured size should be taken into account, thus giving rise to the concept of relative error. Relative error is defined as the ratio of absolute error to the measured true value (usually expressed as a percentage of the absolute value of the ratio), ε represents it. Since the true value is unknown, the measured value x is usually used instead of the true value to approximate the relative error, i.e.

绝对误差（Δ）就是定义所指的测量误差，其大小可以在一定程度上反映测量精度的高低。但当比较大小不同的被测量的测量精度时，用测量的绝对误差就显得不太合理。此时，应把被测量尺寸的因素考虑进去，于是便有了相对误差的概念。相对误差被定义为绝对误差与被测量真值的比值（通常用比值绝对值的百分数），用ε表示。由于真值是未知的，因此通常用测得值x代替真值来近似计算相对误差，即

$$\varepsilon=\frac{|x-x_0|}{x_0}\times 100\%=\frac{\Delta}{x_0}\approx\frac{\Delta}{x_0}\times 100\% \tag{3-2}$$

Based on relative error, the accuracy of multiple measurements of different sizes can be compared. The smaller the relative error of the measurement, the higher the measurement accuracy. For example, the two measured sizes are $x_1=30$ mm and $x_2=50$ mm, respectively, and the absolute errors for measuring these two quantities are $\Delta_1=+0.03$ mm and $\Delta_2=-0.04$ mm. Due to the different sizes of quantities, absolute error cannot be used, but relative error should be used to compare the accuracy of quantities. In this example, the relative errors are $\varepsilon_1=0.1\%$ and $\varepsilon_2=0.08\%$, so the measurement accuracy of the latter is higher than that of the former.

根据相对误差，可以对多个大小不同的被测量的测量精度的高低进行比较。测量的相对误差越小，表明测量精度越高。例如，两个被测量的大小分别为$x_1=30$ mm和$x_2=50$ mm，对这两个被测量进行测量的绝对误差分别为$\Delta_1=+0.03$ mm和$\Delta_2=-0.04$ mm。被测量的大小不同，因此不能用绝对误差，而应该用相对误差来比较测量精度的高低。本例中，相对误差$\varepsilon_1=0.1\%$，$\varepsilon_2=0.08\%$，故后者的测量精度比前者的高。

3.2.2 Sources of Measurement Errors 测量误差的来源

There are many factors that can cause measurement errors in the measurement process of geometric quantities. To improve measurement accuracy, sometimes it is necessary to analyze the causes of measurement errors, calculate the degree of influence of each error factor on the measurement results, and try to reduce these influences. The main sources of measurement errors are as follows.

在几何量的测量过程中，引起测量误差的因素有很多。为提高测量精度，有时要分析测量误差产生的原因，计算各误差因素对测量结果的影响程度，并设法减小这些影响。测量误差的来源主要有以下几个方面。

3.2.2.1 Measurement instrument error 测量器具误差

Measurement instrument error refers to the error caused by the design, manufacturing, and use of the measuring instrument itself, including the following parts.

测量器具误差是指测量器具本身在设计、制造和使用过程中所引起的误差，包括以下几部分。

（1）Principle error 原理误差。

To simplify the design, a considerable number of measuring instruments use approximate measurement transformation principles, and the measurement errors resulting from this are called principle errors. For example, in a lever gear comparator, the linear displacement of the measuring rod is not linearly related to the angular displacement of the pointer, and the error generated by the use of an equal division scale on the dial; Errors caused by neglecting the nonlinearity or incomplete compensation of circuits such as amplification, rectification, and filtering in electric measuring instruments belong to principle errors. There is also a type of principle error caused by the unreasonable structural arrangement of measuring instruments, which is the Abbe error caused by violating the Abbe principle. The Abbe principle states that accurate measurements can only be obtained when the measured axis coincides with or extends from the standard measurement axis. It can be seen that measuring instruments such as micrometers and inner micrometers comply with this principle, while vernier calipers do not.

为了简化设计，许多测量仪器采用了近似测量转换原理，由此产生的测量误差称为原理误差。例如，杠杆齿轮比较仪中，测杆的直线位移与指针的角位移不是线性关系，但表盘上却采用等分刻度所产生的误差即属于此类；电动量仪中忽略放大、整流、滤波等电路的非线性或非线性补偿不完全等所产生的误差，都属于原理误差。还有一类原理误差是由测量器具在结构布置上的不合理所造成的，即违反阿贝原则所引起的阿贝误差。阿贝原则指的是被测量轴线只有与标准量的测量轴线重合或在其延长线上时，测量才会得到精确的结果。由此可见，千分尺、内径千分尺等测量器具符合该原则，游标卡尺不符合。

(2) Reference component error 基准件误差。

The various errors of reference components used in measuring instruments to reflect standard quantities (measurement units), such as measuring blocks, line rulers, light wave lengths, etc., will also be reflected in the measurement results in a certain relationship.

测量器具中用来体现标准量（测量单位）的基准件（如量块、线纹尺、光波波长等）的各种误差，也将以一定的关系反映到测量结果中。

(3) Manufacturing, assembly, and adjustment error 制造、装配、调整误差。

The manufacturing, assembly, and adjustment error of the components of measuring instruments can also result in measurement errors, such as the carving error of vernier calipers, the installation eccentricity of the distribution of indicator dial markings and the rotation axis of pointers.

测量仪器的部件在制造、装配和调整过程中产生的误差也会产生测量误差，例如游标卡尺的刻度误差、指示表盘刻度线分布与指针旋转轴的安装偏心等。

(4) Usage error 使用误差。

After using measuring instruments for a period of time, deformation and wear of their components, as well as changes in electronic component parameters, can also result in measurement errors.

测量器具在使用一段时间后，其零部件的变形、磨损，电子元件参数的改变等，也会产生测量误差。

3.2.2.2 Measurement method error 测量方法误差

Measurement method error refers to the measurement error caused by imperfect measurement methods. For example, surface deformation errors of measuring instruments and parts caused by measuring forces in contact measurement, errors caused by imprecision of calculation formulas in indirect measurements; Errors caused by unreasonable workpiece clamping and positioning during the measurement process.

测量方法误差指的是由测量方法不完善而引起的测量误差。例如，接触式测量中的测量力引起的计量器具和零件表面变形误差、间接测量中计算公式的不精确引起的误差、测量过程中工件装夹定位不合理引起的误差。

3.2.2.3 Measurement environmental error 测量环境误差

Measurement environmental error refers to the measurement error caused by the environmental conditions during measurement not meeting the standard conditions. Environmental conditions refer to temperature, humidity, vibration, air pressure, dust, etc. Among them, temperature has the greatest impact on the measurement results. According to national standards, the measured standard temperature is 20 ℃. If the measurement environment causes the temperature of the measured part and the temperature of the measuring instrument to deviate from the standard temperature, it will cause measurement errors.

测量环境误差指的是由测量时的环境条件不符合标准条件而引起的测量误差。环境条件是指温度、湿度、振动、气压、灰尘等。其中，温度对测量结果的影响最大。根据国家标准的规定，测量的标准温度为20 ℃。若测量环境使被测件的温度及计量器具的温度偏离了标准温度，则会引起测量误差。

3.2.2.4 Personnel error 人员误差

Personnel error refers to measurement errors caused by subjective factors such as emotions, fatigue, technical proficiency, visual discrimination, aiming habits, etc. of the operator.

人员误差指的是由操作者主观上的因素（情绪、疲劳、技术熟练程度、眼睛的分辨能力、瞄准习惯等）所产生的测量误差。

3.2.3 Classification of Measurement Errors 测量误差的分类

According to their characteristics and properties, measurement errors can be divided into the following three categories.

根据特点和性质，测量误差可分为以下三类。

3.2.3.1 System error 系统误差

System error refers to a measurement error that, under certain measurement conditions, remains unchanged in size and sign or varies according to a certain pattern when measuring the same object multiple times. A system error that does not change in size and sign is called a fixed-value system error. For example, measurement error caused by incorrect zero position of a micrometer should be eliminated through the correction method. The correction value is equal to the opposite of the fixed value system error. Adding the correction value to the measured value is equal to subtracting the fixed value system error.

系统误差指的是在一定的测量条件下多次测量同一被测量时，误差的大小和符号均不变或按某一规律变化的测量误差。误差的大小和符号均不变化的系统误差称为定值系统误差。例如，千分尺零位不正确而产生的测量误差。这类误差应通过修正值法予以消除。修正值等于定值系统误差的相反数。给测得值加修正值。将修正值加到测量值上，等于减去定值系统误差。

The system error in which the magnitude and sign of the error vary according to a certain law (such as linear law, sine law, etc.) is called the variable value systematic error, such as the measurement error caused by environmental temperature at different temperatures. The principle for handling such errors is to determine the pattern of change and correct it through appropriate technical means based on the pattern of error variation.

误差的大小和符号按某一规律（如线性规律、正弦规律等）变化的系统误差称为变值系统误差，如环境温度引起的不同温度下的测量误差。对这类误差的处理原则是确定变化的规律后，根据误差的变化规律通过适当的技术手段予以修正。

In addition, based on the mastery of the size, sign, or variation pattern of errors, systematic errors can also be divided into two categories: determined system error and undetermined systematic error. Determined systematic error refers to the systematic error that has been mastered by the measurer in terms of size, sign, or variation pattern, while undetermined systematic error is exactly the opposite. The principles for handling determined system error are the same as before. The principle for handling undetermined system error is to estimate the range of errors and then treat them as random error。

另外，根据对误差的大小、符号或变化规律的掌握情况，系统误差还可分为已定系统误差和未定系统误差两类。已定系统误差指的是大小、符号或变化规律已被测量者掌握的系统误差，未定系统误差则正好相反。对已定系统误差的处理原则同前所述；对未定系统误差的处理原则是估计出误差的范围，然后将其按随机误差处理。

3.2.3.2 Random error 随机误差

Random error refers to a type of measurement error in which the magnitude and sign of the error occur in an unpredictable manner when measuring the same object multiple times under certain conditions. The so-called unpredictability refers to the inability to foresee the magnitude and sign of measurement errors that may occur in the next measurement. Although the size and sign of the random error in a single measurement are unpredictable, if the random error in multiple measurements under the same conditions is considered as a population, then the size and sign of the measurement random error in this population are distributed according to a certain distribution pattern (such as normal distribution, uniform distribution, t-distribution, etc.).

随机误差指的是在一定的条件下多次测量同一被测量时，误差的大小和符号均以不可预测的方式出现的一类测量误差。所谓不可预测，指的是无法预见下一次测量时所出现的测量误差的大小和符号。虽然某一次测量的随机误差的大小和符号不可预测，但若把相同条件下多次测量的测量随机误差看成一个总体，那么这一总体中测量随机误差的大小、符号却是按一定的分布规律（如正态分布、均匀分布、t分布等）分布的。

Random errors cannot be avoided and cannot be completely eliminated in theory. The principle for handling random errors is to reduce their impact using probability theory and mathematical statistics, and to estimate the possible range of errors (limit error) under a certain confidence probability.

随机误差无法避免，在理论上也无法完全消除。对随机误差的处理原则是:用概率论和数理统计的方法减小其影响，估计出在一定置信概率下误差（极限误差）可能出现的范围。

3.2.3.3 Gross error 粗大误差

Gross error refers to a type of measurement error that exceeds expectations and can be understood as an error that significantly distorts the measurement results. This type of error is often caused by subjective operational errors of the measurer (such as reading errors, aiming errors, or significant errors in the installation of the measured part) or unexpected changes in measurement conditions objectively (such as sudden vibrations, electrical pulse interference, etc.).

粗大误差指的是超出预计的一类测量误差，可以理解为明显歪曲测量结果的误差。这类误差往往是由测量者主观上的操作过失（如读数错误、瞄准错误或被测件安装出现较大误差等）或客观上测量条件的意外变化（如突发振动、电脉冲干扰等）导致的。

The presence of gross errors can significantly distort the measurement results, so measurement results containing gross errors cannot be used. The principle for handling such errors is to identify the measured values containing gross errors according to certain discrimination criteria, and then remove them from the measured value sequence (measurement column).

粗大误差的存在会明显歪曲测量结果，所以含有粗大误差的测量结果是不能用的。对这类误差的处理原则是：按一定的判别准则找出含有粗大误差的测得值，然后从测得值序列（测量列）中将它们剔除。

3.2.4 Expression Method of Measurement Accuracy 测量精度的表述方法

The terms commonly mentioned about measurement accuracy include measurement precision, trueness, accuracy, etc., which will be introduced one by one below.

人们经常提到的关于测量精度的术语主要有测量的正（准）确度、精密度、精确度等，下面将对此进行介绍。

3.2.4.1 Trueness 正（准）确度

Trueness reflects the magnitude of the systematic error contained in the measurement results. The smaller systematic error, the higher the trueness (accurate) precision.

正（准）确度反映的是测量结果中所含系统误差的大小，系统误差越小，正（准）确度越高。

3.2.4.2 Precision 精密度

Precision reflects the magnitude of the random error contained in the measurement results. The smaller the random error, the higher the precision.

精密度反映的是测量结果中所含随机误差的大小。随机误差越小，精密度越高。

3.2.4.3 Accuracy 精确度

Accuracy is the combination of positive trueness and precision, which reflects the systematic and random errors contained in measurement results. The accuracy of measurement is only high when both trueness and precision are high. The schematic diagram of the shooting results shown in Fig. 3-3 illustrates the meaning of the three terms mentioned above and their relationship with each other: Fig. 3-3(a) has small random error but large systematic error, that is, high precision but low accuracy; Fig. 3-3(b) shows that the system error is small while the random error is large, indicating high accuracy but low precision; Fig. 3-3(c) shows that both random error and systematic error are small, indicating high accuracy.

精确度（简称精度）是正（准）确度和精密度的综合，是对测量结果中所含系统误差与随机误差的综合反映。只有当正（准）确度、精密度二者都高时，测量的精确度才高。图3-3所示的打靶结果示意图说明了上述三个术语的意义及其相互之间的关系：图3-3(a) 随机误差小而系统误差大，即精密度高而正确度低；图3-3(b) 表示系统误差小而随机误差大，即正确度高而精密度低；图3-3(c) 表示随机误差和系统误差都小，即精确度高。

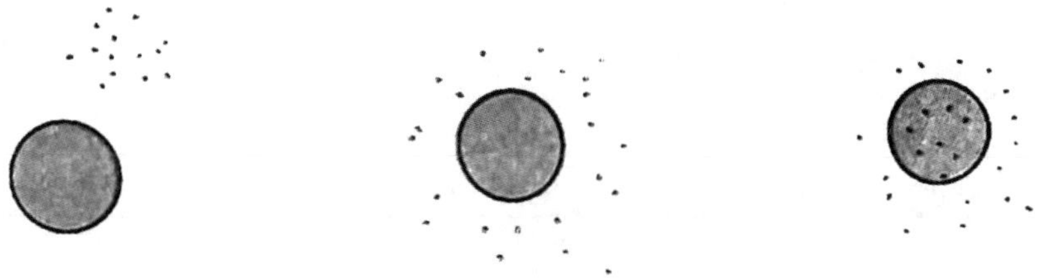

(a) High trueness 精密度高　(b) High precision 正（准）确度高　(c) High accuracy 精确度高

Fig. 3-3 Trueness, precision, accuracy, and their interrelations

图3-3 精密度、正（准）确度、精确度及其关系

3.2.5 Handling of Random Error 随机误差的处理

3.2.5.1 The distribution pattern of random error 随机误差的分布规律

As mentioned earlier, random errors, as a whole, have their inherent patterns. For example, when a certain part of a workpiece is measured 150 times repeatedly under the same measurement conditions using the same method, 150 different measured values are obtained (this series of measured values is often referred to as a measurement sequence). Then, the maximum and minimum measured values are identified, and the difference between the maximum and minimum values gives a dispersion range of the measured values from 7.131 to 7.141 mm. These values are divided into 11 groups with an interval of 0.001 mm, and the number of occurrences n_i in each group is counted. The frequency of each group (the ratio of the number of occurrences *ni* to the total number of measurements N) is calculated, as shown in Tab. 3-4.

前面提到，随机误差就其整体来说是有其内在规律的。例如，在相同测量条件下对一个工件的某一部位用同一方法进行150次重复测量，测得150个不同的读数（这一系列的测得值常称为测量列），然后找出其中的最大测得值和最小测得值，用最大值减去最小值得到测得值的分散范围为7.131～7.141 mm，以每隔0.001 mm为一组分成11组，统计出每一组出现的次数n_i，计算每一组频率（次数n_i与测量总次数N之比），见表3-4。

By plotting the measured value x as the horizontal axis and the frequency n_i/N at which the measured value appears within a certain interval as the vertical axis, a statistical histogram can be obtained as shown in Fig. 3-4. The line connecting the midpoints of each interval is called the actual distribution curve.

以测得值x为横坐标、测得值出现在某一区间内的频率n_i/N为纵坐标作图，可以得到如图3-4所示的统计直方图。各区间中点连成的折线称为实际分布曲线。

Tab. 3-4 Distribution and characteristics of random errors

表3-4 随机误差的分布及其特征

Measurement range 测量值范围	Measurement median 测量中值	Number of occurrences 出现次数(n)	Relative frequency of occurrence 相对出现频率(n_i/N)
7.1305～7.1315	$x_1=7.131$	$n_1=1$	0.007
7.1315～7.1325	$x_2=7.132$	$n_2=3$	0.020
7.1325～7.1335	$x_3=7.133$	$n_3=8$	0.054
7.1335～7.1345	$x_4=7.134$	$n_4=18$	0.120
7.1345～7.1355	$x_5=7.135$	$n_5=28$	0.187
7.1355～7.1365	$x_6=7.136$	$n_6=34$	0.227
7.1365～7.1375	$x_7=7.137$	$n_7=29$	0.193
7.1375～7.1385	$x_8=7.138$	$n_8=17$	0.113
7.1385～7.1395	$x_9=7.139$	$n_9=9$	0.060
7.1395～7.1405	$x_{10}=7.140$	$n_{10}=2$	0.013
7.1405～7.1415	$x_{11}=7.141$	$n_{11}=1$	0.007

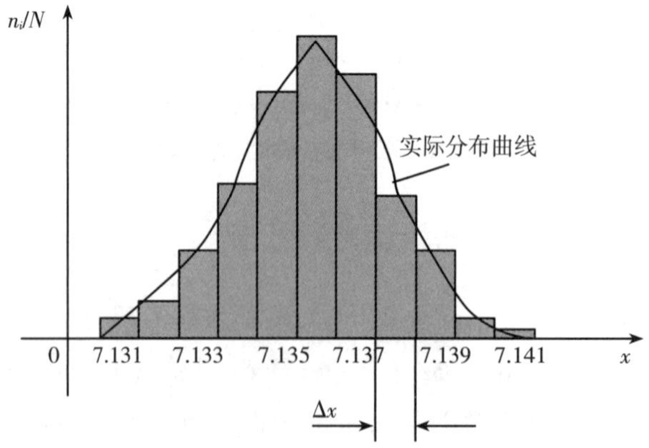

Fig. 3-4 The actual distribution curve of random errors

图3-4 随机误差实际分布曲线

If the number of measurements $n_i \to \infty$, and the interval of the grouping bins $\Delta x \to 0$, a smooth distribution curve can be obtained, as shown in Fig. 3-5. Experiments have shown that, except for a few cases, random measurement errors mostly follow a normal distribution. The distribution curve in Fig. 3-5 is called a normal distribution curve (Gaussian curve), and its vertical coordinate is referred to as probability density.

如果测量次数$n_i \to \infty$，分组区间的间隔$\Delta x \to 0$，则可得到如图3-5所示的光滑分布曲线。实验表明，除少数情况外，测量的随机误差大多服从正态分布。图3-5中的分布曲线称为正态分布曲线（高斯曲线），其纵坐标称为概率密度。

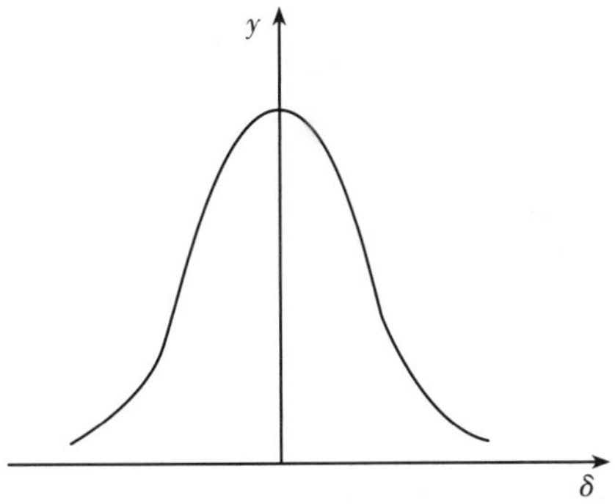

Fig. 3-5 The normal distribution curve

图3-5 正态分布曲线

Random errors that follow a normal distribution have the following characteristics.

服从正态分布的随机误差具有以下几个特性。

(1) unimodality: Small absolute random errors are more likely to occur than large absolute random errors. It can also be seen from the distribution curve and probability density function that there is a peak at $\delta = 0$, indicating that the probability of random error occurring with an absolute value of zero is the highest.

单峰性，绝对值小的随机误差比绝对值大的随机误差出现的概率大。从分布曲线及概率密度函数中也可以看出，在$\delta = 0$处存在一个峰值，表明绝对值为零的随机误差出现的概率最大。

(2) Symmetry means that the probability of random errors with equal absolute values and opposite signs occurring is equal.

对称性，绝对值相等、符号相反的随机误差出现的概率相等。

(3) Compensatory nature: Under the same measurement conditions, multiple repeated measurements are made on the same subject, and the algebraic sum of all random errors is equal to zero.

抵偿性，在相同的测量条件下，对同一被测量进行多次重复测量，所有随机误差的代数和等于零。

(4) Boundedness means that under certain measurement conditions, the distribution range of random errors is limited, that is, the absolute value of random errors will not exceed a certain limit.

有界性，在一定的测量条件下，随机误差的分布范围是有限的，即随机误差的绝对值不会超过一定的界限。

By utilizing the above characteristics, statistical methods can be used to analyze and process measurement random errors, minimizing the impact of random errors on measurement results as much as possible.

利用上述特性，可以用数理统计的方法分析、处理测量的随机误差，尽可能减小随机误差对测量结果的影响。

The mathematical expression for the normal distribution curve is:

正态分布曲线的数学表达式为：

$$y=\frac{1}{\sigma\sqrt{2\pi}}e^{\frac{-\delta^2}{2\sigma^2}} \tag{3-3}$$

In the formula,

where, y—probability density;

σ—standard deviation;

δ—random error;

e—the base of natural logarithms;

式中，y——概率密度；

σ——标准偏差；

δ——随机误差；

e——自然对数的底。

3.2.5.2. Standard deviation of random error σ 随机误差的标准偏差σ

From Eq. (3-3), it can be seen that the probability density is related to the random error δ and standard deviation σ. When $\delta=0$, the probability density y is maximum, ie. $y_{max}=\frac{1}{\sigma\sqrt{2\pi}}$, and it is evident that the maximum probability density (y_{max}) varies with the change of standard deviation σ. The smaller the size of standard deviation σ, the steeper the distribution curve, the more concentrated the distribution of random errors, and the higher the standard measurement accuracy. On the contrary, the larger the size of standard deviation σ, the flatter the distribution curve, and the more dispersed the distribution of random error, indicating lower measurement accuracy. The standard deviation of random error (σ) and arithmetic mean ($\bar{x}$) can be calculated using :

从式（3-3）中可以看出，概率密度大小与随机误差δ、标准偏差σ有关。当$\delta=0$时，概率密度y最大，即$y_{max}=\frac{1}{\sigma\sqrt{2\pi}}$，显然概率密度最大值$y_{max}$是随标准偏差$\sigma$变化的。标准偏差$\sigma$越小，分布曲线越陡，随机误差的分布越集中，标准测量精度越高。反之，标准偏差σ越大，分布曲线就越平坦，随机误差的分布越分散，表示测量精度就越低。随机误差的标准偏差公差σ和算术平均值$\bar{x}$的计算式为：

$$\sigma=\sqrt{\frac{\sum_{i=1}^{n}(x_i-\bar{x})^2}{n-1}} \tag{3-4}$$

$$\bar{x}=\frac{1}{n}\sum_{i=1}^{n}x_i \tag{3-5}$$

Where, n—the number of measurements, where n is adequately large during experimental procedures (commonly, n ranges from 10 to 20);

$\bar{x}$—the arithmetic mean derived from n measurements;

x_i—an individual measurement value.

式中，n——测量次数，实验时n取足够大（一般取$n=10\sim20$）；

$\bar{x}$——n次测量的算术平均值；

x_i——某次测量值。

Standard deviation σ is an indicator that reflects the degree of dispersion of measured values in a measurement column. It represents the standard deviation of a single measurement value (any measured value) in the measurement column. The standard deviation of the arithmetic mean is denoted as $\sigma_{\bar{x}}$, which is related to the number of measurements n.

标准偏差σ是反映测量列中测得值分散程度的一项指标，它表示的是测量列中单次测量值（任一测得值）的标准偏差。用$\sigma_{\bar{x}}$表示算术平均值标准偏差，其数值与测量次数n有关，即

$$\sigma_{\bar{x}} = \frac{\sigma}{\sqrt{n}} \tag{3-6}$$

3.2.5.3 Residual (remaining error) and its application 残差（剩余误差）及其应用

The residual is the error calculated by replacing the true value Q with the arithmetic mean $\bar{x}$, i.e., the difference between a measured value x_i in the measurement series and the arithmetic mean of that series, denoted as v_i.

残差是指用算术平均值$\bar{x}$代替真值Q后计算得到的误差，即测量列中的一个测得值x_i和该测量列的算术平均值$\bar{x}$之差，记为v_i。

$$v_i = x_i - \bar{x} \tag{3-7}$$

Based on the distribution characteristics of random errors that follow the normal curve distribution, residuals have the following two properties:

由符合正态曲线分布规律的随机误差的分布特性可知，残差具有下述两个特性:

(1) When the number of measurements is sufficiently large, the algebraic sum of the residuals approaches zero. i.e., $\sum_{i=1}^{n} v_i \approx 0$.

当测量次数足够多时，残差的代数和趋近于零，即$\sum_{i=1}^{n} v_i \approx 0$。

(2) The sum of the squares of the residuals is minimized. i.e., $\sum_{i=1}^{n} v_i^2$ is the minimum value.

残差的平方和为最小，即$\sum_{i=1}^{n} v_i^2$为最小值。

In practical applications, it is commonly used to verify whether the value $\bar{x}$ obtained in data processing is correct in relation to v_i by checking if $\sum_{i=1}^{n} v_i \approx 0$.

实际应用中，常用$\sum_{i=1}^{n} v_i \approx 0$来验证数据处理中求得的$\bar{x}$与$v_i$是否正确。

3.2.5.4 Limit value of random error ($\delta_{\lim}$) 随机误差的极限值$\delta_{\lim}$

Due to the boundedness of random errors, the size of random errors will not exceed a certain range. The limit value of random error is the measurement limit error.

由于随机误差具有有界性，因此随机误差的大小不会超过一定的范围。随机误差的极限值就是测量极限误差。

According to the knowledge of probability theory, the area between the normal distribution curve and the horizontal axis is equal to the sum of the probabilities of all random errors. If the random error interval falls between $(-\infty \sim +\infty)$, its probability is 1, that is $P=\int_{-\infty}^{+\infty} y\mathrm{d}\delta=\int_{-\infty}^{+\infty}\frac{1}{\sigma\sqrt{2\pi}}\mathrm{e}^{\frac{-\delta^2}{2e^2}}\mathrm{d}\delta=1$. In fact, the random error interval falls within $(-\delta \sim +\delta)$, the probability is less than 1, that is $P=\int_{-\delta}^{+\delta} y\mathrm{d}\delta<1$. For transforming into a standard normal distribution for easy calculation, easy to calculate the integration value (probability value) of $P=\int_{-\delta}^{+\delta} y\mathrm{d}\delta$, the probability integral calculation process is as follows.

由概率论的知识可知，正态分布曲线和横坐标轴间所包含的面积等于所有随机误差出现的概率总和。若随机误差区间为（$-\infty \sim +\infty$），则其概率为1，即 $P=\int_{-\infty}^{+\infty} y\mathrm{d}\delta=\int_{-\infty}^{+\infty}\frac{1}{\sigma\sqrt{2\pi}}\mathrm{e}^{\frac{-\delta^2}{2e^2}}\mathrm{d}\delta=1$。实际上，随机误差区间为（$-\delta \sim +\delta$），其概率小于1，即 $P=\int_{-\delta}^{+\delta} y\mathrm{d}\delta<1$。为化成标准正态分布，便于求出 $P=\int_{-\delta}^{+\delta} y\mathrm{d}\delta$ 的积分值（概率值），其概率积分计算过程如下。

首先引入：

$$t=\frac{\delta}{\sigma},\quad \mathrm{d}t=\frac{\mathrm{d}\delta}{\sigma}\quad(\delta=\sigma t,\quad \mathrm{d}\delta=\sigma\mathrm{d}t)$$

则有

$$P=\int_{-\delta}^{+\delta} y\mathrm{d}\delta=\int_{-\delta}^{+\delta}\frac{1}{\sqrt{2\pi}}\mathrm{e}^{\frac{-t^2}{2}}\mathrm{d}t=\frac{2}{\sqrt{2\pi}}\int_{-\infty}^{+\infty}\mathrm{e}^{\frac{-t^2}{2}}\mathrm{d}t \tag{3-8}$$

再令

$$P=2\phi(t)$$

则有

$$\phi(t)=\frac{1}{\sqrt{2\pi}}\int_{0}^{+t}\mathrm{e}^{\frac{-t^2}{2}}\mathrm{d}t \tag{3-9}$$

This is the Laplace Function (probability integral). The commonly used $\phi(t)$ values are listed in Tab. 3-5. Choosing different t values corresponds to different probabilities, and the credibility of the measurement results also varies.

这就是拉普拉斯函数（概率积分）。常用的$\phi(t)$数值列在表3-5当中。选择不同的t值，就会对应不同的概率，测量结果的可信度也就不一样。

The probability of occurrence of random error within the range of $\pm t\sigma$ is called confidence probability, and t is called confidence factor or confidence coefficient. In geometric measurement, the confidence factor $t=3$ is usually taken, and the confidence probability is $P=2\phi(t)=99.73\%$. That is the probability of δ exceeding $\pm 3\sigma$ is $100\%-99.73\%=0.27\%\approx 1/370$. In actual measurements, the number of measurements is generally not more than a few dozen. The situation of random error exceeding 3σ is actually rare. So the measurement limit error is taken as $\delta_{\lim}=\pm 3\sigma$。$\delta_{\lim}$ also represents the measurement limit error of a single measurement value in the measurement column.

随机误差在$\pm t\sigma$范围内出现的概率称为置信概率，t称为置信因子或置信系数。在几何量测量中，通常取置信因子$t=3$，则置信概率为$P=2\phi(t)=99.73\%$。亦即δ超出$\pm 3\sigma$的概率为$100\%-99.73\%=0.27\%\approx 1/370$。在实际测量中，测量次数一般不会多于几十次。随机误差超出3σ的情况实际上很少出现。所以，取测量极限误差为$\delta_{\lim}=\pm 3\sigma$。$\delta_{\lim}$也表示测量列中单次测量值的测量极限误差。

Tab. 3–5 The Laplace Function (the frequencies associated with four particular *t*-values)

表3–5 拉普拉斯函数（4个特殊t值对应的频率）

t	$\delta=\pm t\sigma$	The probability P of not exceeding $\lvert\delta\rvert$ 不超出$\lvert\delta\rvert$的概率$P=2\phi(t)$	The probability P of exceeding $\lvert\delta\rvert$ 超出$\lvert\delta\rvert$的概率$P=1-2\phi(t)$
1	1σ	0.6826	0.3174
2	2σ	0.9544	0.0456
3	3σ	0.9973	0.0027
4	4σ	0.99936	0.00064

Example 3–2: The measurement values obtained from 10 equal-precision measurements of a certain part's dimension on a certain instrument are shown in Tab. 3–6. It is known that there is no systematic error in the measurements. Calculate the standard deviation σ of a single measurement and the standard deviation of the arithmetic mean $\sigma_{\bar{x}}$, and provide the precision when using a single measurement value as the result and when using the arithmetic mean as the result, respectively.

Tab. 3–6 Measurement data for Example 3–1

表3–6 例3–1的测量数据

Measurement sequence number 测量序号（i）	Measurement value 测量值(x_i)/mm	$x_i-\bar{x}$ /μm	$(x_i-\bar{x})^2$ /μm^2
1	20.008	+1	1
2	20.004	−3	9
3	20.008	+1	1
4	20.009	+2	4
5	20.007	0	0
6	20.008	+1	1
7	20.007	0	0
8	20.006	−1	1
9	20.008	+1	1
10	20.005	−2	4
	$\bar{x}=\frac{1}{10}\sum_{i=1}^{10}x_i=20.007$	$\sum_{i=1}^{10}(x_i-\bar{x})=0$	$\sum_{i=1}^{10}(x_i-\bar{x})^2=22$

例3–2：在某仪器上对某零件尺寸进行10次等精度测量，得到的测量值，见表3–6。已知测量中不存在系统误差、计算单次测量的标准偏差σ和算术平均值的标准偏差$\sigma_{\bar{x}}$，并分别给出以单次测量值作为结果和以算术平均值作为结果的精度。

Solution: From Eq. (3–4), (3–5), and (3–6), we obtain the arithmetic mean of the measurements, the standard deviation of a single measurement, and the standard deviation of multiple measurements, respectively.

解：由式（3–4）、式（3–5）、式（3–6）算术平均值、单次测量的标准偏差和多次测量的标准偏差分别为：

$$\bar{x}=\frac{1}{10}\sum_{i=1}^{10}x_i=20.007\ (\text{mm})$$

$$\sigma = \sqrt{\frac{\sum_{i=1}^{n}(x_i - \bar{x})^2}{n-1}} \approx 1.6\ (\mu m)$$

$$\sigma_{\bar{x}} = \frac{\sigma}{\sqrt{n}} = \frac{1.6}{\sqrt{n10}} \approx 0.5\ (\mu m)$$

Therefore, the precision when using a single measurement value as the result is $\pm 3\sigma \approx \pm 5$ μm; the precision when using the arithmetic mean as the result is $\pm 3\sigma_i = \pm 1.5$ μm。

因此，以单次测量值做结果的精度为$\pm 3\sigma \approx \pm 5$ μm；以算术平均值做结果的精度为$\pm 3\sigma_i = \pm 1.5$ μm。

Exercise 3 习题3

3-1 To composite the following sizes by select suitable gauge blocks from total 83- block set:（1）24.575 mm （2）29.935 mm （3）68.875 mm （4）72.94 mm

试用83块一套的量块，选择组成尺寸为24.545 mm、29.935 mm、68.875 mm和72.94 mm的量块组。

3-2 The measured values of a certain geometric quantity, obtained from 15 measurements, are as follows (unit: millimeters): 30.742, 30.743, 30.740, 30.741, 30.739, 30.740, 30.739, 30.741, 30.742, 30.743, 30.739, 30.740, 30.743, 30.742, 30.741.Calculate the standard deviation and limit error of a single measurement。

对某几何量进行了15次测量，测得值如下（单位：mm）：30.742，30.743，30.740，30.741，30.739，30.740，30.739，30.741，30.742，30.743，30.739，30.740，30.743，30.742，30.741。求单次测量的标准偏差和极限误差。

Chapter 4 Geometric Accuracy Design and Inspection 几何精度设计与检测

4.1 Basic Concept 基本概念

4.1.1 Definition of Geometric Errors 几何误差定义

After machining, there are differences between the actual geometric solids of the parts and their ideal counterparts, which manifest as deviations in the shape, orientation, and position of the geometric features of the parts. All these errors are called geometric errors. Geometric tolerance is given in design, which is used to control the geometric error produced by machining.

加工后零件的实际几何体和理想几何体之间存在差异，即零件几何要素的形状、方向和位置上的偏差，统称为几何误差。几何公差就是设计时给出的，用于控制加工产生的几何误差。

4.1.2 Cause of Geometric Errors and Effects on Performance 几何误差产生的原因及影响

4.1.2.1 Cause of geometric errors 几何误差产生的原因

There are lots of causes for geometric errors of workpiece during manufacturing. Firstly, there are errors of the process system, which consist of machining tools, fixtures, cutting tools and workpieces. Secondly, workpiece deformation caused by force and heat during manufacturing occurs. Thirdly, there is workbench vibration and tool wear. So it is inevitable that there are errors of these machined geometric elements.

在制造过程中，造成工件几何误差的原因有很多。第一，加工刀具、夹具、刀具和工件组成的工艺系统误差；第二，在制造过程中，工件受力、受热变形；第三，工作台振动和刀具的磨损等。这些被加工的几何要素都不可避免地存在误差。

4.1.2.2 Effects on service life and performance of the geometric errors 几何误差对产品的寿命和使用性能的影响

The effects of the geometric errors on service life and performance include the following three aspects:

几何误差对产品的寿命和使用性能的影响可以归纳为以下三个方面：

(1) Effect on assemblability. For example, the location errors of holes will cause difficulty to insert the bolts.

对可装配性的影响。例如，孔的位置误差会导致螺栓难以插入。

(2) Effect on fit properties. For example, the form errors of the contact surface of the hole and the shaft will cause that the gap distribution is non-uniform in clearance fit, then the movement is lopsided if there is

relative motion. In interference fit, the interference value is nonuniform, and then the strength of connection will be decreased.

对配合性质的影响。例如，孔和轴的接触表面存在形状误差，在间隙配合中会导致间隙分布不均匀，进而在相对运动时产生偏斜。在过盈配合中，过盈量不均匀，则会降低连接的强度。

（3） Effect on work accuracy. For example the straightness error of lathe-bed guide rail will influence the movement accuracy of the saddle; the form error and location error of two support journals of the lathe spindle will influence the rotary accuracy of the lathe spindle.

对工作精度的影响。例如，车床床身导轨的直线度误差会影响鞍座的移动精度；车床主轴两个支撑轴颈的形状误差和位置误差会影响车床主轴的旋转精度。

4.2 Geometric Feature and Geometric Tolerance Symbol 零件几何要素和几何公差项目

4.2.1 Geometric Feature 几何要素

4.2.1.1 Size 尺寸

A geometric shape is defined by a linear or angular dimension which is a size. The feature of size can be a cylinder, a sphere, two parallel opposite surfaces, a cone or a wedge.

由一定大小的线性尺寸或角度尺寸确定的几何形状称为尺寸要素，尺寸要素可以是圆柱形、球形、两个平行对应面、圆锥形或楔形。

4.2.1.2 Classificaton 分类

The research object of geometric tolerance is the geometric features of mechanical parts. Geometric features are the general designation of point, lines and surfaces that constitute the geometric features of parts. In order to facilitate the study of geometric tolerances and geometric errors, these geometric features can be classified in different ways.

几何公差的研究对象是机械零件的几何要素。几何要素是构成零件几何特征的点、线、面的统称。为了便于研究几何公差和几何误差，这些几何要素可以从不同角度进行分类。

（1） According to the structural characteristics, they are divided into integral features and derived features. Integral features, also called contour features, the surface and the lines on the surface of a part are called integral features, such as cylinder surface, two end surfaces and plain line on a cylinder surface. Derived features refer to center point, median line or median surface which are derived from integral features. The points, lines and surfaces that make up the symmetric center of integral features are called derived features, also known as center feature, such as the central axis of a cylinder and the center of a sphere, as shown in Fig. 4-1.

几何要素按结构特征分为组成要素和导出要素。组成要素也被称为轮廓要素，指的是构成轮廓的面或面上的线。例如，圆柱体的圆柱面两端面、圆柱面上的素线。导出要素指的是组成要素对称中心的点、线、面各要素，也被称为中心要素。例如，圆柱体的中心轴线，球体的球心，如图4-1所示。

（2） According to the existing state, they are divided into nominal features and actual features. In a drawing, only geometric features without errors are called nominal features. Nominal features have no errors, and the features represented by mechanical drawings are nominal features. The actual existing features of parts are called actual features, which are usually replaced by measured features.

几何要素按存在状态分为公称要素和实际要素。在图样中，只具有几何意义的要素称为公称要素。公称要素没有任何误差, 机械图样所表示的要素均为公称要素。零件实际存在的要素称为实际要素，通常由测量得到的要素替代。

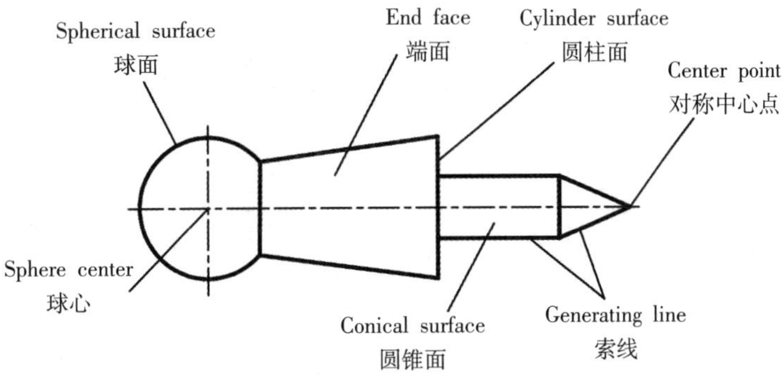

Fig. 4-1 Geometric feature

图4-1 零件的几何要素

4.2.1.3 Single feature 单一要素

The geometric tolerance is unrelated to other features, such as the straightness of the element line of the shaft. And the axis can belong to both a cylinder and a cone.

几何公差与其他特性无关，例如，轴线的直线度，这个轴线可以是圆柱也可以是圆锥。

4.2.1.4 Related feature 关联要素

The geometric tolerance is related to other feature(s), such as a line parallel to another line, the former is a tolerance feature and the latter is datum feature, the former geometric tolerance requirement is related to the latter. Usually, there are four kinds of datum.

几何公差与其他特征有关，如一条线与另一条线平行，前者是公差特征，后者是基准特征，前者的几何公差要求与后者有关。通常，基准有4种。

(1) Single datum: the datum only established by one feature is called single datum. A datum is one axis or one plane.

单一基准：只由单一特征建立的基准称为单一基准。一个基准可以是一个轴或一个平面。

(2) Common datum:one datum established by two or more than two features is called common datum.

公共基准：由两个或两个以上特征建立的基准称为公共基准。

(3) Datum system (three plane datum system): refers to a datum system consisting of three planes perpendicular to each other.

基准体系（三基面体系）：指由互相垂直的三个基准平面组成的基准体系。

(4) Datum target: refers to a datum that is applied only to part of a feature (such as points, lines or local areas that contact with the fixture on the workpiece).

基准目标：指只应用基准的某一特征的一部分（如点、线或局部与夹具接触的工件）。

4.2.2 Symbols and Characteristics of Geometric Tolerances 几何公差符号与特征

There are 14 geometric tolerances in International Standard ISO1101 and National Standard GB/T 1182-2018, the geometric tolerances and tolerance symbols are shown in Tab. 4-1.

国际标准ISO1101和国家标准GB/T 1182—2018将几何公差分为14个项目，它们的名称和符号见表4-1。

Tab. 4-1 Geometric tolerances and tolerance symbols

表4-1 几何公差和几何公差符号

Type of tolerance 公差类型	Geometric feature 几何特征	Symbol 符号	Datum needed 有无基准	Type of tolerance 公差类型	Geometric feature 几何特征	Symbol 符号	Datum needed 有无基准
Tolerance of form 形状公差	Straightness 直线度	⏤	No 无	Tolerance of orientation 方向公差	Parallelism 平行度	∥	Yes 有
	Flatness 平面度	⏥	No 无		Perpendicularity 垂直度	⊥	Yes 有
	Roundness 圆度	○	No 无		Angularity 倾斜度	∠	Yes 有
	Cylindricity 圆柱度	⌭	No 无	Tolerance of location 位置公差	Concentricity or coaxiality 同心度或同轴度	◎	Yes 有
Tolerance of form, orientation or location 形状、方向或位置公差	Profile of a line 线轮廓度	⌒	Yes or No 有或无		Symmetry 对称度	⌯	Yes 有
	Profile of a surface 面轮廓度	⌓	Yes or no 有或无		Position 位置度	⌖	Yes or No 有或无
				Tolerance of run-out 跳动公差	Circular run-out 圆跳动	↗	Yes 有
					Total run-out 全跳动	⌰	Yes 有

4.3 Indication of Geometric Tolerance on Engineer Drawings 几何公差在工程图上的标注

According to the ISO and GB/T1182-2018 standard, in engineering drawings, the geometric tolerance is indicated by using tolerance frames, leader lines, geometric tolerance characteristic symbols, and tolerance values and related symbols. When requirements cannot be indicated with symbols (if unclear or too complex), text is used to describe the requirements in the technical notes.

根据ISO和国家标准GB/T1182—2018的规定，在技术图样中，几何公差应采用几何公差框格、指引线、几何公差特征的符号、公差数值和有关符号等来表示。但是，当要求无法用符号表示（无法清晰表示或表示过于复杂）时，则在技术要求中用文字来描述这些要求。

Geometric tolerance is marked in the form of rectangular tolerance frames on the drawing, as shown in Fig. 4-2 (a). A rectangular tolerance frame is divided into two, three or more compartments. When the measured feature is form tolerance, the tolerance frame has only the first two sections. When the measured feature is orientation, position or run-out tolerance, the tolerance frame is composed of three sections, and the third is optional. The datum section may consist of one to three compartments. The leader with an arrow should point to the corresponding measured feature.

几何公差在图样上用矩形公差框格的形式标注，如图4-2（a）所示。一个公差框格包括两部分、三部分或更多的矩形框格。当被测要素是形状公差时，公差框格只有前面两部分。当被测要素是方向、位置或跳动公差时，公差框格由三个格组成，其中第三个格为可选的基准格，基准格可包含1～3格。带箭头的指引线应指向相应的被测要素。

These compartments contain, from left to right when the tolerance frame is placed in the horizontal direction, in the following order, see Fig. 4-2 (b):

各格公差要求按自左至右顺序水平标注以下内容，如图4-2（b）：

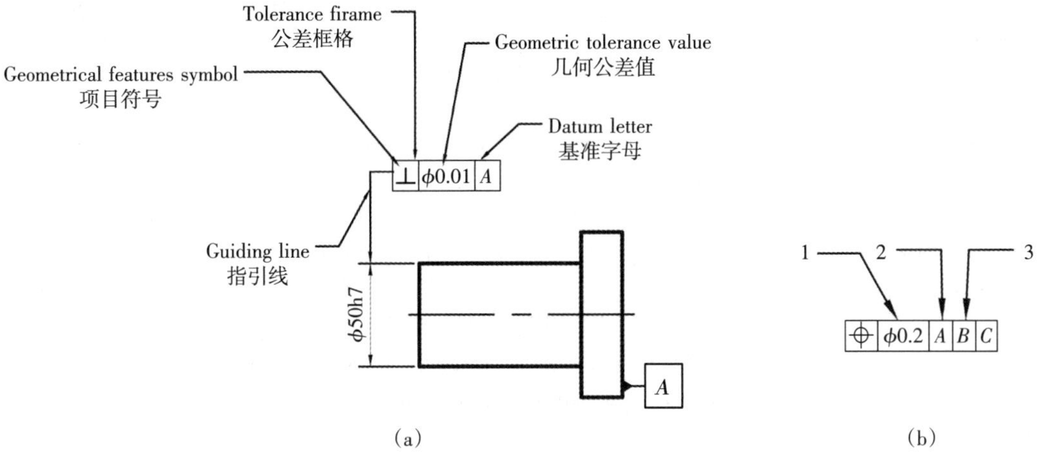

Fig. 4-2 Geometric tolerance frame

图4-2 几何公差框格

(1) First compartment—the symbol for the geometric characteristic.

第一格——几何特征符号。

(2) Second compartment—the geometric tolerance value and its associated requirements. The unit of tolerance value is millimeters (mm). The width of the tolerance zone t is described in linear dimensions (mm) and the pertinent specifications. If the tolerance zone is circular or cylindrical, the value is preceded by the symbol "ϕ", and if the tolerance zone is spherical, the value is preceded by "$S\phi$", Sometimes, more requirements or some tolerance principles are needed for geometric tolerance zone, and the additional symbols should be indicated in the second rectangular frame (being added after the tolerance value). Some additional symbols are shown in Tab. 4-2 and Tab. 4-3.

第二格——几何公差数值和有关要求，公差值的单位为mm。公差带大小用它的宽度或直径表示，由给定的公差值t决定。如果公差带的形状为圆形或圆柱形，则在公差值前应加注符号"ϕ"；如果公差带为圆球形，则在公差值前加注符号"$S\phi$"。有时，几何公差带需要更多的要求或采用一些公差原则，此时应在第二个矩形框中（或公差值后）加注这些附加符号。常用的附加符号见表4-2和表4-3。

Tab. 4-2 Additional symbols (From GB/T 1182-2018)

表4-2 附加符号（摘自GB/T 1182—2018）

Description 说明	Symbol 符号	Description 说明	Symbol 符号
Tolerance feature 被测要素	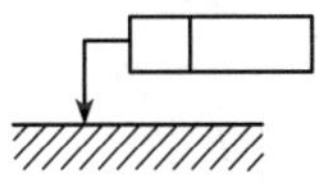	Datum feature 基准要素	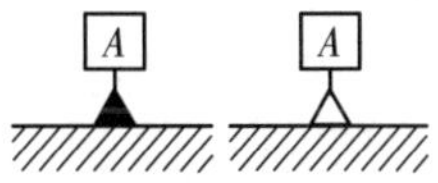
Datum target 基准目标	ϕ2 / A1	Theoretically exact dimension 理论正确尺寸	50

Tab. 4-2（Continued）表4-2（续）

Description 说明	Symbol 符号	Description 说明	Symbol 符号
Projected tolerance zone 延伸公差带	Ⓟ	Maximum material requirement 最大实体要求	Ⓜ
Last material requirement 最小实体要求	Ⓛ	Free status condition（Non-rigid parts） 自由状态条件（非刚性零件）	Ⓕ
All around profile 全周（轮廓）	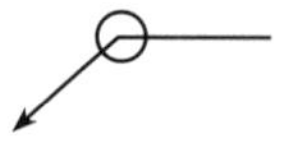	Envelope requirement 包容要求	Ⓔ
Common tolerance zone 公共公差带	CZ	Reciprocity requirement 可逆要求	Ⓡ

Tab. 4-3 Additional symbols in geometric tolerance annotation
表4-3 几何公差标注中的部分附加符号

Explanation 含义	Symbol 符号
The measured feature is only allowed to protrude outward from the material in the middle 被测要素只许中间向材料外凸起	(+)
The measured feature is only allowed to indent inward into the material in the middle 被测要素只许中间向材料内凹下	(−)
The measured feature is only allowed to decrease from left to right in the direction indicated by the symbol 被测要素只许按符号的方向从左至右减小	▷
The measured feature is only allowed to decrease from right to left in the direction indicated by the symbol 被测要素只许按符号的方向从右至左减小	◁

(3) Third and subsequent compartments—Datum (or datum point) letters and related symbols, a single datum is represented by a capital Latin letter, or a datum system or common datum is represented by several letters. Some letters such as E, F, I, J, L, M, O, P, and R are generally not used to avoid misunderstanding. The datum symbol consists of a capital English letter with a small square connected to a small black triangle by a thin solid line, as shown in Fig. 4-3.

第三格和以后各格——基准（或基准点）字母和相关符号，用一个大写拉丁字母表示单个基准或用几个字母表示基准体系或公共基准。代表基准的字母一般不采用E，F，I，J，L，M，O，P，R，以免引起误解。基准符号由带小方格的大写英文字母用细实线与小黑色三角形相连而组成，如图4-3所示。

(a)　　(b)

Fig. 4-3 Symbol of datum

图4-3 基准符号

4.3.1 Marking Geometric Tolerances 几何公差的标注

The marking of the measured feature employs a leader line with an arrow to connect the measured feature and the tolerance frame. The arrow on the leader line points to the measured feature, and its direction indicates the width direction of the tolerance zone. Special attention should be paid to the position and direction of the arrow on the leader line, as different interpretations of the tolerance requirements may arise otherwise. The specific annotation method is as follows:

被测要素的标注采用带箭头的指引线连接被测要素和公差框格。指引线的箭头指向被测要素，箭头的方向为公差带的宽度方向。应特别注意指引线箭头所指的位置和方向，否则公差要求的解释可能不同。具体标注方法如下：

(1) If the measured feature is specified as integral feature, the arrow of the leader line should be placed on the outline of the feature or an extension of the features, and the arrow and the sizeline are obviously staggered. See Fig. 4-4 (a), (b).

当被测要素为组成要素时，指引线的箭头应放置于要素的轮廓线或其延长线上，并与尺寸线明显错开，如图4-4（a）、(b）所示。

(2) When the measurand feature is a derived feature, the arrow of the leader line should be aligned with the size line of the feature, as shown in Fig. 4-4 (c).

当被测要素是导出要素时，指引线的箭头应该与该要素的尺寸线对齐，如图4-4（c）所示。

(3) Use a lead line terminated with an arrow to indicate which feature is a tolerance feature, and the leader line starts from either end of a tolerance frame. The leader line for geometric tolerance can be bent, typically only once, but it should be perpendicular to the frame, see Fig. 4-4 (d).

使用带箭头的指引线来指示公差特征，指引线可从公差框的任意一端开始。几何公差的指引线可以弯曲，通常仅弯曲一次，它应与公差框垂直，如图4-4（d）所示。

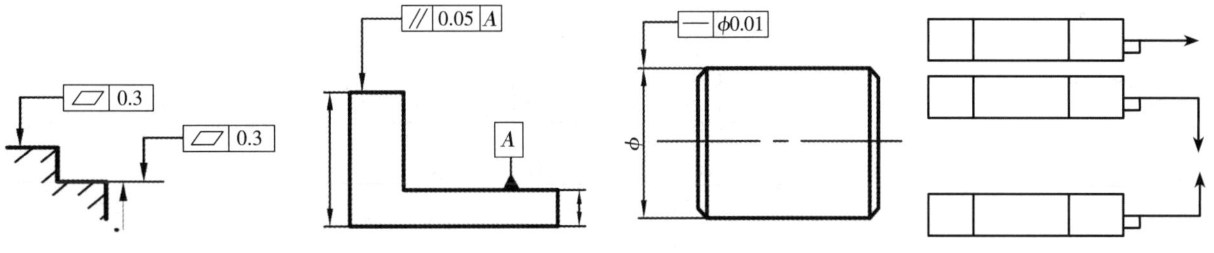

Fig. 4-4 Marking geometric tolerances

图4-4 几何公差的标注

(4) When the measured feature is the axis of a cone, the arrow of the leader line should be aligned with the diameter dimension line of the cone (either the large end or the small end), as shown in Fig. 4-5 (a); If necessary, a blank dimension line can also be drawn inside the cone, and the arrow of the leader line should be aligned with this blank dimension line, as shown in Fig. 4-5 (b); If the cone is marked using an angular dimension, the arrow of the leader line should be aligned with the dimension line of that angle, as shown in Fig. 4-5 (c).

当被测要素为圆锥体的轴线时，指引线的箭头应与圆锥体直径尺寸线（不论大端或小端）对齐，如图4-5（a）所示；必要时也可在圆锥体内画出空白的尺寸线，并将指引线的箭头与该空白的尺寸线对齐，如图4-5（b）所示；如圆锥体采用角度尺寸标注，则指引线的箭头应对着该角度的尺寸线，如图4-5（c）所示。

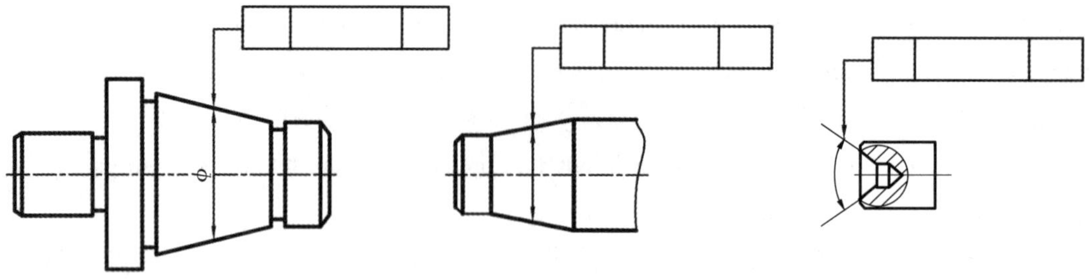

(a) (b) (c)

Fig. 4-5 Marking of the axis of a cone

图4-5 圆锥体轴线的标注

(5) When numerous measured features possess identical geometric tolerance requirements, whether single or multiple, it is permissible to draw multiple indicator arrows along the leader line extending from the frame. Each arrow should be connected to its respective measured feature, as depicted in Fig. 4-6. To maintain control over several measured features within the same tolerance zone, ensuring their coplanarity or collinearity, the common tolerance zone symbol CZ should be appended subsequent to the tolerance value within the tolerance frame, as shown in Fig. 4-7.

当多个被测要素具有相同的几何公差要求时，无论是单项公差还是多项公差，都可以在从框格引出的指引线上绘制多个指示箭头。每个箭头应分别与其各自的被测要素相连，如图4-6所示。用同一公差带控制多个被测要素，以保证这些要素共面或共线时，应在公差框格内公差值的后面加注公共公差带符号CZ，如图4-7 所示。

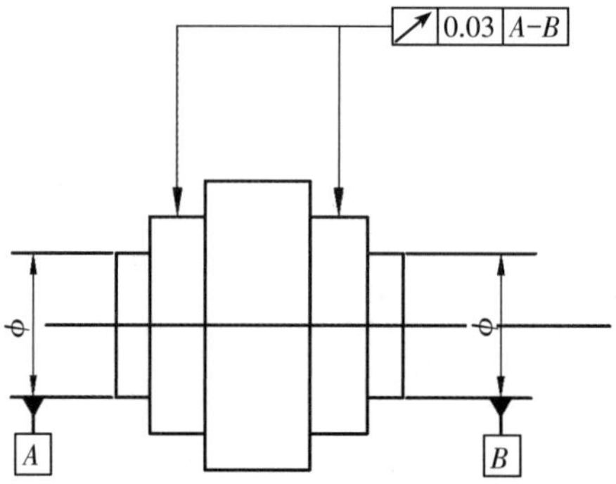

Fig. 4-6 Simplified marking for multiple features with the same requirements

图4-6 多要素同要求的简化标注

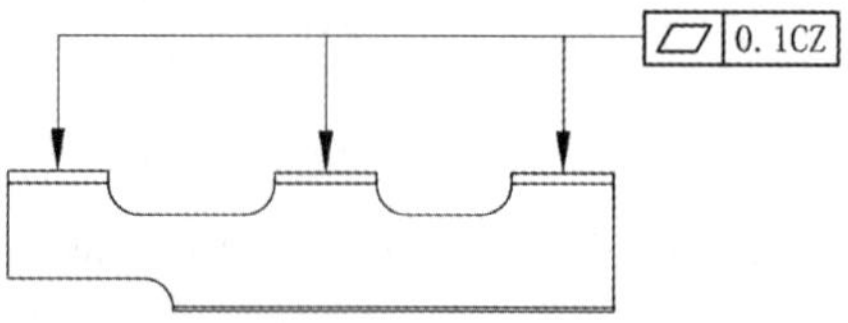

Fig. 4-7 Marking when multiple features share the same tolerance zone

图4-7 多要素用同一公差带时的标注

(6) When there are multiple geometric tolerance requirement for the same measured feature and the marking methods are consistent, for simplification, these frames can be drawn together, with tolerance frames one under another and a single leader line, as shown in Fig. 4-8.

当同一个被测要素有多项几何公差要求，其标注方法又一致时，为了简化，可以将这些框格绘制在一起，公差框格一个在另一个下面并引用一根指引线，如图4-8所示。

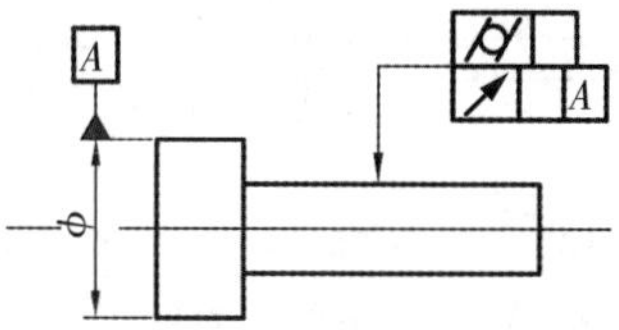

Fig. 4-8 Simplified marking multiple geometric characteristics on the same feature

图4-8 同一要素多项要求的简化标注

4.3.2 Marking Datum Feature 基准要素的标注

(1) The single datum is represented by one letter; the common datum is represented by two letters separated by horizontal lines, as shown in Fig. 4-9 (a); the datum system is represented by two or three letters, arranged from left to right according to the sequence of datum Ⅰ, Ⅱ and Ⅲ, as shown in Fig. 4-9 (b).

(a) (b)

Fig. 4-9 Frame of datum

图4-9 基准的公差框格

单一基准由一个字母表示；公共基准采用由横线隔开的两个字母表示，如图4-9（a）所示；基准体系用两个或三个字母表示，按基准的先后顺序从左至右排列，分别为第Ⅰ基准、第Ⅱ基准和第Ⅲ基准，如图4-9（b）所示。

(2) When the datum feature is a contour line or a surface, the small black isosceles triangle of the datum symbol is close to the contour line or its extension line of the datum feature, and the datum line and the size line of the contour are obviously staggered, as shown in Fig. 4-10. When the datum feature is an axis or a central plane is determined by a feature with sizes, the datum line of the datum symbol is aligned with the size line, as shown in Fig. 4-11. Regardless of the orientation of the datum symbol on the drawing, the letters in the small square of the datum symbol shall be written horizontally.

当基准要素是轮廓线或表面时，基准符号的小黑色等腰三角形靠近基准要素的轮廓线或其延长线，且基准连线与轮廓的尺寸线明显错开，如图4-10所示。当基准要素是轴线、中心平面或由带尺寸的要素确定时，则基准符号的基准连线应与尺寸线对齐，如图4 11所示。无论基准符号在图纸上的方向如何，基准符号小方格中的字母都应水平书写。

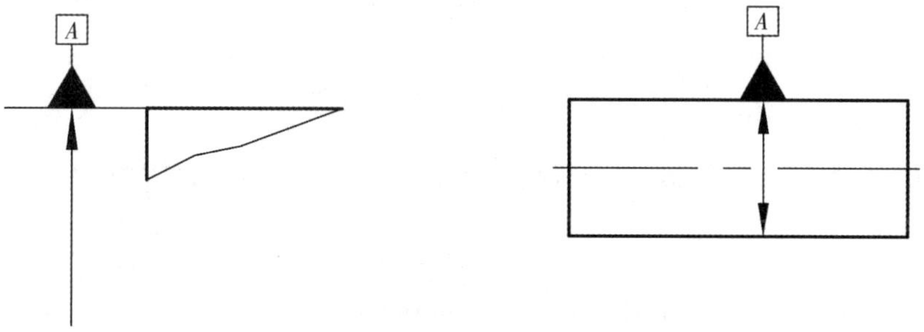

Fig. 4-10 Datum marking method of integral features

图4-10 组成要素的基准标注

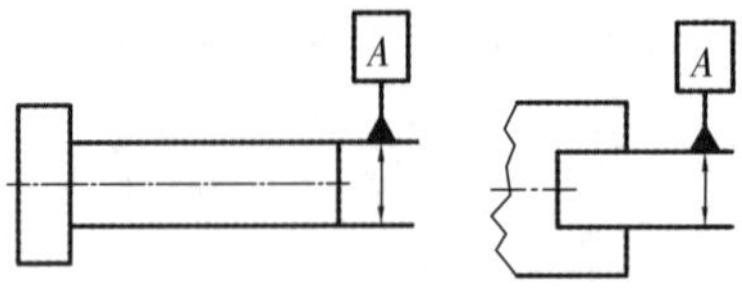

Fig. 4-11 Datum marking method of derived features

图4-11 导出要素的基准标注

(3) When the datum feature is the axis of a center hole or a cone, it should be marked according to the method shown in Fig. 4-12.

当基准要素为中心孔或圆锥体的轴线时，则按图4-12所示方法标注。

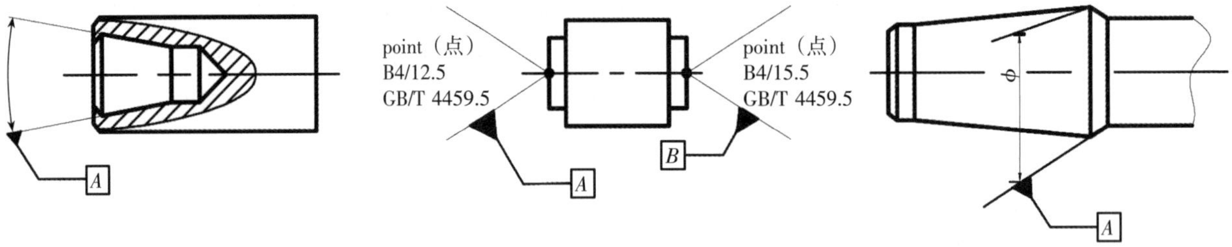

Fig. 4-12 Marking with the axis of a center hole or cone as the datum feature

图4-12 中心孔和圆锥体轴线为基准要素的标注

The valve stem shown in Fig. 4-13 serves as a typical example of geometric tolerance annotation. The meanings of the annotated geometric tolerances are as follows: The cylindricity tolerance of the ϕ16 mm rod body is 0.005 mm; The coaxiality tolerance of the M8×1 threaded hole axis relative to the ϕ16 mm axis of the rod body is ϕ0.1 mm; The circular run-out tolerance of the SR750 mm spherical surface relative to the ϕ16 mm axis is 0.003 mm.

以图4-13所示的气门阀杆作为几何公差标注的典型示例。其中所注各几何公差的含义为：杆身ϕ16 mm的圆柱度公差为0.005 mm；M8×1的螺孔轴线对于杆身ϕ16 mm轴线的同轴度公差为ϕ0.1 mm；SR750 mm的球面对于ϕ16 mm轴线的圆跳动公差为0.003 mm。

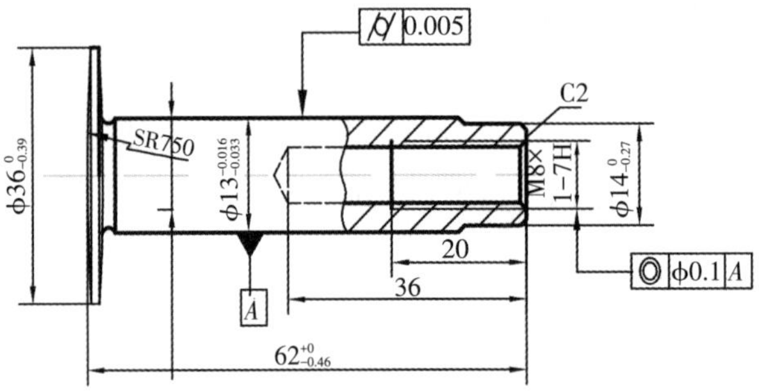

Fig. 4-13 Example of geometric tolerance annotation

图4-13 几何公差标注示例

4.3.3 Characteristics of Geometric Tolerance Zone 几何公差带的特征

Geometric tolerance zones encompass four essential characteristics: shape, size, orientation, and position. When all measured elements conform to the specified tolerance zone, they are deemed qualified. The configuration of the geometric tolerance zone is dependent on the ideal shape of the measured elements, as well as the prescribed tolerance characteristics. Based on these geometric tolerance characteristics, they can be categorized into nine primary forms, as detailed in Tab. 4-4.

几何公差带具有形状、大小、方向和位置4个特征要素。只要被测要素全部落在给定的公差带内，就表示该被测要素合格。几何公差带的形状取决于被测要素的理想形状和给定的公差特征。根据几何公差的特征，可将其分为9种主要形式，见表4-4。

Tab. 4-4 The main shapes of tolerance zones

表4-4 几何公差带的主要形状

Planar region 平面区域		Spatial region 空间区域	
Two parallel lines 两平行直线	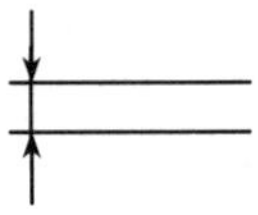	Sphere 球	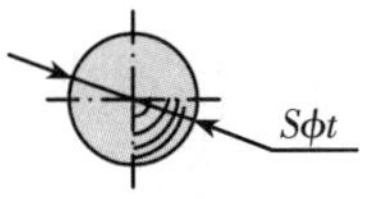
Two equidistant curves 两等距曲线	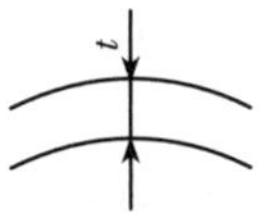	Cylindrical surface 圆柱面	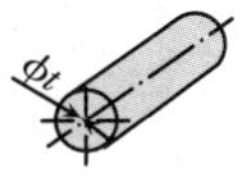
Two concentric circles 两同心圆	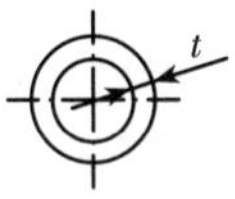	Two coaxial cylindrical surfaces 两同轴圆柱面	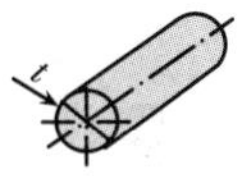
Circle 圆	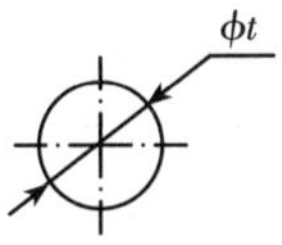	Two parallel planes 两平行平面	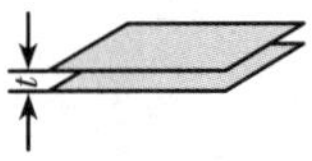
		Two equidistant surfaces 两等距曲面	

4.4 Definition of Geometric Tolerance 几何公差的定义

4.4.1 Definition and Characteristics of Form Tolerances 形状公差的定义及特征

Form tolerance limits the deviation of a single feature from its geometric ideal line or surface. A form tolerance is defined by its tolerance zone, which is the region within which any shape deviation is permissible. Form tolerance value defines the width of this zone as the maximum allowable value of the form deviation. The feature may have any form, within this zone, unless otherwise specified, and the location of the tolerance zone is floating.

形状公差限制了单一实际要素偏离其几何理想线或表面的程度。形状公差通过形状公差带定义，形状公差带是允许形状偏差存在的任何形状的区域。形状公差是实际形状偏离其理想形状所允许的最大值。如果没有特殊规定，形状公差带可以是任意形状的，零件实际要素在该区域内为合格。公差带的位置是浮动的。

Form tolerances include straightness, fatness, circularity, and cylindricity, and also include profile of any line and profile of any surface in the case without related datum. The measured features are straight lines, planes, circles, cylindrical surfaces, curves, and curved surfaces, respectively. The definitions, some examples of indications and explanations are shown in Tab. 4-5.

形状公差包括直线度、平面度、圆度、圆柱度及与基准无关的任意线轮廓度和面轮廓度。被测要素分别为直线、平面、圆、圆柱面、曲线和曲面。其定义及标注示例和解释见表4-5 。

Tab. 4-5 Definition of form tolerances zone and its indication examples and explanations

表4-5 形状公差带定义及其标注示例和解释 单位：mm

Feature 几何特征	Symbol 符号	The definition of tolerance zone 公差带定义	Example of indication and explanation 标注示例和解释	Datum needed 有无基准
		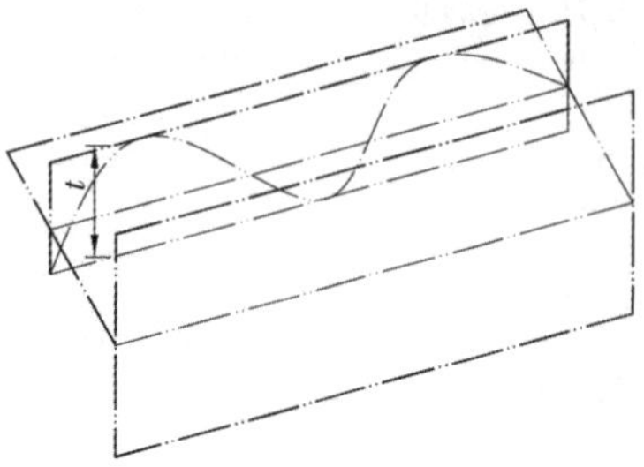 In the specific plane or the considered plane, the tolerance zone is limited by two parallel straight lines with a distance of *t* and in the specified plane and direction 给定平面内的直线度，公差带为在给定平面内和给定方向上，间距等于公差值的两个平行直线所限定的区域	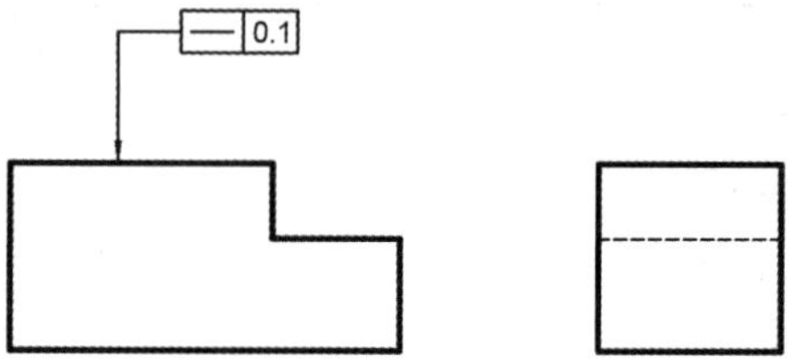 It is qualified if any extracted line on the upper surface parallel to the plane of projection is contained between two parallel straight lines 0.1 mm apart 在任一平行于图示投影面的平面内，上表面的实际投影线应限定在间距等于0.1 mm的平行直线之间	No 否
Straightness 直线度	—	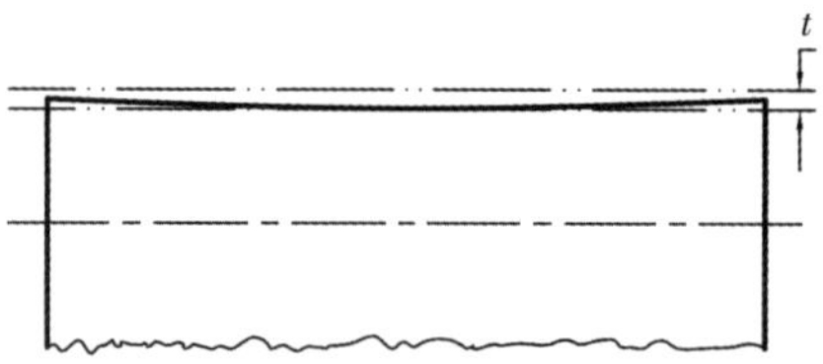 In the particular direction, the tolerance zone is limited by two parallel planes with a distance of *t* 在给定方向上为间距等于公差值*t*的两平行直线所限定的区域	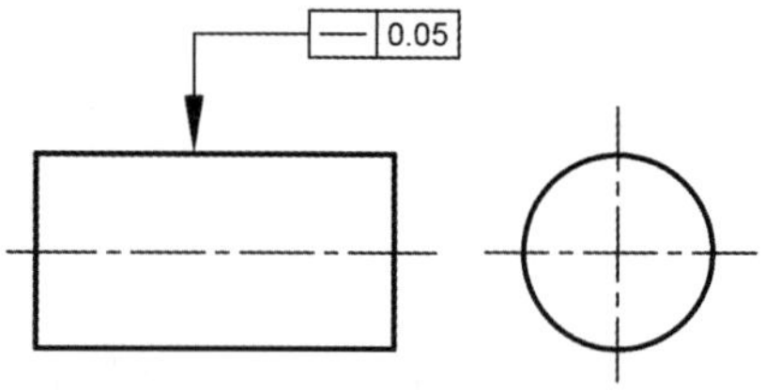 It is qualified if any extracted generating line on the cylindrical surface is contained between two parallel straight lines 0.05 mm apart 实际轮廓线或素线应限定在间距等于0.1 mm的两个平行直线之间	No 否
		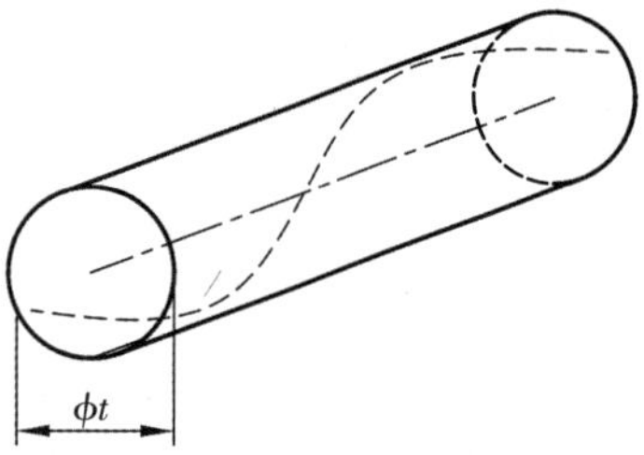 Straightness is in the random direction.The tolerance zone is limited by *t*, When a straightness control is applied to a cylinder of diameter *ϕt*, if the tolerance value is preceded by the symbol *ϕ* 直线度适用于任意方向时，其公差带为由公差值*t*所限定的区域。当对直径为*ϕt*的圆柱体应用直线度控制且公差值前标注符号*ϕ*	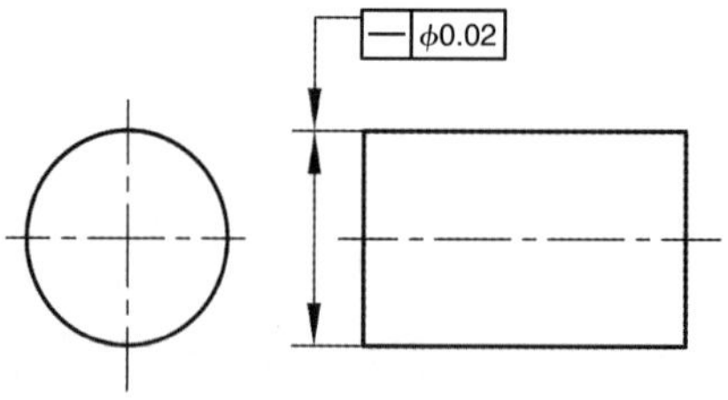 The axis must lie within the cylindrical tolerance zone.The tolerance zone is a cylinder, and the diameter of the cylinder is equal to the straightness tolerance value（*ϕ*0.02） 外圆柱面的实际轴线应限定在直径等于*ϕ*0.02 mm的圆柱面内	No 否

Tab. 4-5（Continued）表4-5（续）

Feature 几何特征	Symbol 符号	The definition of tolerance zone 公差带定义	Example of indication and explanation 标注示例和解释	Datum needed 有无基准
Flatness 平面度	⏥	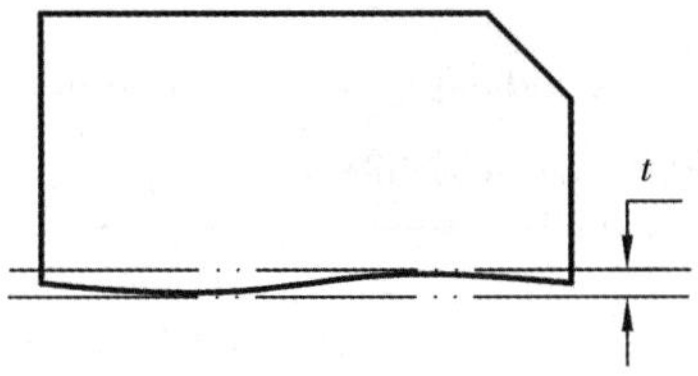 All of the measured surface must be within two parallel planes *t* apart 被测表面应限定在间距为*t*的两个平行平面之间	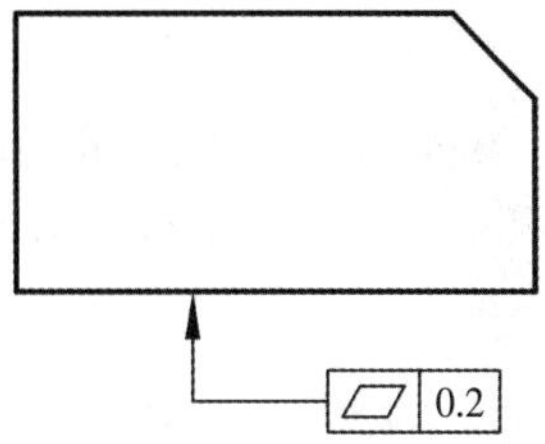 It is qualified if the extracted（actual）surface is contained between two parallel planes 0.2 mm apart 提取（实际）表面位于间距为0.2毫米的两个平行平面之间	No 否
Circularity 圆度	○	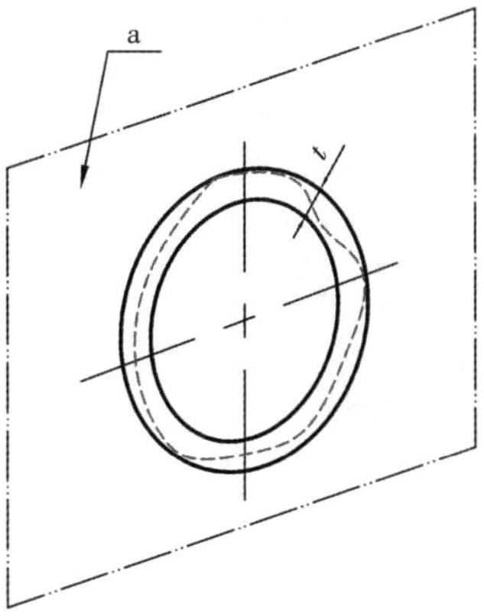 Circularity tolerance zone is two coaxial circles *t* apart 圆度公差带是间距为*t*两个共面同心圆之间	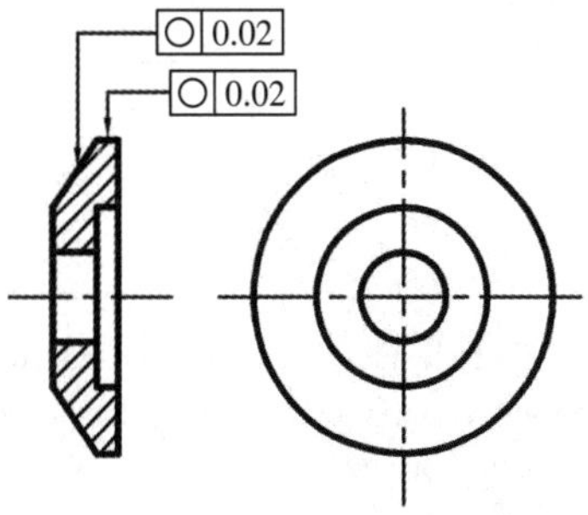 It is qualified if the extracted circumferential line, in any cross-section of the cylinder, and is contained between two co-planar concentric circles with a different in radii of 0.02 mm 在圆柱面的任一横截面内, 实际圆周应限定在半径差等于0.02 mm 的两个共面同心圆之间; It is qualified if any extracted circumferential line, resulting in the intersection of the revolute surface and a section cone is contained in a conical zone limited by two circles 0.02 mm apart. 在圆锥面的任一横截面内，实际圆周应限定在半径差等于0.02 mm的两个共面同心圆之间	No 否
Cylindricity 圆柱度	⌭	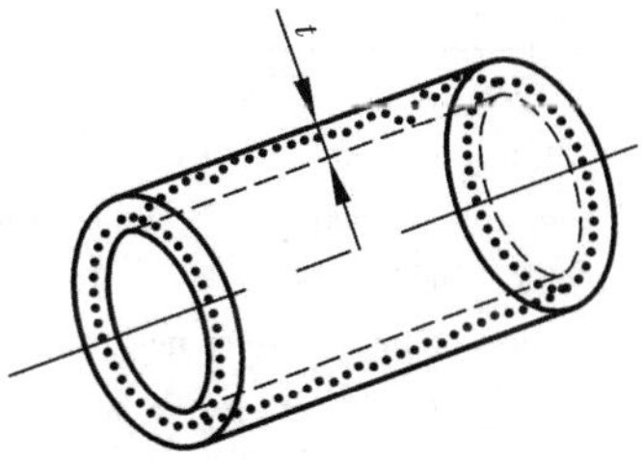 Tolerance zone is limited by two coaxial cylinders with a difference in radii of *t* 被测表面应限定在间距为*t*的两个平行圆柱面之间	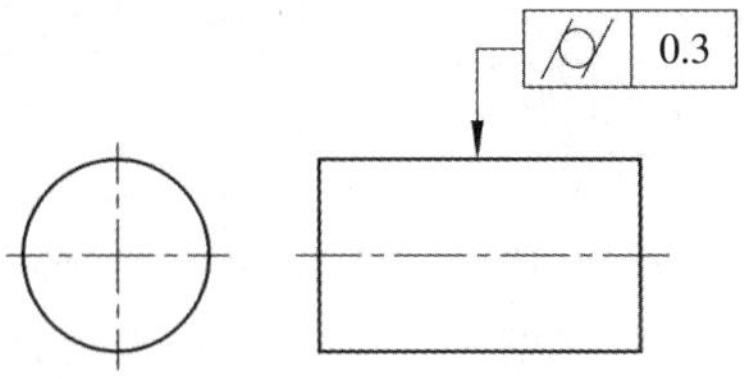 It is qualified if the extracted (actual) cylindrical surface is contained between two coaxial cylinders with a different in radii of 0.3 mm 它是合格条件提取的（实际的）圆柱面被包含在两个半径相差0.3 mm的同轴圆柱之间	No 否

4.4.2 Profile Tolerances and the Tolerance Zone 轮廓度公差与公差带

The profile tolerance includes line profile tolerance and surface profile tolerance. When there is no datum requirement, it is form tolerance; When there is a datum requirement, it is orientation or location tolerance, which are shown in Tab. 4-6.

轮廓度公差包括线轮廓度公差和面轮廓度公差。无基准要求时为形状公差，有基准要求时为方向公差、位置公差。

Tab. 4-6 Definition of form tolerances zone and its indication examples and explanations

表4-6 形状公差带定义及其标注示例和解释 单位：mm

Feature 几何特征	Symbol 符号	The tolerance zone 公差带定义	Example of indication and explanation 标注示例和解释	Datum needed 有无基准
Profile tolerances of any line 线轮廓度公差	⌒	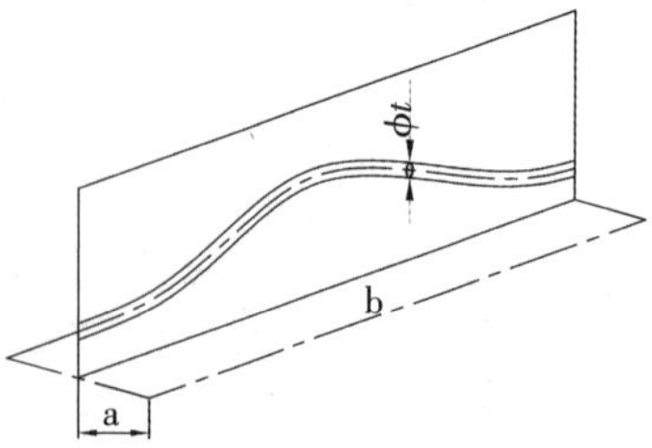 The tolerance zone is limited by two lines enveloping circles of diameter t, the centers of situated on a line having the which ares theoretically exact geometric form 公差带为直径等于公差值t，圆心位于被测要素理论正确几何形状上的一系列圆的两包络线所限定的区域	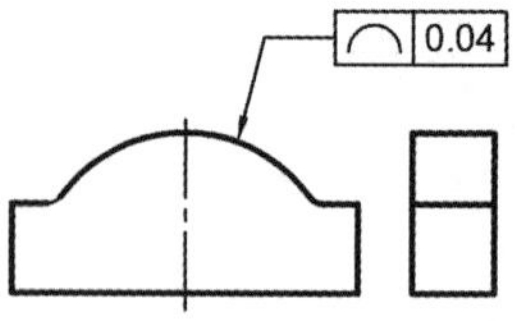 It is qualified if the extracted (actual) profile line, in any section parallel to the projection, is contained between two equidistant lines enveloping circles of diameter 0.04 mm, the centers of which are situated on a line having the theoretically exact geometric form 在提取的（实际）轮廓线上，任何与投影平行的截面所包含的实际轮廓均位于两等距线之间，这两条等距线所包围的圆的直径为0.04 mm，圆心位于被测要素理论正确几何形状上的一系列圆的两包络线之间	No 否
		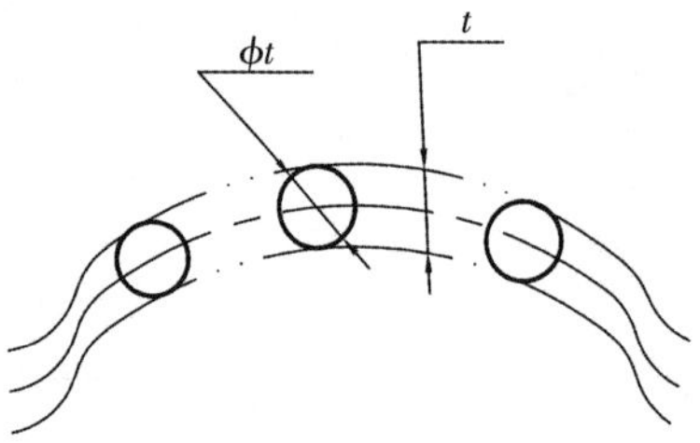 When there is a datum, the tolerance zone is limited by two lines enveloping circles of diameter t, the centers of which are situated on a line having the theoretically exact geometric form with respect to datum 有基准时，公差带为直径等于公差值t，圆心位于由基准确定的被测要素理论正确几何形状上的一系列圆的两包络线所限定的区域	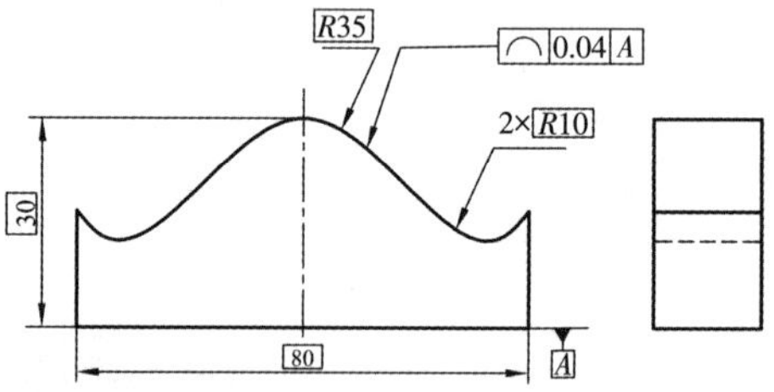 It is qualified if in each section, parallel to the plane of projection or datum plane A, extracted profile line shall be contained between two equidistant lines enveloping circles of diameter 0.04 mm, the centers of which are situated on a line having the theoretically exact geometric form with respect to datum plane A 在任一平行于图示投影面的截面内，实际轮廓线应限定在直径等于0.04 mm、圆心位于由基准A确定的被测要素理论正确几何形状上的一系列圆的两包络线之间	Yes 是

Tab. 4-6 (Continued) 表4-6 (续)

Feature 几何特征	Symbol 符号	公差带定义 The tolerance zone	标注示例和解释 Example of indication and explanation	Datum needed 有无基准
Profile tolerances of any surface 面轮廓度公差	⌓	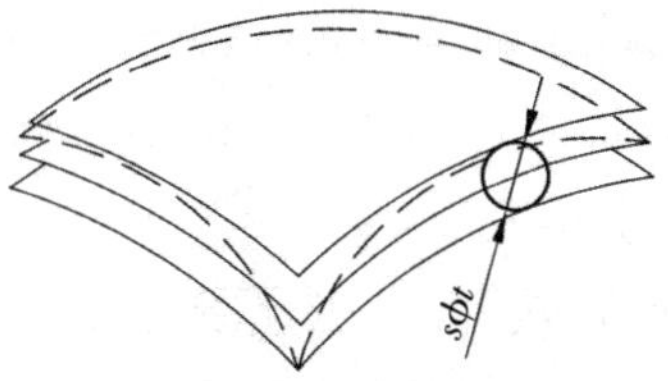 When there is no datum, the tolerance zone is limited by two surfaces enveloping spheres of diameter t, the centers of which are situated on a surface having the theoretically exact geometric form 没有基准时，公差带为直径等于公差值t，球心位于被测要素理论正确几何形状上的一系列圆球的两包络面所限定的区域	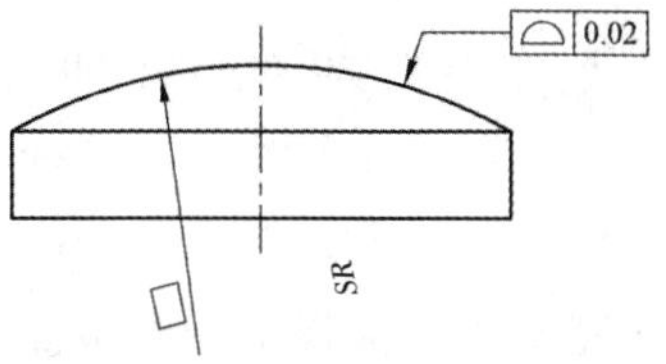 It is qualified if the extracted (actual) surface is contained between two equidistant surfaces enveloping spheres of diameter 0.02 mm, the centers of which are situated on a surface having the theoretically exact geometric form 所提取的(实际)表面位于直径等于0.02 mm、球心位于被测要素理论正确几何形状上的一系列圆球的两包络面之间	No 否
		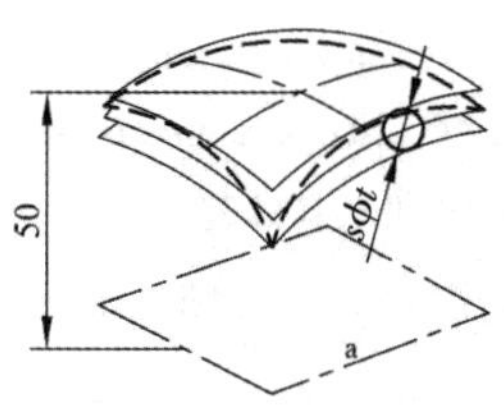 The tolerance zone is limited by two surfaces enveloping spheres of diameter t, the centres of which are situated on a surface having the theoretically exact geometric form 有基准时,公差带为直径等于公差值t球心位于由基准确定的被测要素理论正确几何形状上的一系列圆球的两包络面所限定的区域	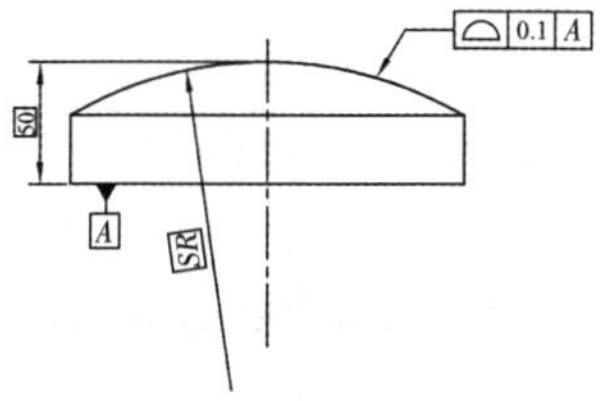 It is qualified if the extracted (actual) surface is contained between two equidistant surfaces enveloping spheres of diameter 0.1 mm, the centers of which are situated on a surface having the theoretically exact geometric form with respect to datum A 实际轮廓面应限定在直径等于0.1 mm,球心位于由基准平面A确定的被测要素理论正确几何形状上的一系列圆球的两包络面之间	Yes 是

4.4.3 Directional Tolerance, Positional Tolerance and Run-out Tolerance 方向公差、位置公差和跳动公差

The national standard (GB/T 1182-2020) stipulates that orientation tolerances, location tolerance and run -out tolcrancc havc datum rcquircmcnts.

国家标准（GB/T 1182—2020）规定方向公差、位置公差和跳动公差都有基准要求。

4.4.3.1 Datum 基准

A datum is a theoretically exact plane, point or axis from which a dimensional measurement is made. Datum features are selected on the basis of part function and assembly requirements.

理论上，基准可以是一个精确的平面、点或轴，几何尺寸检测过程是基于基准。基准要素是基于零件功能和装配要求来进行选择的。

4.4.3.2 Orientation tolerances 方向公差

The orientation tolerance is the allowable variation of the related actual features to the datum in a direction. The orientation tolerance zone has a clear direction relative to the datum and has the function of controlling the direction and shape of the measured feature at the same time. There is a datum requirement for the orientation tolerance.

方向公差是关联实际要素对基准在方向上的允许变动量。方向公差带相对基准具有明确的方向，具有同时控制被测要素方向和形状的功能。方向公差有基准要求。

On the premise of ensuring functional requirements, its form tolerance is usually not required additionally after an orientation tolerance is given for a feature. If there are further requirements for shape accuracy, form tolerance and orientation tolerance can be given at the same time, but the tolerance value of form tolerance should be less than that of orientation tolerance, The parallelism tolerance and flatness tolerance are 0.03 mm and 0.01 mm respectively, as shown in Fig. 4-14.

在保证功能要求的前提下，对被测要素给出方向公差后，一般不再给出其形状公差。如果对形状精度有进一步要求时，可同时给出形状公差和方向公差，但是形状公差的公差值应该小于方向公差的公差值。如图4-14所示，对被测平面给出0.03 mm的平行度公差和0.01 mm的平面度公差。

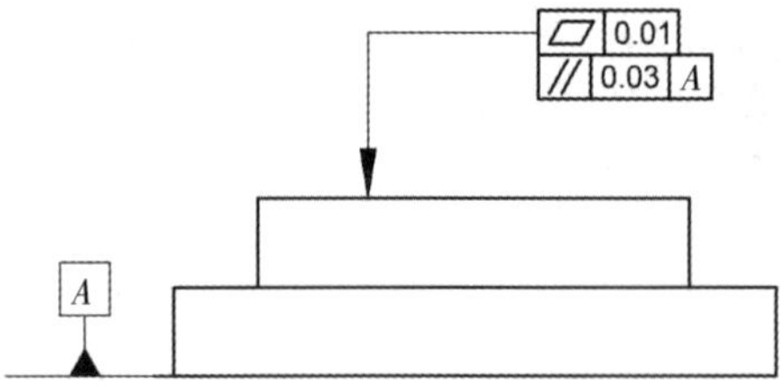

Fig. 4-14 Parallelism tolerance and flatness tolerance are given at the same time

图4-14 同时给出方向公差和形状公差

Orientation tolerances include parallelism, perpendicularity and angularity. Orientation control must contain a datum reference. The definitions of the orientation tolerance zones some examples of annotation and explanation of the three orientation tolerances are shown in Tab. 4-7.

方向公差包括平行度、垂直度和倾斜度，方向公差。方向控制必须以基准为参考方向。方向公差的公差带定义、标注样例和解释见表4-7。

Tab. 4-7 Definition of orientation tolerances zone and its indication examples and explanations

表4-7 方向公差的公差带定义、标注样例和解释

Feature 几何特征	Symbol 符号	The tolerance zone 公差带定义	Example of indication and explanation 标注示例和解释	Datum needed 有无基准
Parallelism 平行度	//	Parallelism tolerance of a line related to datum system 被测直线对基准系统的平行度公差		Yes 是
		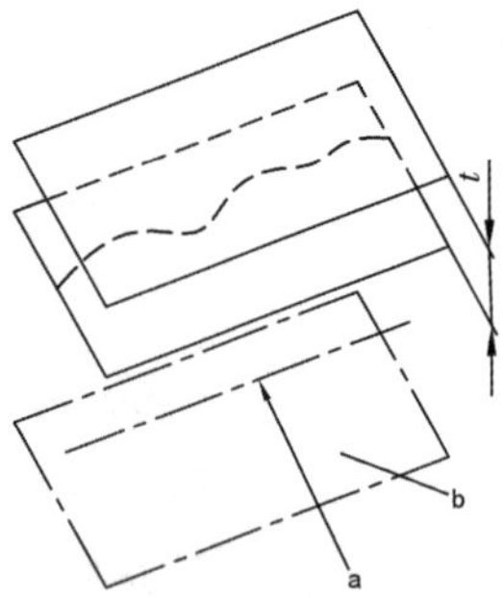 The tolerance zone is limited by two parallel planes a distance t apart. Two parallel planes are parallel to the datums and oriented along a specified direction 公差带为间距等于公差值t，平行于两基准且沿规定方向的两平行平面所限定的区域	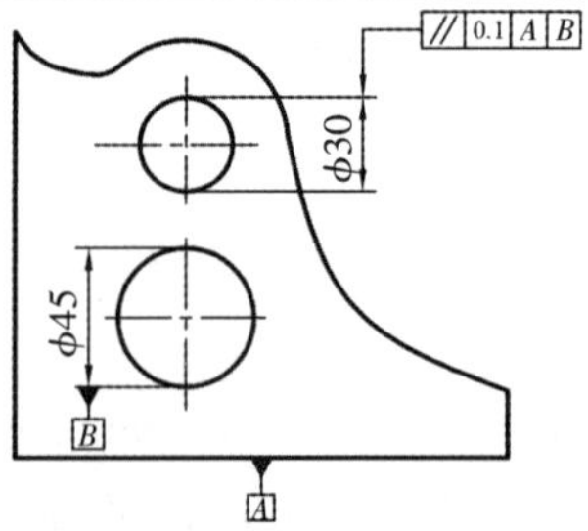 It is qualified during the extracted (actual) median line is contained between two parallel planes 0. 1 mm apart which parallel to datum axis A.The planes defining the tolerance zone are all parallel to the datum plane B specified by the orientation plane. Datum B serves as an auxiliary datum for datum A 提取(实际)中心线应限定在间距等于0.1mm、平行于基准轴线A的两平行平面之间。限定公差带的平面均平行于由定向平面规定的基准平面B。基准B为基准A的辅助基准	

Tab. 4-7 (Continued) 表4-7（续）

Feature 几何特征	Symbol 符号	The tolerance zone 公差带定义	Example of indication and explanation 标注示例和解释	Datum needed 有无基准
Parallelism 平行度	//	Parallelism tolerance of measured plane to datum plane 被测平面对基准平面的平行度公差		
		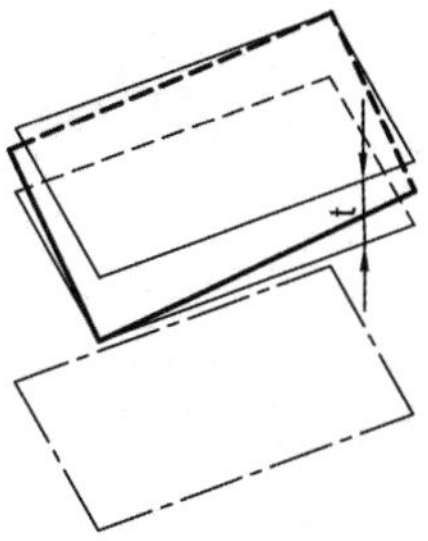 It is qualifed during the extracted median line shall be contained between two parallel planes 0.01 mm apart, which are parallel to the datum plane A 被测直线应限定在间距等于0.01 mm且平行于基准平面A的两平行平面之间	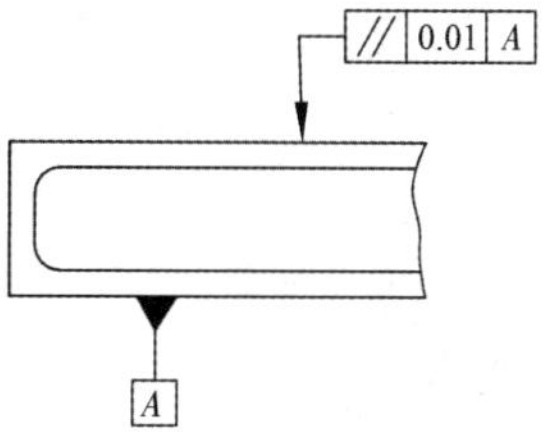 Parallelism tolerance of a line related to a datum plane, the tolerance zone is limited by two parallel planes with a distance of t and parallel to the datum plane. The direction of tolerance zone is fixed 被测直线对基准平面的平行度公差，公差带为间距等于公差值t且平行于基准平面的两平行平面所限定区域，公差带的方向是固定的	Yes 是
		Parallelism tolerance of measured line to datum line 被测直线对基准直线的平行度公差		
		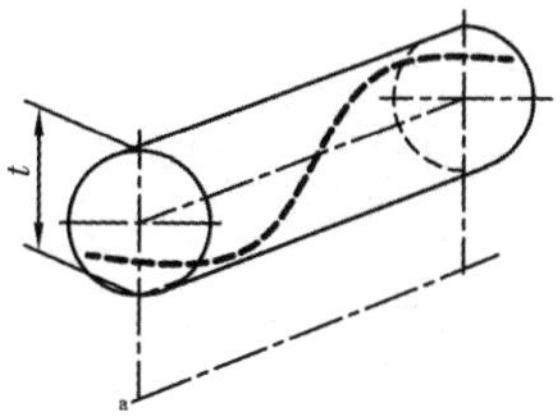 Parallelism tolerance of a line related to a datum axis, the tolerance zone is limited by a cylinder of diameter t, parallel to the datum, if the tolerance value is preceded by the symbol ϕ. The direction of tolerance zone is fixed 被测直线对基准直线的平行度，如果公差值前加了ϕ，其公差带为直径等于公差值t，且轴线平行于基准直线的圆柱面所限定区域，公差带的方向是固定的	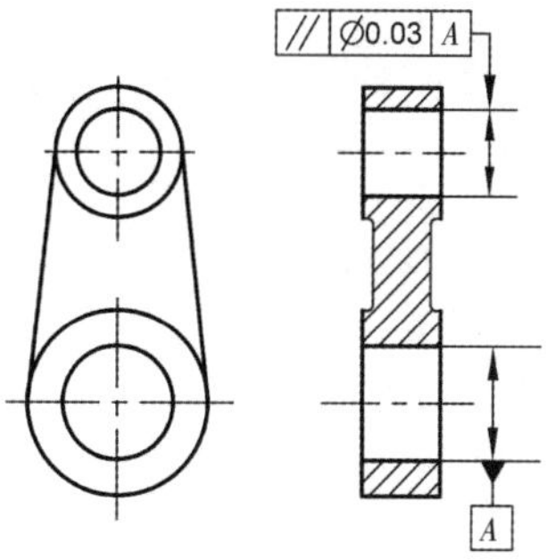 It is qualified if the extracted axis is within a cylindrical zone of diameter 0.03 mm, parallel to the datum axis A 被测孔的实际轴线应限定在直径等于0.03 mm且平行于基准轴线A的圆柱面内	Yes 是
Perpendicularity 垂直度	⊥	Perpendicularity tolerance of measured plane to datum plane 被测平面对基准平面的垂直公差		
		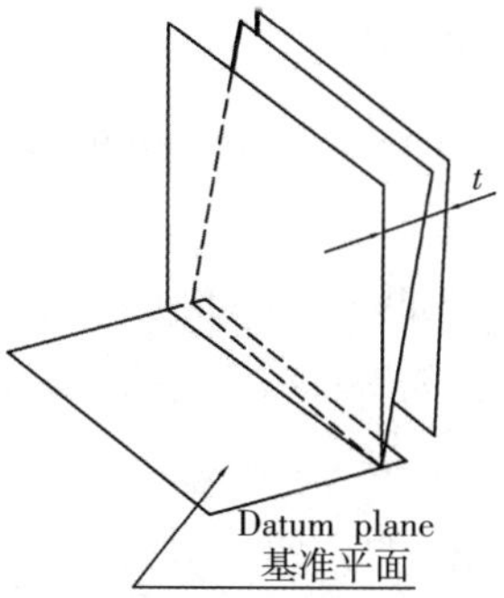 The tolerance zone is limited by two parallel planes a distance t apart and perpendicular to the datum plane 公差带为公差值t，且垂直于基准平面A的两平行平面所限定的区域	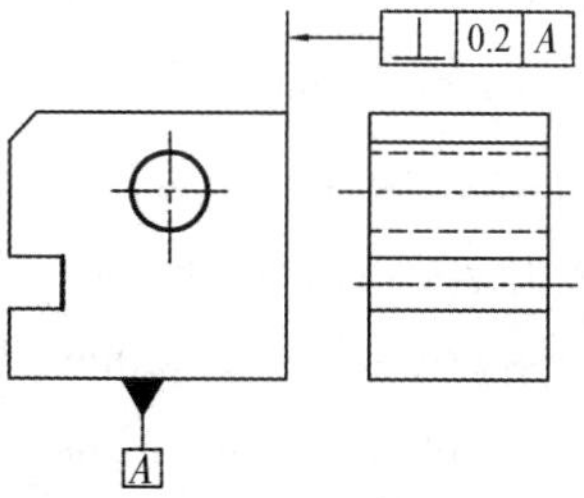 It is qualified if the extracted (actual) surface is contained between two parallel planes 0.2 mm apart, which are perpendicular to datum plane A 实际表面应限定在间距等于0.2 mm且垂直于基准平面A的两平行平面之间	Yes 是

Tab. 4-7（Continued）表4-7（续）

Feature 几何特征	Symbol 符号	The tolerance zone 公差带定义	Example of indication and explanation 标注示例和解释	Datum needed 有无基准
Perpendicularity 垂直度	⊥	Perpendicularity tolerance of measured line to datum plane 被测直线对基准平面的垂直度公差		Yes 是
		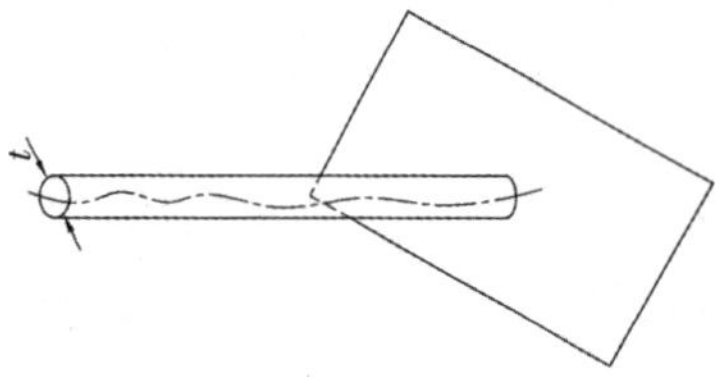 Perpendicularity tolerance of a line related to datum plane, the tolerance zone is limited by a eylinder of diameter t perpendicular to the datum, if the tolerance value is preceded by the symbol ϕ. The direction of tolerance zone is fixed 被测直线对基准平面的垂直度，如果公差值前加了ϕ，其公差带为直径等于公差值ϕt且轴线垂直于基准平面的圆柱面所限定区域，公差带的方向是固定的	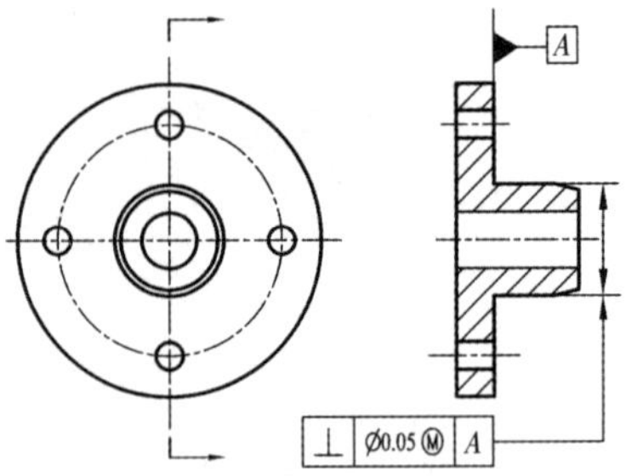 It is qualified if the extracted median line of the cylinder is within a cylindrical zone of diameter 0.05 mm, which is perpendicular to the datum plane A 被测孔的提取（实际）轴线应限定在直径等于0.05 mm且垂直于基准平面A的圆柱面内	
Angularity 倾斜度	∠	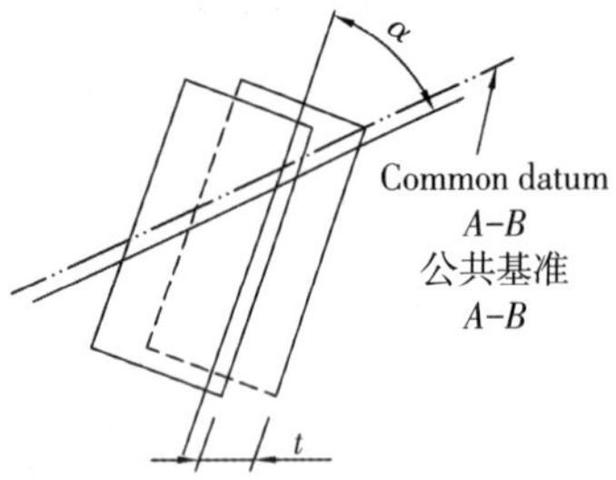 Line and datum line in the same plane: the tolerance zone is limited by two parallel planes a distance t apart and inclined at the specified angle to the datum 公差带为间距等于公差值t的两平行平面所限定的区域。该两平行平面按规定角度倾斜于基准轴线。被测线与基准线在不同的平面内	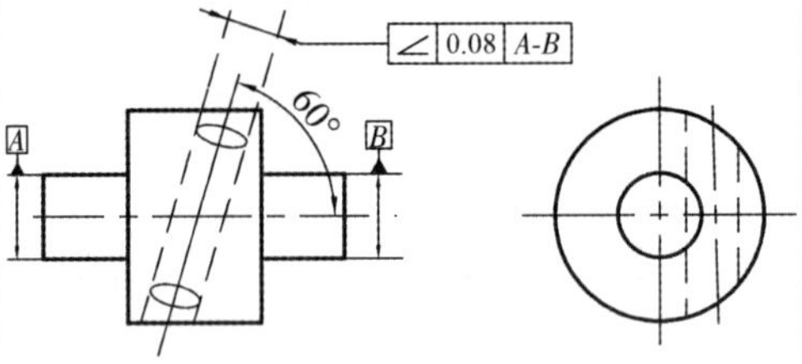 It is qualified if the extracted（actual）median line is contained between two parallel planes 0.08 mm apart that are inclined at a theoretically exact angle of 60° to the common datum straight line $A-B$ 提取（实际）中心线应限定在间距等于0.08 mm的两平行平面之间。该两平行平面按理论正确角度60°倾斜于公共基准轴线$A-B$	Yes 是

4.4.3.3 Characteristics of orientation tolerances 方向公差的特点

All orientation tolerances have datums. The tolerance zones have certain orientation to the datum, however,（the position is floating. Besides orientation deviation, the orientation tolerances als、o limit the form deviation of the tolerance feature.）For example, the parallelism tolerance of a plane can control the flatness and straightness errors of that plane. If necessary, a form tolerance is specified to the same tolerance feature, the value of form tolerance should be smaller than the value of orientation tolerance, i.e. $T_f < T_o$.

所有方向公差都有基准。方向公差带相对于基准有确定的方向，而其位置往往是浮动的。除了方向偏差，方向公差还限制被测要素形状偏差。如平面的平行度公差，可以控制该平面的平面度和直线度误差。如果需要对同一被测被测要素有进一步要求时，那么形状公差数值应小于方向公差值，即$T_f < T_o$。

4.4.3.4 Location tolerance

Location tolerance is the allowable variation of the location of the related feature relative to the datum. The position tolerance zone has a correct location relative to the datum, determined by the theoretically exact dimension（TED）, with TED for coaxiality and symmetry being zero. It controls the tolerance feature's location, direction, and shape. Location tolerances（coaxiality, symmetry, location）limit location deviations relative to datum, with tolerance features（points, lines, planes）and datum features（lines, planes）.

位置公差是关联要素相对于基准的位置允许变动量。位置公差带的位置由理论正确尺寸（TED）确定（同轴度、对称度的TED为0)，控制公差要素的位置、方向和形状。位置公差（同轴度、对称度、位置度）限制关联要素相对于基准的位置偏差，公差要素包括点、直线、平面，基准要素主要为直线、平面。

Coaxiality（concentricity）and symmetry are special location tolerances. They adopt the same tolerance concept but apply to different geometric shapes. The measured feature characteristics of coaxiality tolerance are axes with rotational features such as cylinders, cones, and center points, and the datum is also an axis. Coaxiality tolerance is used to control the error of coincidence between the measured axis（center point）and the datum axis. Symmetry tolerance controls the feature errors constructed around an axis or an intermediate plane.The measured feature using location tolerance should maintain the exact positional relationship with the datum feature specified on the engineering drawing. The accurate positional relationship between the measured feature and the datum feature is determined by the theoretical exact dimension（abbreviated as TED）. The TED for coaxiality and symmetry tolerances is 0, and it does not need to be indicated on the engineering drawing.

同轴度（同心度）和对称度是特殊位置公差，基于相同概念但适用于不同几何。同轴度的公差要素为带有旋转特征（如圆柱、圆锥、中心点）的轴线，基准为同轴轴线（控制公差轴线与基准轴线的重合误差)。对称度公差控制围绕中心点、轴或中间平面的特征误差。应用位置公差的公差要素需与图纸基准要素保持精确位置关系（TED）确定，同轴度和对称度的理论正确尺寸（TED）等于零，在图纸上无需标注。

Location tolerance is related to datum(s), the orientation and position of tolerance zones are determined by datum. Location tolerance has the function to limit the deviation of location, orientation and form. For example, the position tolerance of a plane can limit the flatness error and orientation error related to datum of this plane; The coaxiality tolerance can limit the straightness error and parallelism error related to datum axis. So usually, under the premise of functional requirements, the orientation tolerances and form tolerances are not applied when the location tolerances are applied. If necessary, the form tolerances or orientation tolerances with high precision are needed, then the form tolerances or orientation tolerances can be applied, however, the tolerance value of form tolerance should be smaller than the location tolerance value, as shown in Tab. 4-8.

位置公差是与基准相关联的，公差带的方向和位置依据基准来确定。位置公差具有综合控制被测要素的位置、方向和形状误差的功能。如平面的位置公差，可限制与该平面对应基准相关的平面度误差和方向误差；同轴度公差可控制被测轴线的直线度误差及相对于基准轴线的平行度误差。因此，在满足功能要求时，通常应用位置公差后不再加方向或形状公差。若需对方向或形状有更高精度要求时，可另行标注方向或形状公差，但形状公差值需小于位置公差值（见表4-8)。

Tab. 4-8 Definition of location tolerances zone and its indication examples and explanations

表4-8 位置公差的公差带定义及其标注示例和解释

Feature 几何特征	Symbol 符号	The tolerance zone 公差带定义	Example of indication and explanation 标注示例和解释	Datum needed 有无基准
Concentricity 同心度	◎	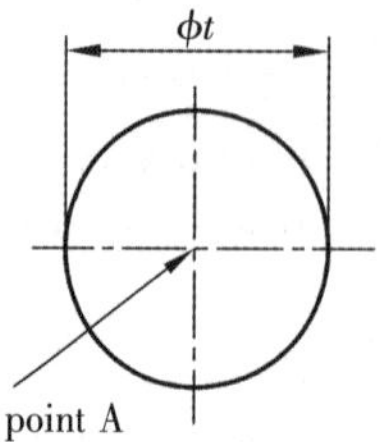 Datum point A 基准点A Concentricity tolerance of a point, the tolerance zone is limited by a circle of diameter t, the tolerance value shall be preceded by the symbol ϕ, the center of the circular tolerance zone coincides with the datum point. The location of tolerance zone is fixed 点的同心度，公差带为直径等于公差值t的圆周所限定的区域，公差值前应加ϕ，圆周的圆心与基准点重合。公差带的位置是固定的	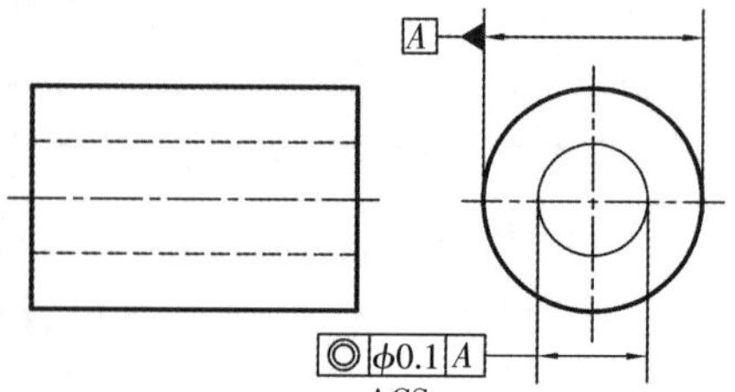 ACS It is qualified if the extracted center of the inner circle in any cross-section is be within a circle of diameter 0.1 mm, concentric with datum point A defined in the same acrosssection 在任一截面内，外圆的实际中心点应限定在直径等于ϕ0.1 mm 且以基准点A为圆心的圆周内	Yes 是
Coaxiality 同轴度	◎	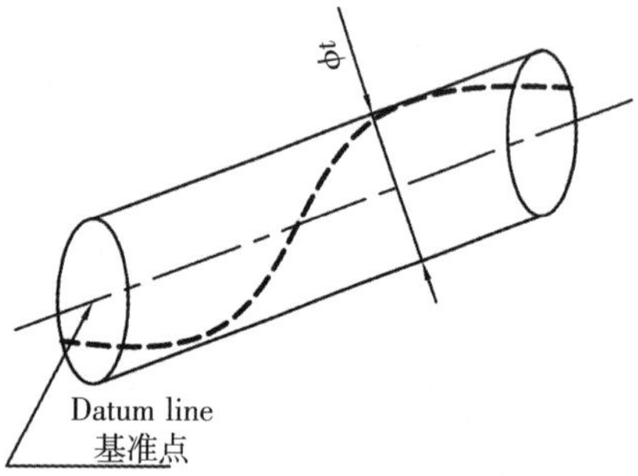 Coaxiality tolerance of a axis, the tolerance zone is limited by a cylinder of diameter ϕt, the tolerance value is preceded by the symbol ϕ, the axis of the cylindrical tolerance zone coincides with the datum, the location of tolerance zone is fixed 轴线的同轴度，其公差带为直径等于公差值ϕt，公差值前应加ϕ，且轴线与基准轴线重合的圆柱面所限定的区域。公差带的位置是固定的	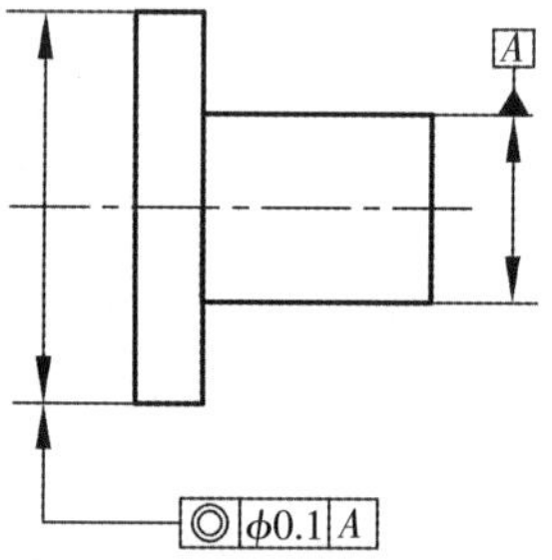 It is qualified if the extracted median line of the tolerance cylinder is within a cylindrical zone of diameter ϕ0.1 mm, the shaft of which is the datum A 圆柱面的提取中心线应限定在直径等于ϕ0.1 mm 且轴线与基准轴线A重合的圆柱面内	Yes 是
Symmetry 对称度	≡	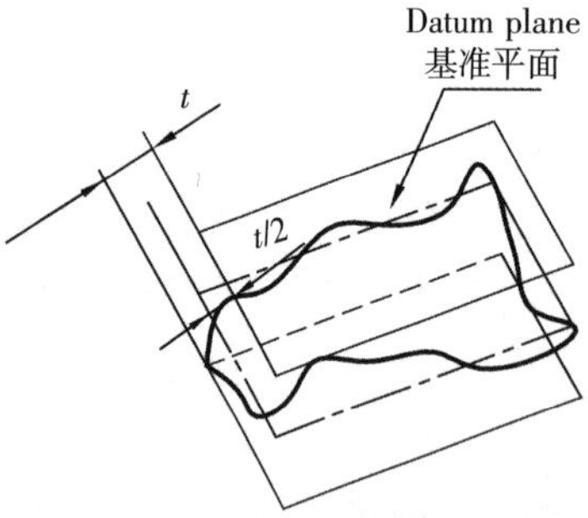 Symmetry tolerance of a median plane. The tolerance zone is limited by two parallel planes a distance t apart, symmetrically disposed about the median plane, with respect to the datum. The location of tolerance zone is fixed 被测平面对基准平面的对称度，公差带为间距等于公差值t且对称于基准中心平面的两平行平面所限定的区域。公差带位置是固定的	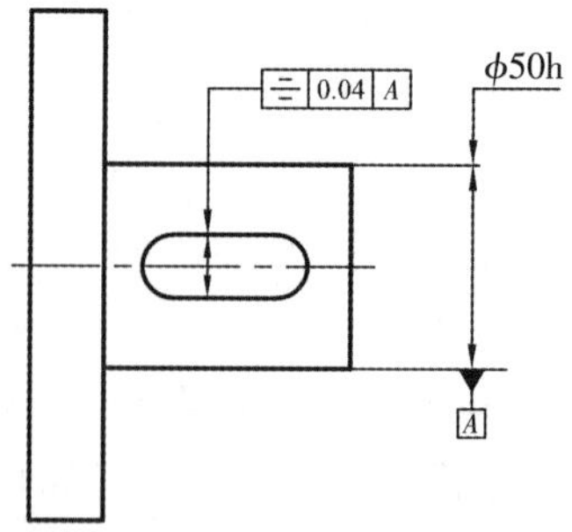 It is qualified if the extracted (actual) median line is contained between two parallel planes 0. 04 mm apart, which are symmetrically disposed about datum plane A 被测实际中心平面应限定在间距等于0.04 mm且对称于基准中心平面A的两平行平面之间	Yes 是

Tab. 4-8 (Continued) 表4-8（续）

Feature 几何特征	Symbol 符号	The tolerance zone 公差带定义	Example of indication and explanation 标注示例和解释	Datum needed 有无基准
Position 位置度	⌖	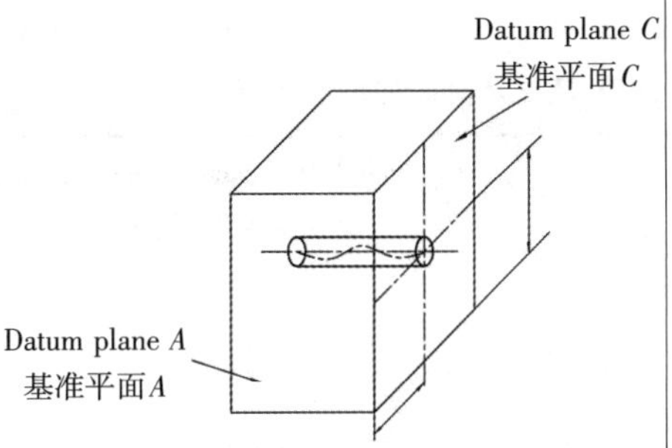 Position tolerance of a line, the tolerance zone is limited by a cylinder of diameter *t*, the tolerance value is preceded by the symbol ϕ. The axis of the tolerance cylinder is fixed by theoretically exact sizes with respect to datum. The location of tolerance zone is fixed 线的位置度，公差带为直径等于公差值*t*的圆柱面所限定的区域，公差值前应加ϕ，该圆柱面的轴线的理论正确位置由基准和理论正确尺寸确定。公差带的位置是固定的	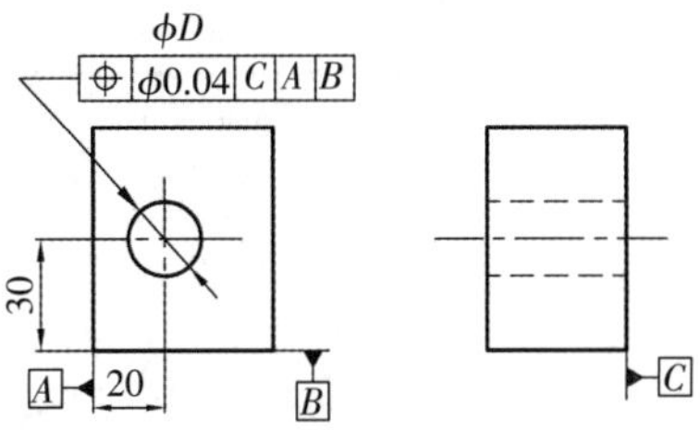 It is qualified if the extracted median line is within a cylindrical zone of diameter ϕ0.04 mm, the shaft of which coincides with the theoretically exact position of the considered hole, with respect to datum planes A, B and C 当提取的中线位于直径为ϕ0.04 mm的圆柱面区域内。该圆柱面的轴线与所考虑孔的理论正确位置重合，且相对于基准平面A，B和C时，即为合格	Yes 是

4.4.3.5 Run-out tolerance 跳动公差

Run-out tolerance is the maximum allowable run-out when the actual feature rotates around the datum shaft for one cycle or rotates continuously. The tolerance features of run-out tolerance usually are cylindrical surfaces, cone surfaces and end planes. The datum feature is a shaft. Run-out is a composite tolerance used to control the functional relationship of one or more features of a part to a datum axis. According to the measuring direction, the circular run-out is divided into radial run-out, axial run-out and slant run-out.

跳动公差是关联实际要环绕基准轴线回转一周或连续回转时所允许的最大跳动量。它的被测要素为圆柱面、圆锥面和端平面等组成要素。基准要素为轴线。跳动公差是一个综合公差，用于控制零件上一个或多个特征相对于基准轴线的功能关系。根据测量方向的不同，圆跳动分为径向圆跳动、轴向圆跳动和斜向圆跳动（应用较少）。典型的圆跳动、全跳动公差的公差带定义及其标注示例和解释见表4-9。

Run - out tolerances are comprehensive tolerances, which can limit form errors, orientation errors and location errors. Radial circular run - out tolerance can limit circularity errors and concentricity errors; Radial circular run-out tolerance can limit orientation errors of a plane relative to the axis; Radial total run-out tolerance can limit cylindricity errors and coaxiality errors. Axial total run-out can limit perpendicular error of a plane to datum axis and flatness error (axial total run-out is equivalent to perpendicularity of a plane to a datum axis). As run - out is easily measured, usually run - out tolerance is preferably specified under the same functional condition in designing. Definition of run - out tolerance zone and its indication examples and explanations are shown in Tab. 4-9.

跳动公差具有综合控制被测要素的位置、方向和形状的作用。径向全跳动公差带可综合控制同轴度和圆柱度误差；轴向圆跳动公差可以限制平面相对于轴的方向误差；径向全跳动公差可以限制圆柱度误差和同轴度误差。轴向全跳动可以限制平面相对于基准轴的垂直度误差和平面度误差（轴向全跳动相当于平面相对于基准轴的垂直度)。由于跳动易于测量，因此在设计过程中，通常在相同的功能条件下优先规定跳动公差。典型的圆跳动、全跳动公差的公差带定义及其标注示例和解释见表4-9。

Tab. 4-9 Definition of run-out tolerances zone and its indication examples and explanations

表4-9 圆跳动公差的公差带定义及其标注示例和解释

Feature 几何特征	Symbol 符号	The tolerance zone 公差带定义	Example of indication and explanation 标注示例和解释	Datum needed 有无基准
Circular run-out 圆跳动	↗	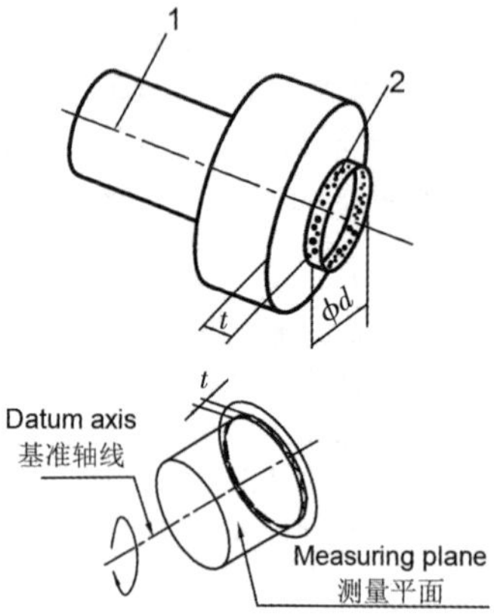 Circular run-out tolerance-axial, the tolerance zone is limited to any cylindrical section by circles a distance *t* apart lying in the cylindrical section, the shaft of which coincides with the datum, The central location of tolerance zone is fixed, 轴向圆跳动公差带为与基准轴线同轴的任一直径的圆柱截面上,间距等于公差值*t*的两个等径圆所限定的圆柱面区域。公差带的中心位置固定	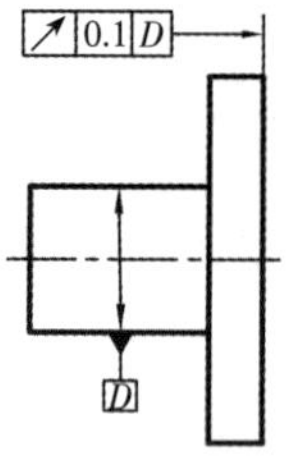 It is qualified during line in any cylindrical section , the shaft of which coincides with datum shaft D, shall be contained between two concentric circles with a distance of 0.1 mm 在任意圆柱截面上，其轴线与基准轴A重合的线应被包络在间距为0.1毫米的两个同心圆之间的，即为合格	Yes 是
		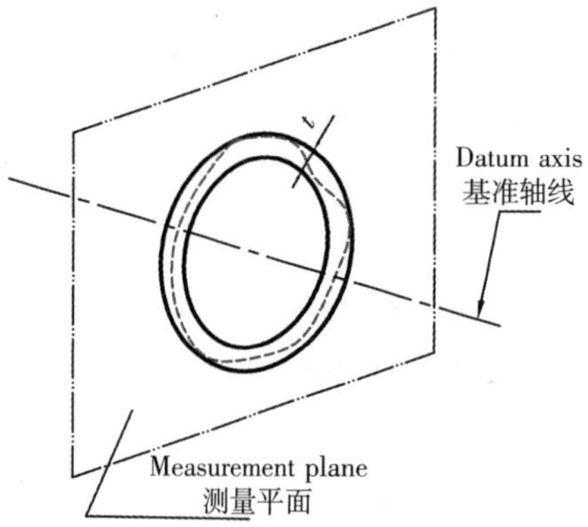 Circular run-out tolerance-radial, the tolerance zone is limited within any cross-section perpendicular to the datum shalt by two concentric circles with a difference in radii of *t*, the centers of which coincide with the datum. The central location of tolerance zone is fixed 径向圆跳动公差带为在任一垂直于基准轴线的横截面内、半径差等于公差值*t*圆心在基准轴线上的两同心圆所限定的区域。公差带的中心位置固定	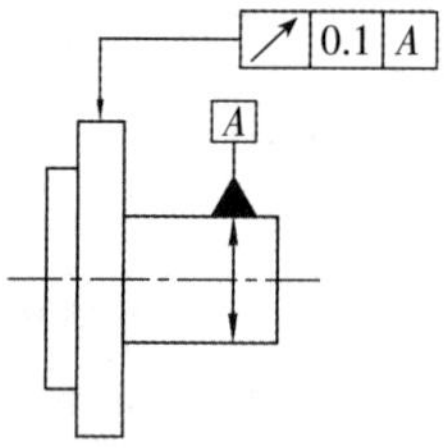 It is qualified during the extracted line in any cross - section plane perpendicular to datum shaft A shall be contained between two concentric circles with difference in radii of 0. 1 mm 在任一垂直于基准轴线*A*的横截面内，被测圆柱面的实际圆应限定在半径差等于0.1 mm，且圆心在基准轴线*A*上的两个同心圆之间	Yes 是

Tab. 4-9（Continued）表4-9（续）

Feature 几何特征	Symbol 符号	The tolerance zone 公差带定义	Example of indication and Explanation 标注示例和解释	Datum needed 有无基准
Circular run-out 圆跳动	↗	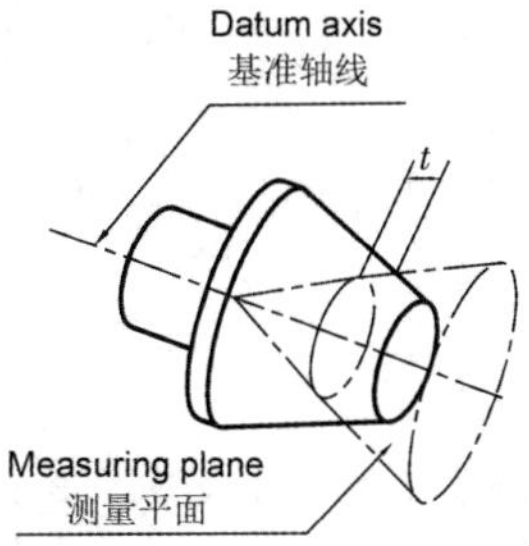 Slant run-out tolerance, the tolerance zone is limited within any conical section by two circles a distance t apart, the axes of which coincide with the datum. The width of the tolerance zone is normal to the specified geometry unless otherwise indicated.The central location of tolerance zone is fixed 斜向圆跳动公差带为与基准轴线同轴的某一圆锥截面，间距等于公差值t的直径不相等的两个圆所限定的圆锥面区域。除非另有规定，测量方向应垂直于被测表面。公差带的中心位置固定	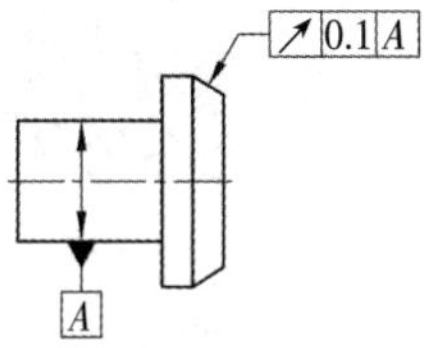 It is qualified during the extracted line in any conical section, the shaft of which coincides with datum shalt A, shall be contained between two circles within the conical section with a distance of 0.1 mm 在与基准轴线A同轴的任一圆锥面，实际轮廓线应限定在素线方向间距等于0.1 mm的直径不相等的两个圆之间	Yes 是
		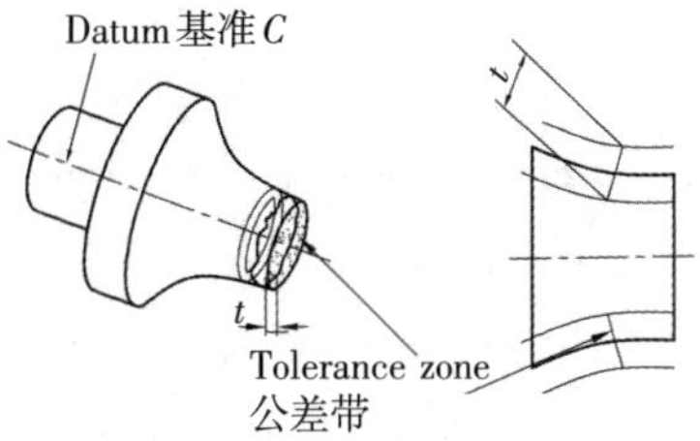 The tolerance zone is limited within any conical section by two circles a distance t apart, the axes of which coincide with the datum The width of the tolerance zone is normal to the specified geometry unless otherwise indicated.When the generator line for the tolerance feature is not straight, the apex angle of the conical section will change depending on the actual position 公差带为与基准轴线同轴的某一圆锥截面，间距等于公差值t的直径不相等的两个圆所限定的圆锥面区域。当公差特征的母线不是直线时，圆锥截面的顶角将根据实际位置而变化	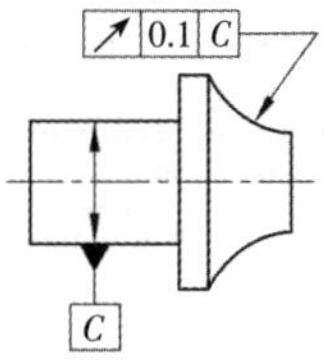 It is qualified during the extracted line in any conical section, the shaft of which coincides with datum shalt A, shall be contained between two circles within the conical section with a distance of 0.1 mm 在与基准轴线A同轴的任一圆锥面，实际轮廓线应限定在素线方向间距等于0.1 mm 的直径不相等的两个圆之间	Yes 是
Total run-out 全跳动	⌰	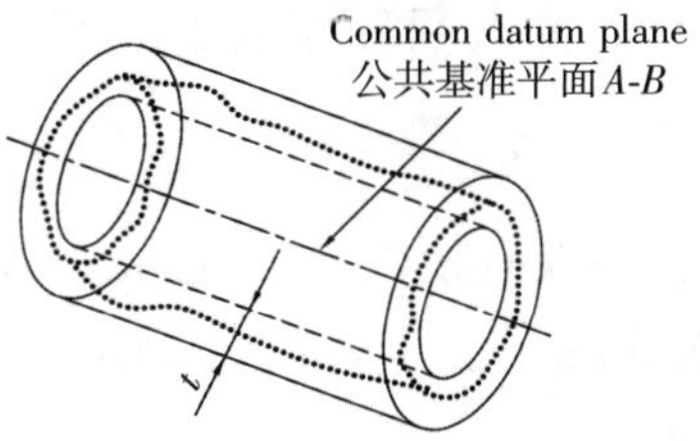 Radial total run-out tolerance, the tolerance zone is limited by two coaxial cylinders with a difference in radii of t, the axes of which coincide with the datum. The central location of tolerance zone is fixed 径向全跳动公差带为半径差等于公差值t且轴线与基准轴线重合的两个圆柱面所限定的区域。公差带的中心位置固定	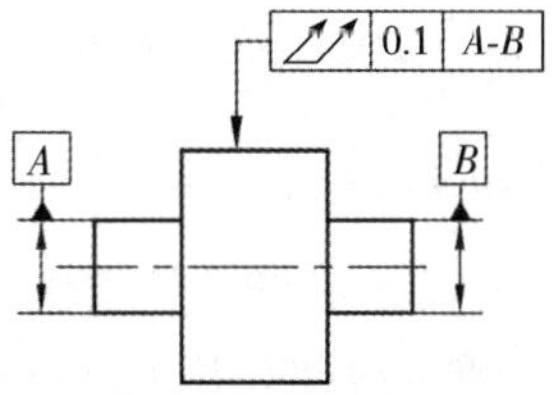 It is qualified during the extracted surface shall be contained between two coaxial cylinders with a difference in radii of 0.1 mm and the axes coincident with the common datum straight line A–B 被测圆柱面的整个实际表面应限定在半径差等于0.1 mm 且轴线与公共基准轴线A–B重合的两个圆柱面之间	Yes 是

Tab. 4-9（Continued）表4-9（续）

Feature 几何特征	Symbol 符号	The tolerance zone 公差带定义	Example of indication and Explanation 标注示例和解释	Datum needed 有无基准
Total run-out 全跳动	↗↗	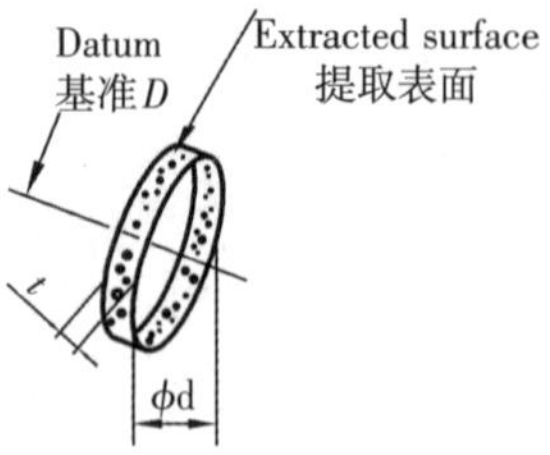 Axial total run-out tolerance, the tolerance zone is limited by two parallel planes a distance t apart and perpendicular to the datum.The central location of tolerance zone is fixed 轴向全跳动公差带为半径差等于公差值t且垂直于基准轴线的两个平行平面所限定的区域。公差带的中心位置固定	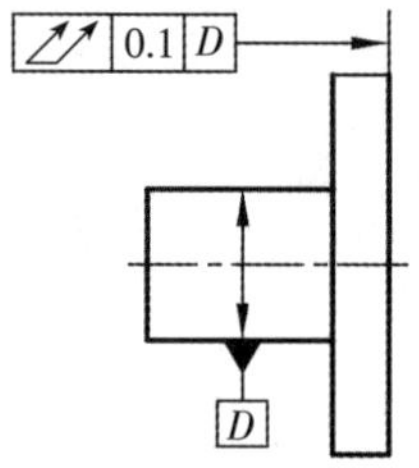 It is qualified during the extracted surface shall be contained between two parallel planes 0.1 mm apart, which are perpendicular to datum shaft D 被测实际表面应限定在间距等于0.1 mm且垂直于基准轴线D的两个平行平面之间	Yes 是

4.5 Tolerance Principle 公差原则

During designing mechanical components, both the size tolerance and geometric tolerance are specified simultaneously for the key geometric features based on functional requirements. The actual states of a workpiece's geometric features result from the combined effect of the size deviations and geometric deviations, both of which influence the fitting performance. Therefore, during design and inspection, it is essential to clarify the relationship between geometric tolerances and dimensional tolerances. The principle govering such interrelations is called the tolerance principle. Tolerance principle is divided into independency principle and mutuality requirement. This Book introduces independency principle, envelope material requirement, maximum material requirement and minimum meterial requirement.

在设计零件时，根据零件的功能要求，对于零件上重要的几何要素，常常需要同时给定尺寸公差、几何公差等。零件上几何要素的实际状态是要素的尺寸误差和几何误差综合作用的结果，两者都会影响零件的配合性能。因此，在设计和检测时需要明确几何公差与尺寸公差之间的关系。确定这种相互关系的原则称为公差原则。公差原则分为独立原则和相关要求两大类。本教材将介绍独立原则、包容要求、最大实体要求、最小实体要求。

4.5.1 Relevant Terminology and Definitions 有关术语及定义

4.5.1.1 External function size 体外作用尺寸（D_{fe}，d_{fe}）

External function size of hole D_{fe} refers to over the given length of the measured feature, the diameter or the width of the largest ideal surface that can externally contact the actual inner surface or the smallest ideal surface that can externally contact the actual external surface of the workpiece. External function size of shaft d_{fe} refers to over the given length of tolerance feature, the width or diameter of the smallest perfect feature that can be circumscribed about the feature so that it just contacts the surface at the highest points outside the workpiece.

在被测要素的给定长度上，与实际内表面体外相接的最大理想面的直径或宽度，或与实际外表面体外相接的最小理想面的直径或宽度，称为孔的体外作用尺寸，用D_{fe}表示。在被测要素的给定长度上，实际外表面体外相接的最小理想面的直径或宽度，称为轴的体外作用尺寸，用d_{fe}表示。

For the related features, the axis or central plane of the ideal plane must maintain the orientation relationship with the datum specified in the drawing, as shown in Fig. 4–15 (a)。

对于关联要素，该理想面的轴线或中心平面，必须与基准保持图样给定的方位关系。如图4–15（a）所示。

Since the external function size of a hole is smaller than its local size and the external function size of the shaft is larger than its local size, the external function size can be calculated using Eq.(4–1) and Eq.(4–2).

由于孔的体外作用尺寸小于局部尺寸，轴的体外作用尺寸大于局部尺寸，孔和轴的体外作用尺寸可以通过式（4–1）和（4–2）求得：

$$D_{fe} = D_a - f_G \tag{4-1}$$

$$D_{fe} = d_a + f_G \tag{4-2}$$

Where，f_G—the form, orientation or location deviation of derived feature.

式中，f_G——导出要素的形状、方向或位置误差。

4.5.1.2 Internal function size 体内作用尺寸（D_{fi}, d_{fi}）

Internal function size of hole D_{fi} refers to the width or diameter of the smallest perfect feature that can be inscribed within the actual inner surface and contacts it at the highest points over the given length of tolerance feature. Internal function size of shaft d_{fi} refers to over the given length of tolerance feature, the width or diameter of the largest perfect feature that can be inscribed within the feature so that it just contacts the surface at the highest points inside the workpiece.

在被测要素的给定长度上，与实际内表面体内相接并在最高点接触的最小理想面的直径或宽度，称为孔的体内作用尺寸，用D_{fi}表示。在被测要素的给定长度上，与实际外表面体内相接且工件内部最高点接触的最大理想面的直径或宽度，称为轴的体内作用尺寸，用d_{fi}表示。

For the related features, the axis or central plane of the ideal plane must maintain the geometric relationship specified in the drawing relative to the datum, as shown in Fig. 4–15 (b)。

对于关联要素，该理想面的轴线或中心平面必须与基准保持图样给定的几何关系，如图4–15（b）所示。

The internal function size of an actual hole is larger than the local size and the internal function size of a shaft is smaller than the actual size, and the internal function size can be calculated by Eq.(4–3) and Eq.(4–4).

实际孔的体内作用尺寸大于该孔的局部尺寸，实际轴的体内作用尺寸小于该轴的实际尺寸。因此，轴的体内作用尺寸和孔的体内作用尺寸可以通过式（4–3）和（4–4）计算。

$$D_{fi} = D_a + f_G \tag{4-3}$$

$$D_{fi} = d_a - f_G \tag{4-4}$$

Where，f_G—the form, orientation or location deviation of derived feature.

式中，f_G——导出要素的形状、方向或位置误差。

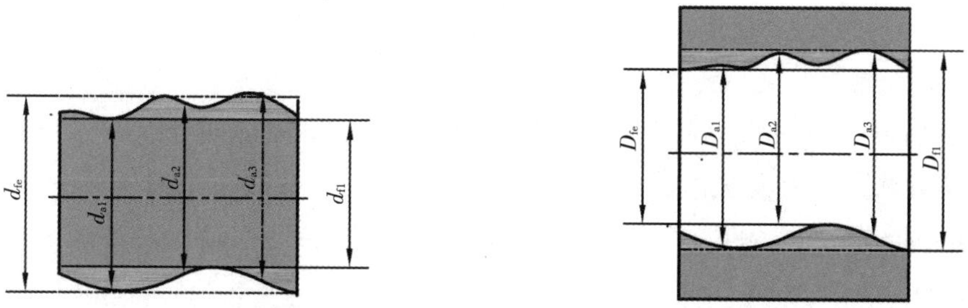

(a) External surface (shaft) 外表面（轴）　　(b) Internal surface (hole) 内表面（孔）

Fig. 4–15 Actual size and function size

图4–15 实际尺寸和作用尺寸

4.5.1.3 Maximum material condition（MMC）and maximum material size（MMS）最大实体状态（MMC）与最大实体尺寸（MMS）

The maximum material condition（MMC）is the state in which a size feature contains the maximum amount of material within the specified size limits, for example, the largest shaft diameter or smallest hole diameter.

最大实体状态（MMC）是指在规定的尺寸范围内，某尺寸要素的各部位材料量均达到最大值的条件，例如轴的最大直径或孔的最小直径。

Maximum material size（MMS）is the size defining the maximum material condition of a feature, i.e. the limit size where the material is at the maximum, e.g. maximum limit of size of a shaft or minimum limit size of a hole. The MMS of shaft is represented by d_M and the MMS of hole is represented by D_M. The relationship between limit size and maximum material size is shown in Eq.（4-5）and Eq.（4-6）.

最大实体尺寸（MMS）是用于定义实际要素最大实体状态下的极限尺寸，即材料处于最大值时的极限尺寸，例如，轴的最大极限尺寸或孔的最小极限尺寸。轴的最大实体尺寸用d_M表示，即轴的最大极限尺寸；孔的最大实体尺寸用D_M表示，即孔的最小极限尺寸。两者分别如式（4-5）和式（4-6）所示。

$$d_M = d_{max} \tag{4-5}$$

$$D_M = D_{min} \tag{4-6}$$

4.5.1.4 Least material condition（LMC）and least material size（LMS）最小实体状态（LMC）与最小实体尺寸（LMS）

Least material condition（LMC）is the condition in which a feature of size contains the least amount of material everywhere within the stated limits of size, for example, the smallest shaft diameter or the largest hole diameter.

最小实体条件（LMC）是指在规定的尺寸限值范围内，某尺寸要素的各部位均处于材料量最小的极限状态。例如，轴的最小直径或孔的最大直径。

Least material size（LMS）is the size defining the least material condition of a feature, i.e. the limit size where the material is at the minimum, e.g. minimum limit size of a shaft or maximum limit size of a hole. The LMS of shaft is represented by d_L, and the LMS of the hole is represented by D_L. The relationship of limit size and maximum material size is shown in Eq.（4-7）and Eq.（4-8）.

最小实体尺寸（LMS）是用于定义实际要素最小实体状态下的极限尺寸，即材料处于最小值时的极限尺寸，例如，轴的最小极限尺寸或孔的最大极限尺寸。轴的最小实体尺寸用d_L表示，即轴的最小极限尺寸；孔的最小实体尺寸用D_L表示，即孔的最大极限尺寸。两者分别如式（4-7）和式（4-8）所示。

$$d_L = d_{min} \tag{4-7}$$

$$D_L = D_{max} \tag{4-8}$$

4.5.1.5 Maximum material virtual condition（MMVC）and maximum material virtual size（MMVS）最大实体实效状态与最大实体实效尺寸

Maximum material virtual condition（MMVC）is a collective state of the considered feature of size when the feature is at MMC and its derived feature's geometric deviation is equal to the specified geometric tolerance on engineering drawing at a given length. Maximum material virtual size（MMVS）is the external function size of the considered feature of size in which at MMVC, the MMVS of hole and shaft is represented by D_{MV}

and d_{MV}, shown in Eq.（4-9）and Eq.（4-10）.

在给定长度上，最大实体实效状态（MMVC）是指被测要素处于最大实体状态，且导出要素的几何偏差等于工程图样上规定的几何公差值的综合状态。最大实体实效尺寸（MMVS）是指被测要素在MMVC状态下，被测尺寸要素的体外作用尺寸。D_{MV}表示孔的最大实体实效尺寸，d_{MV}表示轴的最大实体实效尺寸代号。关系如式（4-9）和式（4-10）所示。

$$D_{MV} = D_M - t = D_{min} - t \text{ Ⓜ} \tag{4-9}$$

$$d_{MV} = d_M + t = d_{max} + t \text{ Ⓜ} \tag{4-10}$$

Where，t—the geometric tolerance of derived feature；

Ⓜ—symbol of maximum material requirement.

式中，t——导出要素的几何公差；

Ⓜ——最大实体要求。

4.5.1.6 Least material virtual condition（LMVC）and least material virtual size（LMVS）最小实体实效状态（LMVC）和最小实体实效尺寸（LMVS）

Least material virtual condition（LMVC）is a collective state of the considered feature of size in which the feature is at its LMC and its derived feature's geometric deviation is equal to the specified geometric tolerance specified on engineering drawing. Least material virtual size（LMVS）is the internal function size of the considered feature of size in which at LMVC, the LMVS of hole and shaft is represented by D_{LV} and d_{LV}, shown in Eq.（4-11）and Eq.（4-12）.

最小实体实效状态（LMVC）是指被测要素处于最小实体状态且导出要素的几何偏差等于工程图样上规定的几何公差的综合状态。最小实体实效尺寸（LMVS）是指被测要素在LMVC状态下被测要素的体内作用尺寸，其中，D_{LV}表示孔的最小实体实效尺寸，d_{LV}表示轴的最小实体实效尺寸。关系如式（4-11）和式（4-12）所示。

$$D_{LV} = D_L + t \text{ Ⓛ} \tag{4-11}$$

$$d_{LV} = d_L - t \text{ Ⓛ} \tag{4-12}$$

Where，t—the geometric tolerance of derived feature；

Ⓛ—symbol of minimum material requirement.

式中，t——导出要素的几何公差；

Ⓛ——最小实体要求。

4.5.1.7 Boundary 边界

Boundary is the perfect limit envelope surface（cylinder or two parallel planes）specified in designing. According to the functional and economy requirements, different boundaries such as maximum material boundary（MMB), least material boundary（LMB), maximum material virtual boundary（MMVB）and least material virtual boundary（LMVB）are defined. The calculation formulas for various ideal boundary dimensions are shown in Tab. 4-10.

由设计给定的具有理想形状的极限包容面（极限圆柱或两平行平面）称为边界。根据零件的功能和经济性要求,可以给出最大实体边界（MMB)、最小实体边界（LMB)、最大实体实效边界（MMVB)、最小实体实效边界（LMVB)。各种理想边界尺寸的计算公式见表4-10。

Tab. 4-10 Formulas for calculating various ideal boundary dimensions

表4-10 各种理想边界尺寸的计算公式

Terms 术语	Symbol and formula 符号和公式
Maximum material boundary size of hole 孔的最大实体边界尺寸	$MMB_D = D_M = D_{min}$
Maximum material boundary size of shaft 轴的最大实体边界尺寸	$MMB_d = d_M = d_{max}$
Minimum material boundary size of hole 孔的最小实体边界尺寸	$LMB_D = D_L = D_{max}$
Minimum material boundary size of shaft 轴的最小实体边界尺寸	$LMB_d = d_L = d_{min}$
Maximum material virtual size of hole 孔的最大实体实效边界尺寸	$MMVB_D = D_{MV} = D_M - t = D_{min} - t$ Ⓜ
Maximum material virtual size of shaft 轴的最大实体实效边界尺寸	$MMVB_d = d_{MV} = d_M + t = d_{max} + t$ Ⓜ
Least material virtual size of hole 孔的最小实体实效边界尺寸	$LMVB_D = D_{LV} = D_L + t = D_{max} + t$Ⓛ
Least material virtual size of shaft 轴的最小实体实效边界尺寸	$LMVB_d = d_{LV} = d_L - t = d_{min} - t$ Ⓛ

4.5.2 Independency Principle 独立原则

Independency Principle stipulates that dimensional or geometric requirements on a drawing are met independently. It is the basic principle that the relationship of size tolerances and geometric tolerances should conform to. When independency principle is adopted, its characteristics are shown below.

独立原则是指图样上给定的几何公差与尺寸公差相互无关，应分别满足要求的公差原则。它是尺寸公差与几何公差关系应遵循的基本原则。当采用独立性原则时，其特点如下。

(1) The size tolerances specified on drawing are only to control the size deviation and do not to control the geometric deviations; and at the same time, the geometric tolerances are only to control the geometric deviations and do not control the size deviations.

图纸上规定的尺寸公差只控制尺寸误差，不控制几何误差；同时，几何公差只控制几何误差，不控制尺寸误差。

(2) There are no additional symbols indicated on drawings when independency principle is applied.

当采用独立性原则时，图纸上无附加符号。

Fig. 4-16 is an example of independency principle used for individual feature. After machining, the local size of extracted feature should be kept between $\phi19.967$ mm and $\phi20$ mm, and the straightness deviation of axis should not be more than $\phi0.02$ mm.

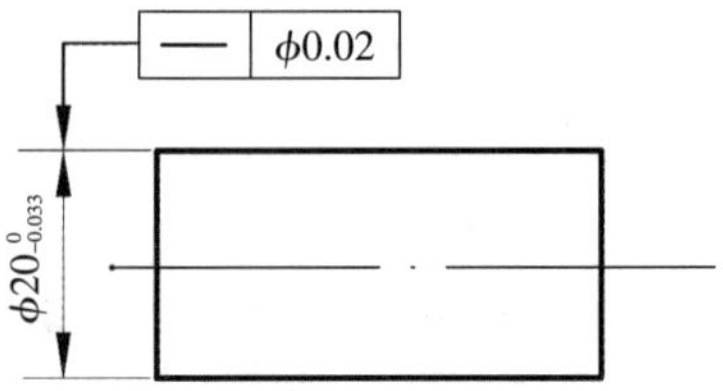

Fig. 4-16 Independency principle is used for single feature

图4-16 独立原则应用于单一要素

图4-16是用于单个特征的独立性原则的示例。加工后，提取特征的局部尺寸应保持在ϕ19.967 ~ ϕ20 mm，其轴线的直线度误差不允许大于ϕ0.02 mm。

Independency principle is the essential tolerance principle, so it is widely used. Unless there is an obvious advantage in adopting mutuality principles, usually independency principle is used to specify the size tolerances and geometric tolerances.

独立原则是最基本的公差原则，因而被广泛应用。除了采用特殊原则具有明显的优势，通常采用独立原则来规定尺寸公差和几何公差。

4.5.3 Relevant Requirements 相关要求

The relevant requirements refer to the tolerance requirements in which the given geometric tolerances and dimensional tolerances on the drawing are interrelated. The relevant requirements include envelope requirement, maximum material requirement（MMR), least material requirement（LMR）and reversibility requirement. In this book, only envelope requirement, maximum material requirement（MMR), least material requirement（LMR）are introduced.

相关要求是指图样上给定的几何公差与尺寸公差相互有关联的公差要求。相关要求包括包容要求、最大实体要求、最小实体要求和可逆要求。本书中，仅介绍包容要求、最大实体要求、最小实体要求。

4.5.3.1 Envelope requirement 包容要求

The envelope requirement, also called Taylor Principle, specifies that the surface of a single feature of size should not exceed the imaginary envelope of perfect（geometric ideal）form at maximum material size. That is, the extracted integral feature should not exceed the maximum material boundary, and the extracted local size should not exceed the least material size. For a hole（internal feature of size), the external function size should be equal to or more than the maximum material size and the extracted local size should be equal to or less than the least material size. For a shaft（external feature of size), the external function size should be equal to or less than the maximum material size and the extracted local size should be equal to or more than the least material size. The envelope requirement is indicated by the symbol Ⓔ placed after the linear tolerance。

包容要求，也称为泰勒原则，其规定单一尺寸要素的表面在最大实体尺寸时不应超出理想边界(几何理想）形状的假想包容面，即提取的整体特征尺寸不应超出最大实体边界，且其提取的局部实际尺寸不应超出最小实体尺寸。对于孔（内部尺寸特征）而言，体外作用尺寸应等于或大于最大实体尺寸，且局部尺寸应等于或小于最小实体尺寸。对于轴（外部尺寸特征）而言，体外作用尺寸应等于或小于最大实体尺寸，且提取的局部尺寸应等于或大于最小实体尺寸。包容要求可以通过在线性公差尺寸后加注符号Ⓔ来表示。

When an individual feature adopts the envelope requirement, the actual size and shape error of the feature depend on each other within the maximum solid boundary, and the allowable shape error value completely depends on the actual size. Therefore, if the actual size of the shaft or hole is the maximum physical size everywhere, the shape error must be zero to be qualified.

单一要素采用包容要求时，在最大实体边界范围内，该要素的实际尺寸和形状误差相互依赖，所允许的形状误差值完全取决于实际尺寸的大小。因此，若轴或孔的实际尺寸处处皆为最大实体尺寸，则其形状误差必须为零才是合格的。

Example 4-1: Drawing annotation in Fig. 4-17 indicates that the actual contour of an individual feature shaft shall not exceed the maximum solid boundary with a boundary size of ϕ20 mm, that is, the external functional size of the shaft shall not be greater than the maximum material size of ϕ20 mm (the maximum limit size of the shaft). The actual size of the shaft shall not be less than the minimum material size of ϕ19.987 mm (the minimum limit size of the shaft). Because the shaft is limited by the maximum material boundary, when the shaft is in the maximum material state, there is no shape error; When the shaft is in the minimum material state in Fig. 4-17, the allowable value of the shaft straightness error can reach ϕ0.013 mm. However, in most cases, the shaft is in a certain actual size and has shape error (including shaft straightness). In this case, because the actual size deviates from the maximum material state, the shape tolerance value of the shaft can be derived from this deviation, and the deviation can be compensated as much as possible. In other words, the shape tolerance value of the shaft is equal to the compensation value. At present, the actual size of the shaft is ϕ19.998 mm, and the shape tolerance values of the shaft are listed in Tab. 4-11.

例4-1：图4-17的图样标注，表示单一要素轴的实际轮廓不得超出边界尺寸为ϕ20 mm 的最大实体边界，即轴的体外作用尺寸应不大于ϕ20 mm 的最大实体尺寸（轴的最大极限尺寸）。轴的实际尺寸应不小于ϕ19.987 mm的最小实体尺寸（轴的最小极限尺寸）。由于轴受到最大实体边界的限制，当轴处于最大实体状态时，不允许存在形状误差；图4.17中，当轴处于最小实体状态时，其轴线直线度误差允许值可达到ϕ0.013 mm。但是，在大部分情况下，轴是有某一实际尺寸的，而且具有形状误差（包括轴线直线度）。这时，由于实际尺寸偏离了最大实体状态，轴的形状公差值可以从这个偏离中可得到最大程度的补偿。也就是说，轴的形状公差值等于补偿值。现假设轴的实际尺寸为ϕ19.998 mm，将轴的形状公差值列于表4-11。

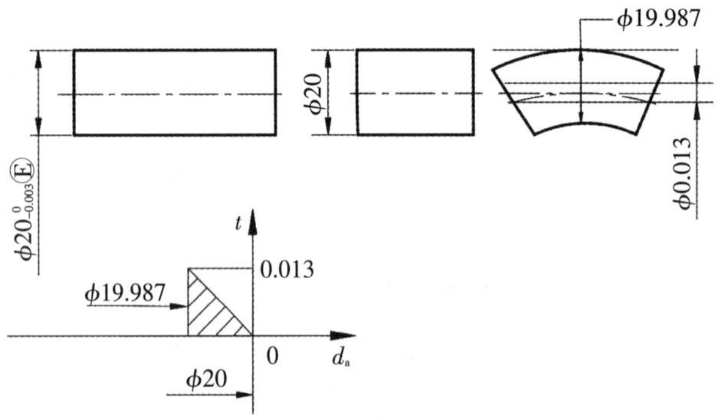

Fig. 4-17 Examples of application of independency principle

图4-17 包容要求应用举例

Tab. 4-11 Tolerance value of shape required by inclusion

表4-11 包容要求的形状公差值

Maximum material size 最大实体尺寸	ϕ20	Tolerance value 形状公差值	0
Actual size 实际尺寸	ϕ19.998		Compensation value: 补偿值： ϕ20 − ϕ19.998 = ϕ0.002
Least material size 最小实体状态	ϕ19.987		ϕ0.013
Maximum material boundary 最大实体边界	$MMB_d = \phi 20$		——

When the actual feature deviates from the maximum material state, the tolerance of size is allowed to compensate the form tolerance, and the compensation amount depends on the deviation from the maximum material state. In Tab. 4-11,when the actual size is $\phi19.998$ mm, the compensation tolerance value is equal to $\phi20$ mm minus $\phi19.998$ mm, that is, $\phi0.002$ mm. The form tolerance referred to may be the straightness of the shaft, the straightness of the plain line, or the roundness, and the tolerance value is equal to the compensation value,which is 0.002 mm.

当实际要素偏离最大实体状态时，包容要求允许尺寸公差补偿给形状公差，补偿量取决于偏离最大实体状态的多少。在表4-11中，当实际尺寸为$\phi19.998$ mm时, 此时补偿形状公差值等于$\phi20$ mm减去$\phi19.998$ mm即$\phi0.002$ mm，所指的形状公差可能是轴线直线度，也可能是素线直线度，还可能是圆度，公差值等于补偿值，同为0.002 mm。

Based on the above mentioned meaning, when judging the qualification conditions of parts, the following parts qualification conditions can be obtained,such as Eq. (4-13)to Eq. (4-15)respectively for holes and shafts.

从含义上判断零件的合格条件，可以得出孔和轴的零件合格条件，分别如式（4-13）至式（4-15）所示。

$$\begin{cases} D_k \geqslant D_M \\ D_a \leqslant D_L \end{cases} \text{即} \begin{cases} D_a - f \geqslant D_{min} \\ D_a \leqslant D_{max} \end{cases} \tag{4-13}$$

$$\begin{cases} D_{fe} \leqslant D_M \\ D_a \geqslant D_L \end{cases} \text{即} \begin{cases} D_a - f \leqslant D_{min} \\ D_a \geqslant D_{max} \end{cases} \tag{4-14}$$

$$\begin{cases} D_{min} \leqslant D_a \leqslant D_{max} \text{或} D_{min} \leqslant D_a \leqslant D_{max} \\ f \leqslant t = \text{补偿值 (Compensation value)} \end{cases} \tag{4-15}$$

Where, for the shaft（轴）: $t = d_M = d_a$; for the hole（孔）: $t = D_a - D_M$;

Example 4-2: A shaft is processed according to the size $\phi50^{0}_{-0.039}$ Ⓔ, and the size in the pattern is processed according to the envelope requirement. After processing, the actual size of the shaft is measured $d_a = \phi49.97$ mm, and the shaft straightness error $f_G = \phi0.02$ mm. Judge whether the part is qualified.

例4-2： 按尺寸 $\phi50^{0}_{-0.039}$ Ⓔ加工一个轴，图样上该尺寸按包容要求加工。加工后测得该轴的实际尺寸$d_a = \phi49.97$ mm，其轴线直线度误差$f_G = \phi0.02$ mm，判断该零件是否合格。

Solution: The first method: judge from the meaning,

$d_{max} = \phi50$ mm, $d_{min} = \phi49.95$ mm

$$\begin{cases} d_{fe} = d_a + f_G = \phi49.97 + \phi0.02 = \phi49.99 < d_M = d_{max} = \phi50 \\ d_a = \phi49.97 > d_L = d_{min} = \phi49.9561 \end{cases}$$

So the part is qualified.

解： 方法一：从含义上判断，

$d_{max} = \phi50$ mm, $d_{min} = \phi49.95$ mm

$$\begin{cases} d_{fe} = d_a + f_G = \phi49.97 + \phi0.02 = \phi49.99 < d_M = d_{max} = \phi50 \\ d_a = \phi49.97 > d_L = d_{min} = \phi49.9561 \end{cases}$$

故零件合格。

The second method：judge from the deviation，

$D_{max} = \phi50$ mm, $d_{min} = \phi49.95$ mm

$D_{min} = \phi49.95 < d_a = \phi49.97 < d_{max} = \phi50$

$f_G = \phi0.02 < t = d_M - d_a = \phi50 - \phi49.97 = \phi0.03$

So the part is qualified.

方法二：从偏离状态判断，

$D_{max}=\phi50$ mm, $d_{min}=\phi49.95$ mm

$D_{min}=\phi49.95<d_a=\phi49.97<d_{max}=\phi50$

$f_G=\phi0.02<t=d_M-d_a=\phi50-\phi49.97=\phi0.03$

故零件合格。

4.5.3.2 Maximum material requirement 最大实体要求

When the clearances of holes and shafts fit together, whether they can be assembled freely and meet the functional requirements usually depends on the combined effect of local actual size and geometric error. For example, the bolt holes on two flanges are assembled with the bolts holding them together, when the local actual sizes of the bolt holes and bolts reach the maximum material sizes and their geometric errors also reach the given geometric tolerance value, their assembly clearance is minimum; When their local actual sizes deviate from the maximum material size and reach the least material size and the geometric error is zero, their assembly clearance is the maximum. Accordingly, if the local actual sizes of bolt holes and bolts deviate from their maximum material sizes in the direction of the minimum material sizes,they are free to be assembled even if their geometric error exceeds the given geometric position tolerance (but not beyond a certain limit).

孔与轴间隙配合时，它们能否自由装配和保证功能要求，通常取决于局部实际尺寸和几何误差的综合效应。例如，两个法兰盘上的螺栓孔与固紧它们的螺栓相装配，当螺栓孔和螺栓的局部实际尺寸都达到最大实体尺寸，且它们的几何误差也都达到给定几何公差值时，它们的装配间隙为最小值；当它们的局部实际尺寸偏离最大实体尺寸而达到最小实体尺寸和几何误差为零时，它们的装配间隙为最大值。据此，如果螺栓孔和螺栓的局部实际尺寸向最小实体尺寸方向偏离其最大实体尺寸，即使它们的几何误差超出给定几何公差值（但不超出某一限度）。它们也能自由装配。

Only interchangeable features are required for assembly, usually the maximum material requirement is adopted. Therefore, the maximum material requirement is generally used in situations where ensuring assemblability is prioritized over other functions. In this way,dimensional tolerance can be fully used to compensate geometric tolerances, which is beneficial to manufacturing and inspection. Maximum material requirements can only be used for the axis, form, direction and position tolerances of central surfaces. If this principle can be applied correctly in design, it will bring beneficial economic effect to production. For example, maximum material requirements are widely applied to the positional tolerances of through holes arranged on the circumference of a bolted or screwed - connected disk part in order to make full use of the through - hole dimensional tolerances given in the drawing and obtain the best technical and economic benefits.

只要求装配互换关系的要素，通常采用最大实体要求。因此，最大实体要求一般主要用于保证可装配性、对其他功能要求较低的场合。这样可以充分利用尺寸公差补偿几何公差，有利于制造和检验。最大实体要求只能用于轴线及中心面的形状、方向与位置公差。设计时若能正确地应用此原则，将给生产带来经济效益。例如，用螺栓或螺钉连接的圆盘零件，圆周布置的通孔的位置度公差多采用最大实体要求，以便充分利用图样上给出的通孔尺寸公差，获得最佳的技术和经济效益。

(1) The meaning of the maximum material requirement and the method of marking on the drawing 最大实体要求的含义和在图样上的标注方法。

The maximum material requirements apply to derived features (central features). Adhere to the maximum material virtual boundary. When the actual size deviates from the maximum material size, the geometric error value is allowed to exceed the given tolerance value. The maximum material requirement applies to both the measured features and the datum feature. The symbol required for the maximum features is Ⓜ. When applied to the measured features,it shall be marked after the tolerance value of the geometric tolerance frame; When applied to the datum feature, it shall be marked after the datum letter code of the geometric tolerance box.

最大实体要求适用于导出要素（中心要素）。遵守最大实体实效边界，遵守最大实体实效边界（原则),当实际尺寸偏离最大实体尺寸时，允许几何误差值超出给定的公差值。最大实体要求既适用于被测要素也适用于基准要素。最大实体要求的符号为Ⓜ。应用于被测要素时，在几何公差框格的公差值后标注；应用于基准要素时，在几何公差框格的基准字母代号后标注。

（2） The maximum material requirement applies to the measured feature 最大实体要求应用于被测要素。

When the maximum material requirement is applied to the measured feature, the actual contour of the measured feature should not exceed the maximum material virtual boundary at any given length, that is, its external functional size should not exceed the maximum material effect size, and its local actual size should not exceed the maximum material size and the least material size.

最大实体要求应用于被测要素时，被测要素的实际轮廓在任何给定的长度上处处不得超出最大实体实效边界，即其体外作用尺寸不应超出最大实体实效尺寸，且其局部实际尺寸不超出最大实体尺寸和最小实体尺寸。

Maximum material requirement is applied to the measured feature, the geometric tolerance values of measured features are given in the state of the features in the maximum material. When the actual contour of the measured features deviate from its maximum material state, namely the deviation from its maximum material size, the geometric error values related to the actual size can exceed the geometric tolerance value given under the maximum material condition, that is, the geometric tolerance values can be increased.

最大实体要求应用于被测要素时，被测要素的几何公差值是在该要素处于最大实体状态时给出的。当被测要素的实际轮廓偏离其最大实体状态，即其实际尺寸偏离最大实体尺寸时，几何误差值可超出在最大实体状态下给出的几何公差值，即此时的几何公差值可以增大。

When the given geometric tolerance value is zero,it is zero geometric tolerance. At this time, the maximum material virtual boundary of the measured feature is equal to the maximum material boundary. The qualification conditions of the parts are described below.

当给定几何公差为0时，即为零几何公差。此时，被测要素的最大实体实效边界等于最大实体边界。零件的合格条件如下所述。

（1） Based on the meaning, the qualification conditions of parts can be deduced as follows, such as Eq.（4-16） for holes and Eq. （4-17） for shafts respectively.

从含义上推断孔和轴的零件合格条件，分别得到如式（4-16）和式（4-17）。

$$\begin{cases} D_{fe} \geqslant D_{MV} \\ D_M \leqslant D_a \leqslant D_L \end{cases} \text{即} \begin{cases} D_a - f \geqslant D_{min} + t \\ D_{min} \leqslant D_a \leqslant D_{max} \end{cases} \tag{4-16}$$

$$\begin{cases} d_{fe} \leqslant d_{MV} \\ d_L \leqslant d_a \leqslant d_M \end{cases} \text{即} \begin{cases} d_a + f \leqslant d_{max} + t \\ d_{min} \leqslant d_a \leqslant d_{max} \end{cases} \tag{4-17}$$

Where, f = given value + compensation value.

式中，f = 给定值 + 补偿值。

（2） Judging from the deviation from the maximum material condition, the compensation value of tolerance can be calculated by this degree of deviation from the maximum material condition, as shown in Eq. (4-18).

从偏离最大实体状态上判断，按偏离最大实体状态的程度可计算出公差的补偿值，如式（4-18）所示。

$$d_{min} \leqslant d_a \leqslant d_{max} \text{ 或 } D_{min} \leqslant D_a \leqslant D_{max} \tag{4-18}$$

The given value is the tolerance value in the tolerance frame. The compensation value is the deviation value from the maximum material condition. For the shaft, the compensation value is equal to the maximum material size minus the actual size, i.e. $t = d_M - d_a$; for the hole, the compensation value is equal to the actual size of the hole minus the maximum material size, i.e, $t = D_a - D_M$.

式中的给定值是几何公差框格中给定的公差值。补偿值是偏离最大实体状态的偏离值。对于轴来说，补偿值等于最大实体尺寸减去实际尺寸，即$t = d_M - d_a$；对于孔来说，补偿值等于孔的实际尺寸减去最大实体尺寸，即$t = D_a - D_M$。

Maximum material requirements are applied to the measured features, as shown in Fig. 4-18. Let the actual size be ϕ50.12 mm, and the shaft perpendicularity tolerance allowed is shown in Tab. 4-12.

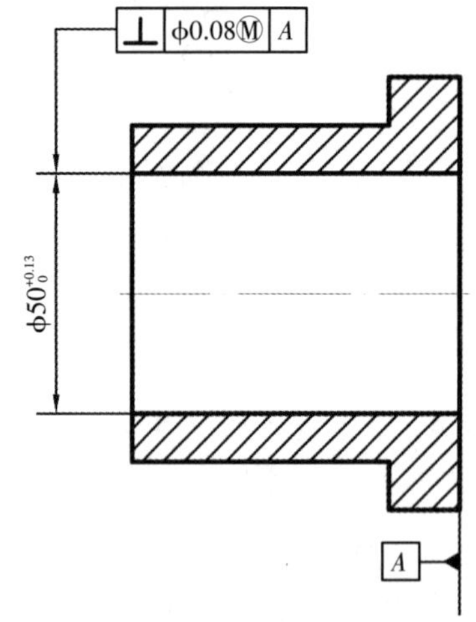

Fig. 4-18 Application example of maximum material requirement (MMR)

图4-18 最大实体要求应用举例

最大实体要求应用于被测要素，如图4-18所示。设实际尺寸为ϕ50.12 mm，允许的轴线垂直度公差见表4-12。

Tab. 4-12 Tolerance value of the maximum material requirement

表4-12 最大实体要求的几何公差值

Maximum material size 最大实体尺寸	ϕ50	Tolerance value 形状公差值	0.08
Actual size 实际尺寸	ϕ50.12		Given value+Compensation value: 给定值＋补偿值: $\phi0.08 + (\phi50.12 - \phi50) = \phi0.2$
Least material size 最小实体状态	ϕ50.13		Given value + Compensation value: 给定值＋补偿值: $\phi0.08+(\phi50.13 - \phi50) = \phi0.2$
Maximum material boundary 最大实体边界	$d_{MV} = \phi49.92$		—

Example 4-3: A hole is processed. The actual size of the hole measured after processing is $\phi50.02$ mm, and the error value of perpendicularity of the shaft is $\phi0.09$ mm. Judge whether the part is qualified.

例4-3：加工一个孔，加工后测得孔的实际尺寸为$\phi50.02$，轴的垂直度误差值f为$\phi0.09$，判断该零件是否合格。

Solution: Judge from the deviation:

$D_{max} = \phi50.13$ mm, $D_{min} = \phi50$ mm

$D_{min} = \phi50$ mm $< D_a = \phi50.02$ mm $< D_{max} = \phi50.13$ mm

$f = \phi0.09$ mm $< t$ = given value + compensation value

$= \phi0.08 + (\phi50.02 - \phi50) = \phi0.10$ mm

So the parts are qualified.

解：从偏离状态判断：

$D_{max} = \phi50.13$ mm, $D_{min} = \phi50$ mm

$D_{min} = \phi50$ mm $< D_a = \phi50.02$ mm $< D_{max} = \phi50.13$ mm

$f = \phi0.09$ mm $< t$ = 给定值 + 补偿值 $= \phi0.08 + (\phi50.02 - \phi50) = \phi0.10$ mm

故零件合格。

4.5.3.3 Minimum material requirement 最小实体要求

Minimum material requirement is applicable to extracted derived features. Minimum material requirements are widely used to ensure the minimum wall thickness and control the maximum distance from the surface to the central features under the premise of obtaining the best technical and economic benefits. It is mainly used to limit the positional variation of features, mostly used for position tolerance, and is suitable for derived (central) features.

最小实体要求适用于提取导出要素。在获得最佳技术和经济效益的前提下，最小实体要求广泛应用于保证最小壁厚和控制表面至中心要素的最大距离等功能要求。它主要用于限制要素的位置变动，多用于位置公差，适用于导出（中心）要素。

For example, the wall thickness between two adjacent holes on a part reaches its minimum when the local actual size of both holes reaches the least material size and the position error between them also attains the specified position tolerance. Accordingly, if the local actual sizes of two holes deviate from their least material size toward the maximum material size, they can still guarantee the minimum wall thickness even if their positional error exceeds the given location tolerance value (but within a certain limit).

例如，对于零件上相邻两孔之间的壁厚，当两孔的局部实际尺寸都达到最小实体尺寸，且它们之间的位置误差也达到给定的位置度公差时，它们之间的壁厚为最小值。据此，如果两孔的局部实际尺寸向最大实体尺寸方向偏离其最小实体尺寸，即使它们的位置误差超出给定位置度公差值（但不超出某一限度),它们也能保证最小壁厚。

(1) Definition of minimum material requirements and method of marking on drawings 最小实体要求的含义和在图样上的标注方法。

Minimum material requirements apply to derived (central) features. Adhering to the minimum material virtual boundary, the actual contour of the measured feature must lie within the minimum material virtual boundary. When the actual size deviates from the least material size, the geometric error is allowed to exceed the given tolerance value. The minimum material requirement applies to both the measured features and the datum features. The symbol required for the minimum material is Ⓛ. When applied to a measured features, the tolerance value of the geometric tolerance box of the features to be measured shall be marked. When applied to the datum feature, it is marked after the datum letter in the geometric tolerance box.

最小实体要求适用于导出（中心）要素。被测要素遵守最小实体实效边界，其实际轮廓处于最小实体实效边界之内。当实际尺寸偏离最小实体尺寸时，允许几何误差值超出给定的公差值。最小实体要求既适用于被测要素，也适用于基准要素。最小实体要求的符号为 Ⓛ。应用于被测要素时，在被测要素几何公差框格的公差值后标注；应用于基准要素时，在几何公差框格的基准字母代号后标注。

（2）Minimum material requirement applies to the measured feature 最小实体要求应用于被测要素。

When the minimum material requirement is applied to the measured feature, the actual contour of the measured feature should not exceed the minimum material virtual boundary at any given length. The measured features of geometric tolerance values are given in least material condition. When the actual contour of measured features deviates from its least material condition, namely its actual size deviates from least material sizes, geometric error values can exceed the geometric tolerance given under the least material condition, at this moment, the geometric tolerance values can be increased.

最小实体要求应用于被测要素时，被测要素的实际轮廓在任何给定的长度上处处不得超出最小实体实效边界。被测要素的几何公差值是在该要素处于最小实体状态时给出的，当被测要素的实际轮廓偏离其最小实体状态，即其实际尺寸偏离最小实体尺寸时，几何误差值可超出在最小实体状态下给出的几何公差值，此时的几何公差值可以增大。

When the given geometric tolerance value is zero, it is zero geometric tolerance. The minimum material virtual boundary of the feature under test is equal to the minimum material boundary, and the minimum material virtual size is equal to the least material size. Minimum material requirements are primarily applied to related features but can also be used for individual features. The qualification conditions of the parts are as follows.

当给出的几何公差值为零时，则为零几何公差。被测要素的最小实体实效边界等于最小实体边界，最小实体实效尺寸等于最小实体尺寸。最小实体要求主要应用于关联要素，也可用于单一要素。零件的合格条件如下所述。

（1）Judging the qualification conditions of parts from the meaning, the following parts qualification conditions can be derived as shown in Eq.（4–19）and Eq.（4–20）respectively for holes and shafts.

从含义上推断孔和轴合格条件，分别如式（4–19）和式（4–20）所示。

$$\begin{cases} D_{fe} \leqslant D_{LV} \\ D_M \leqslant D_a \leqslant D_L \end{cases} \text{即} \begin{cases} D_a + f \leqslant D_{max} + t \\ D_{min} \leqslant D_a \leqslant D_{max} \end{cases} \tag{4-19}$$

$$\begin{cases} d_{fe} \geqslant d_{LV} \\ d_L \leqslant d_a \leqslant d_M \end{cases} \text{即} \begin{cases} d_a - f \geqslant d_{max} - t \\ d_{min} \leqslant d_a \leqslant d_{max} \end{cases} \tag{4-20}$$

（2）Judging from the deviation from the least material condition, the compensation value of tolerance can be calculated according to the deviation from the maximum material condition, as shown in Eq.（4–21）.

从偏离最小实体状态上判断零件的合格条件，按偏离最小实体状态的程度可计算出公差的补偿值，如式（4–21）所示。

$$d_{min} \leqslant d_a \leqslant d_{max} \text{ 或 } D_{min} \leqslant D_a \leqslant D_{max} \tag{4-21}$$

The “given value” refers to the tolerance value in the tolerance frame. The compensation value is the deviation value from the least material condition. For a shaft,the compensation value is equal to the actual size minus the least material size, i.e. $t = d_a - d_L$; for a hole, the compensation value is equal to the least material size of the hole minus the actual size,that is, $t = D_M - D_a$.

式中的给定值是公差框格中给定的公差值。补偿值是偏离最小实体状态的偏离值。对于轴来说，补偿值等于实际尺寸减去最小实体尺寸，即$t = d_a - d_L$；对于孔来说，补偿值等于孔的最小实体尺寸减去实际尺寸，即$t = D_M - D_a$。

Example 4-4: A shaft is processed as shown in Fig. 4-19. The actual size of the hole measured after processing is ϕ19.7 mm, and the error value(f) of straightness of the axis is ϕ0.1 mm. Judge whether the part is qualified.

例4-4: 按图4-19加工一个孔，加工后测得孔的实际尺寸为ϕ19.7 mm，轴的垂直度误差值(f) 为ϕ 0.1，判断该零件是否合格。

Solution: The shaft adopts the minimum material requirement, as shown in Fig. 4-19 (a). When the actual size of the shaft is the minimum material size of ϕ19.7 mm, the straightness tolerance of the axis is given as ϕ0.1 mm, see Figure 4-19 (b). The minimum material virtual size of the shaft is $d_{LV} = d_{min} - t$ Ⓛ = ϕ(19.7-0.1) mm = ϕ19. 6 mm

解: 图4-19 (a) 所示的轴采用了最小实体要求，当轴的实体尺寸为最小实体尺寸ϕ19.7 mm时，轴线的直线度公差为给定值ϕ0.1 mm，见图4-19 (b)，轴的最小实体实效尺寸为 $d_{LV} = d_{min} - t$ Ⓛ = ϕ (19.7 - 0.1) mm = ϕ19. 6 mm。

When the actual size of the shaft deviates from the minimum material size, the straightness error is allowed to increase,that is to say, the dimensional tolerance compensates for the geometric tolerance. When the actual size of the shaft is the maximum material size of ϕ20 mm, the maximum allowable straightness error reaches, t_{max} = the given value + the additional value = ϕ0.1 mm + ϕ0.3 mm = ϕ0.4 mm. The dynamic tolerance chart for the compensation of straightness error is shown in Fig. 4-19 (c).

当轴的实际尺寸偏离最小实体尺寸时，直线度误差允许增大，即尺寸公差补偿给几何公差。当轴的实际尺寸为最大实体尺寸ϕ20 mm时，直线度误差允许达到的最大值t_{max} = 给定值 + 补充值 = ϕ0.1 mm + ϕ0.3 mm = ϕ0.4 mm。图4-19 (c) 为其补偿的动态公差图。

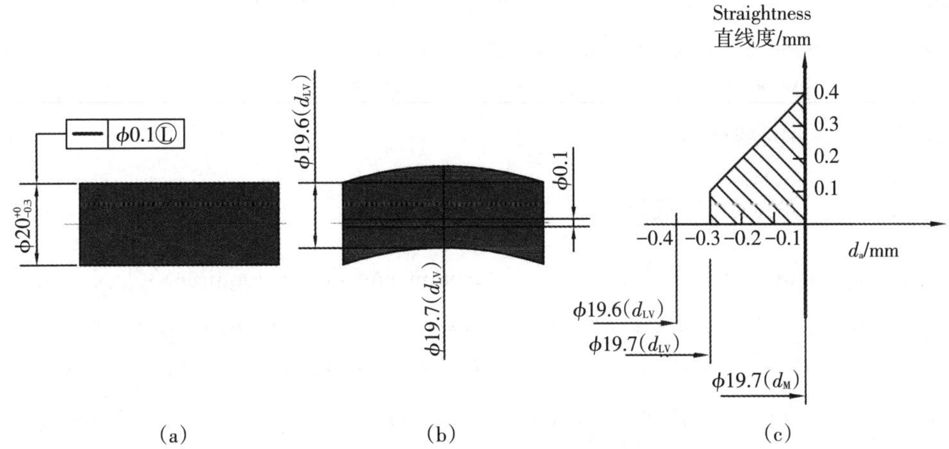

Fig. 4-19 Application example of minimum material requirement (MMR)

图4-19 最小实体要求应用举例

4.6 Selection of Geometric Tolerance 几何公差的选用

4.6.1 Selection of Geometric Characteristics 几何公差特征的选择

Following aspects should be considered when geometric characteristics are selected:

在选择几何特征时，应考虑以下几个方面:

(1) Structure properties of workpieces: to analyse the geometric errors that maybe exist in the machining of workpieces. For example, cylindrical surface workpieces have cylindrical error, stepped shafts and holes have coaxial errors.

工件的结构特性：分析被加工工件可能存在的几何误差。例如，圆柱面工件存在圆柱度误差，阶梯轴和孔存在同轴度误差。

(2) Functional requirements of workpieces: Analyse the geometric characteristics that affect the functional requirement. For example, the main errors affecting lathe spindle accuracy are the cylindrical error and coaxial error of the shaft neck in the front and the rear. The movement accuracy of lathe guide rail is mainly affected by straightness error of the guide rail.

工件的功能要求：分析影响功能要求的几何特征。例如，影响车床主轴精度的主要误差是前后轴颈的圆柱度误差和同轴度误差。机床导轨的运动精度主要受导轨直线度误差的影响。

(3) Properties of every geometric characteristics: there are individual characteristics (straightness, flatness, circularity) and composite characteristics (cylindricity, location) in the 14 geometric characteristics. It is preferred to select the composited characteristics to reduce the amount of characteristics on the engineering drawings.

每一个几何特征的性质：在14个几何特征中有各自的特征（直线度、平面度、圆度）和综合特征（圆柱度、位置）。最好选择复合特征，以减少工程图纸上的特征数量。

(4) Measurement conditions: the measurement conditions include the availability of measurement instrument, the measurement difficulty, measurement efficiency, etc. In the case of satisfying the function requirements, the characteristics that are easier to should be selected first.

测量条件因素：测量条件包括有无测量仪器、测量难度、测量效率等。在满足功能要求的情况下，应首先选择易于测量的特性。

4.6.2 Selection of Datum 基准的选择

The selection of datum is mainly based on the function and design requirements of workpiece, and the aspects should be considered at the same time, such as the principle of unifying datum and the structure properties of workpiece. Dadum principle unification: the datum features should be the same with features of designing datum, location datum and the assembly datum. Among the three datum planes, the plane which has the largest effect on function or the location is most stable to the tolerance feature should be selected as primary datum.

基准的选择主要依据工件的功能和设计要求，同时考虑基准统一原则和工件的结构特性等因素。基准统一原则：基准特征应与设计基准、定位基准和装配基准相同。在三个基准面中，对功能影响最大或位置对公差特性最稳定的平面应作为第一基准面。

The envelope requirement is selected when the fit properties are very important, and maximum material requirement is selected when ensuring free assembly, and minimum material requirement is selected when ensuring the minimum wall thickness or the maximum distance. Otherwise, the independence principle is selected.

当配合性能非常重要时，选择包容要求。在保证自由装配时，选择最大实体要求。在保证最小壁厚和最大距离时，选择最小实体要求。否则，选择独立原则。

4.6.3 Selection of Geometric Tolerance Value 几何公差值的选择

4.6.3.1 The principles of geometric tolerance selection 几何公差值的选择原则

For the economic reason,the geometric tolerance value as large as possible is selected in the case of satisfying the functional requirements. Since there are no reliable methods to calculate geometric tolerance values, the analogy method is usually used to select geometrical tolerance values. The following aspects should be considered when the geometrical tolerance values are selected.

从经济角度考虑，在满足功能要求的情况下，尽可能选取较大的几何公差值。到目前为止，还没有可靠的方法来计算几何公差值，所以一般都用类比法来选择几何公差值。在选择几何公差值时，应考虑以下几个方面：

（1）The relationship of size tolerance and geometric tolerance: typically, the values of form tolerance T_f, orientation tolerance $T_{0,}$ocation tolerance T_L and size tolerance T_S, should satisfy the Eq.（4–22）.

$$T_F < T_0 < T_L < T_S \tag{4–22}$$

尺寸公差与几何公差的关系：形状公差 T_F、方向公差 T_0、位置公差 T_L和尺寸公差 T_S的值通常应满足式（4–22）所示的关系。

形状公差 T_F < 方向公差 T_0 < 位置公差 T_L < 尺寸公差 T_S

For example, for two parallel planes with geometric tolerances, the flatness tolerance value should be smaller than the parallelism tolerance value, and the parallelism tolerance value should be smaller than the distance size tolerance value between them.

例如，两个平行平面的几何公差，平面度公差值应小于平行度公差值，平行度公差值应小于它们之间的距离尺寸公差值。

（2）The properties of the workpiece structure: When meeting the function requirements, the geometric tolerance grades can be reduced by 1 ~ 2 grades for the workpiece with complex, structure, poor rigidity（such as thin and long spindles, thin walled workpieces）or difficult to machine and measure, such as：

工件结构的特点：在满足功能要求的前提下，对于结构复杂、刚性较差（如细长轴、薄壁件）或加工和测量难度较大的工件，其几何公差等级可降低1 ~ 2级。具体包括：

① The hole or shaft of which the rate of length to diameter is large;

长径比较大的孔或轴；

② The hole or shaft of which the distance is large;

孔或轴的跨距较大时；

③ The plane surface of work piece of which the width is large, the width is generally 1/2 times greater than length;

平面宽度较大的工件（一般宽度为长度的1/2及以上）；

④ The perpendicularity or parallelism of the line to line, line to plane and the plane to plane;

涉及线对线、线对面、平面对平面的垂直度或平行度；

⑤ The workpiece should follow the specified standards which have been standardized, for example,the cylindricity tolerances of shafts and holes in housings fitted with rolling bearings, the straightness tolerance of the railway guides in the machine tools,the parallelism of holes in the gear boxes.

已标准化的特定部件应遵循专用标准。例如：滚动轴承配合部位的轴颈与轴承孔的圆柱度公差；机床导轨的直线度公差；齿轮箱中孔系的平行度要求。

4.6.3.2 Geometric tolerance grades with specified tolerance values 注出公差值的几何公差等级

In GB/T 1184-1996, the tolerance grades and values are specified for 11 geometric tolerance characteristics, and the value series of position tolerance are specified. However, the tolerance grades for position tolerance are not specified. The two geometric tolerance characteristics, i.e. profile of any line and profile of any surface, are not specified for tolerance grades and values. There are 12 grades for the 11 geometric characteristics for which the tolerance grades are specified, shown in Tab. 4-13 to Tab. 4-16.

GB/T 1184—1996中规定了11个未注几何公差特征的公差等级和公差值，规定了位置公差的公差值，但未规定公差等级。任意线的轮廓度和表面轮廓度这两项几何公差特征均未规定公差等级和公差值。公差等级的11个几何特征共有12个等级，见表4-13至表4-16。

Tab. 4-13 Geometric tolerance on straightness and flatness

表4-13 直线度和平面度的公差值

Main parameter 主参数（L）/mm	Tolerance grades 公差等级											
	1	2	3	4	5	6	7	8	9	10	11	12
	Tolerance values 公差值/μm											
≤10	0.2	0.4	0.8	1.2	2	3	5	8	12	20	30	60
>10 ~ 16	0.25	0.5	1.0	1.5	2.5	4	6	10	15	25	40	80
>16 ~ 25	0.3	0.6	1.2	2	3	5	8	12	20	30	50	100
>25 ~ 40	0.4	0.8	1.5	2.5	4	6	10	15	25	40	60	120
>40 ~ 63	0.5	1.0	2	3	5	8	12	20	30	50	80	150
>63 ~ 100	0.6	1.2	2.5	4	6	10	15	25	40	60	100	200
>100 ~ 160	0.8	1.5	3	5	8	12	20	30	50	80	120	250
>160 ~ 250	1.0	2	4	6	10	15	25	40	60	100	150	300
>250 ~ 400	1.2	2.5	5	8	12	20	30	50	80	120	200	400
>400 ~ 630	1.5	3	6	10	15	25	40	60	100	150	250	500
>630 ~ 1000	2	4	8	12	20	30	50	80	120	200	300	600
>1000 ~ 1600	2.5	5	10	15	25	40	60	100	150	250	400	800
>1600 ~ 2500	3	6	12	20	30	50	80	120	200	300	500	1000
>2500 ~ 4000	4	8	15	25	40	60	100	150	250	400	600	1200
>4000 ~ 6300	5	10	20	30	50	80	120	200	300	500	800	1500
>6300～10000	6	12	25	40	60	100	150	250	400	600	1000	2000

Tab. 4-14 Geometric tolerance on circularity and cylindricity

表4-14 圆度和圆柱度的公差值

Main parameter 主参数[$d(D)$]/mm	Tolerance grades 公差等级											
	1	2	3	4	5	6	7	8	9	10	11	12
	Tolerance values 公差值/μm											
≤3	0.1	0.2	0.3	0.8	1.2	2	3	4	6	10	14	25
>3 ~ 6	0.1	0.2	0.4	1.0	1.5	2.5	4	5	8	12	18	30
>6 ~ 10	0.12	0.25	0.4	1.0	1.5	2.5	4	6	9	15	22	36
>10 ~ 18	0.15	0.25	0.5	1.2	2	3	5	8	11	18	27	43
>18 ~ 30	0.2	0.3	0.6	1.5	2.5	4	6	9	13	21	33	52
>30 ~ 50	0.25	0.4	0.6	1.5	2.5	4	7	11	16	25	39	62
>50 ~ 80	0.3	0.5	0.8	2	3	5	8	13	19	30	46	74
>80 ~ 120	0.4	0.6	1.0	2.3	4	6	10	15	22	35	54	87
>120 ~ 180	0.6	1.0	1.2	3.5	5	8	12	18	25	40	63	100
>180 ~ 250	0.8	1.2	2	4.5	7	10	14	20	29	46	72	115
>250 ~ 315	1.0	1.6	2.5	6	8	12	16	23	32	52	81	130
>315 ~ 400	1.2	2	3	7	9	13	18	25	36	57	89	140
>400 ~ 500	1.5	2.5	4	8	10	15	20	27	40	63	97	155

Tab. 4-15 Geometric tolerance on parallelism, perpendicularity, and angularity

表4-15 平行度、垂直度、倾斜度的公差值

Main paramete 主参数(L)/mm	Tolerance grades 公差等级											
	1	2	3	4	5	6	7	8	9	10	11	12
	Tolerance values 公差值/μm											
≤10	0.4	0.8	1.5	3	5	8	12	20	30	50	80	120
>10 ~ 16	0.5	1.0	2	4	6	10	15	25	40	60	100	150
>16 ~ 25	0.6	1.2	2.5	5	8	12	20	30	50	80	120	200
>25 ~ 40	0.8	1.5	3	6	10	15	25	40	60	100	150	250
>40 ~ 63	1.0	2	4	8	12	20	30	50	80	120	200	300
>63 ~ 100	1.2	2.5	5	10	15	25	40	60	100	150	250	400
>100 ~ 160	1.5	3	6	12	20	30	50	80	120	200	300	500
>160 ~ 250	2	4	8	15	25	40	60	100	150	250	400	600
>250 ~ 400	2.5	5	10	20	30	50	80	120	200	300	500	800
>400 ~ 630	3	6	12	25	40	60	100	150	250	400	600	1000
>630 ~ 1000	4	8	15	30	50	80	120	200	300	500	800	1200
>1000 ~ 1600	5	10	20	40	60	100	150	250	400	600	1000	1500
>1600 ~ 2500	6	12	25	50	80	120	200	300	500	800	1200	2000
>2500 ~ 4000	8	15	30	60	100	150	250	400	600	1000	1500	2500
>4000 ~ 6300	10	20	40	80	120	200	300	500	800	1200	2000	3000
>6300 ~ 10000	12	25	50	100	150	250	400	600	1000	1500	2500	4000

Tab. 4-16 Geometric tolerance on coaxial, symmetry and run-out

表4-16 同轴度、对称度和跳动的公差值

Main parameter 主参数(L)/mm	Tolerance grades 公差等级											
	1	2	3	4	5	6	7	8	9	10	11	12
	Tolerance values 公差值/μm											
≤1	0.4	0.6	1.0	1.5	2.5	4	6	10	15	25	40	60
>1～3	0.4	0.6	1.0	1.5	2.5	4	6	10	20	40	60	120
>3～6	0.5	0,8	1.2	2	3	5	8	12	25	50	80	150
>6～10	0.6	1,0	1.5	2.5	4	6	10	15	30	60	100	200
>10～18	0.8	1.2	2	3	5	8	12	20	40	80	120	250
>18～30	1.0	1.5	2.5	4	6	10	15	25	50	100	150	300
>30～50	1.2	2	3	5	8	12	20	30	60	120	200	400
>50～120	1.5	2.5	4	6	10	15	25	40	80	150	250	500
>120～250	2	3	5	8	12	20	30	50	100	200	300	600
>250～500	2.5	4	6	10	15	25	40	60	120	250	400	800
>500～800	3	5	8	12	20	30	50	80	150	300	500	1000
>800～1250	4	6	10	15	25	40	60	100	200	400	600	1200
>1250～2000	5	8	12	20	30	50	80	120	250	500	800	1500
>2000～3150	6	10	15	25	40	60	100	150	300	600	1000	2000
>3150～5000	8	12	20	30	50	80	120	200	400	800	1200	2500
>5000～8000	10	15	25	40	60	100	150	250	500	1000	1500	3000
>8000～10000	12	20	30	50	80	120	200	300	600	1200	2000	4000

4.7 General Geometric Tolerances 未注几何公差

The general geometric tolerances are specified for machined workpieces in GB/T1184-1996. These general geometric tolerances are derived from measurements on workpieces formed by metal removal. Usually,only such features that have geometric tolerance indication on the drawing are measured.

GB/T1184—1996规定了未注几何公差。未注几何公差是由金属加工表面的工件测量结果得出的。通常只有图纸上标有几何公差指示的特征项目才会被测量。

General geometric tolerance refers to the tolerance which is not marked separately and can be guaranteed under the normal processing conditions in the workshop. For sizes with generalgeometric tolerance, it is not necessary to indicate the limit deviation after the size.

未注几何公差是指未单独标注出的公差，指在车间通常加工条件下可保证的公差。采用公差的尺寸，在该尺寸后不需要注出其极限偏差数值。

The general tolerances of profile, angularity, position and total run-out are specified, since they are limited by angular tolerance or general size tolerances. Special annotations are not required for these unspecified tolerances.

对于轮廓度、角度、位置度和总跳动的一般公差均有明确规定，这些规定受到角度公差或一般尺寸公差的严格制约。对这些项目的未注公差不必做特殊的标注。

General tolerances on circularity are equal to diameter tolerance value, but they shall not exceed the tolerance for circular runout specified in Tab. 4-20.

圆度的未注公差值等于直径尺寸公差值，但不能大于表4-20中的圆跳动公差值。

The general tolerance on cylindricity is not provided.

圆柱度的未注公差一般不提供。

General tolerances on angularity in the definition are not specified. For angles indicated in theoretically exact sizes, the angularity tolerance has to be indicated on the drawing individually.

倾斜度的未注公差并未给出明确定义。对于理论正确尺寸所指示的倾斜度，必须在图纸上单独注明角度的理论正确值。

General tolerance on coaxiality is not specified, and it can be limited by general tolerances on radial circular run-out.

同轴度公差，当未规定同轴度的未注公差时，可由径向圆度跳动的未注公差限制。

For general tolerances of straightness, flatness, perpendicularity, symmetry, and circular runout, the standard specifies three tolerance grades: H, K, and L, with grade H being the highest and grade L the lowest. When applying these tolerances, the standard number and tolerance grade code shall be indicated in the technical requirements, such as: "General tolerances GB/T1184-K".

对于直线度、平面度、垂直度、对称度和圆跳动的未注公差，标准中规定了H、K、L三个公差等级，其中H级最高，L级最低。选用时应在技术要求中注出标准号及公差等级代号，如：未注几何公差按GB/T1184-K"

The general tolerances on straightness, flatness, perpendicularity, symmetry, circular run-out are given in Tab. 4-17 to 4-20.

直线度、平面度、垂直度、对称度和圆跳动的未注公差,见表4-17至表4-20。

Tab. 4-17 General tolerances on straightness and flatness

表4-17 直线度和平面度的未注公差值

Accuracy grades 精度等级	Ranges of nominal length 公称长度范围/mm					
	~10	>10 ~ 30	>30 ~ 100	>100 ~ 300	>300 ~ 1000	>1000 ~ 3000
H	0.02	0.05	0.1	0.2	0.3	0.4
K	0.05	0.1	0.2	0.4	0.6	0.8
L	0.1	0.2	0.4	0.8	1.2	1.6

Note: When selecting a straightness tolerance from the table,the length of the corresponding line must be taken; In the case of a flatness tolerance,the longer length of the surface or the diameter of the circular surface must be taken.

注：从表中选择直线度公差时，必须取相应直线的长度为参考量。而在选取平面度公差时，则必须取较长的表面长度或圆形的直径。

Tab. 4-18 General tolerances on perpendicularity

表4-18 垂直度的未注公差值

Accuracy grades 精度等级	Ranges of nominal length 公称长度范围/mm			
	~100	>100 ~ 300	>300 ~ 1000	>1000 ~ 3000
H	0.2	0.3	0.4	0.5
K	0.4	0.6	0.8	1.0
L	0.6	1.0	1.5	2

Note: The longer of the two sides forming the rectangular angle is to be taken as the datum.If the nominal lengths of the two sides are equal,either of them may apply as the datum.

注：以形成直角的两边中较长的一个作为基准。如果两侧的公称长度相等，则可将任意一个作为基准。

Tab. 4-19 General tolerances on symmetry

表4-19 对称度的未注几何公差值

Accuracy grades 精度等级	Ranges of nominal length 公称长度范围/mm			
	~100	>100 ~ 300	>300 ~ 1000	>1000~3000
H	0.5			
K	0.6		0.8	1.0
L	0.6	1.0	1.5	2

Note: They apply to symmetrical features. If one of the two features is symmetrical and the other cylindrical,the longer feature is to be taken as the datum.If the nominal lengths of the two features are equal,either of them may apply as the datum.

注：在特定情况下，若一个特征为对称形状，而另一个为圆柱形，那么较长的特征应被选定为基准。如果两个特征的公称长度相等，则它们中的任何一个都可作为基准。

Tab. 4-20 General tolerances on radial, axial and inclined circula run-out

表4-20 径向、轴向和倾斜圆跳动的未注几何公差值

Accurary grades 精度等级	Run-out tolerance 公差值
H	0.1
K	0.2
L	0.5

Note: The bearing surfaces are to be taken as the datum if they are designated as bearing surfaces.In the other case, for radial circular run-out, the longer feature is to be taken as the datum. If the nominal lengths of the two features are equal, either of them may apply as the datum.

注：如果支承面被指定，则以支承面为基准。在其他情况下，对于径向圆跳动，以较长的特征作为基准。如果两个特征的公称长度相等，则它们中的任何一个都可以作为基准。

4.8 Assessment and Measurement of Geometric Deviations 几何偏差评定与检测

Geometric deviation is the variation quantity of extracted tolerance feature from ideal associated feature (ideal feature). It is qualified when all of the extracted tolerance features are within the geometric tolerance zone, otherwise it is unqualified.

几何偏差是指被测提取要素相对于其理想关联要素（理想要素）的变动量。当所有被测提取要素均位于几何公差带内时判定为合格，否则为不合格。

4.8.1 Assessment Criteria of Geometric Deviation 几何偏差评定准则

Minimum envelope zone method is an assessment method to assess geometric deviation, and it serves as the basis to arbitrate the assessment result. Minimum envelope zone is called minimum zone for short. Minimum zone refers to the enveloping zone is with minimum width or minimum diameter when enveloping the extracted tolerance features. The geometric deviation value is the diameter or width of minimum zone. The minimum zone forms of every geometric tolerance are consistent with the geometric tolerance zone form, the width or diameter is determined by the measured extracted feature itself, and its orientation and position should be kept in the relationship with datum（s）indicted on engineering drawing.

最小包容区域法（简称最小区域法）是几何偏差评定的基准方法，并作为判断评定结果的依据。最小包容区域是指包容被测要素时具有最小宽度或最小直径的区域。几何偏差值为该最小区域的宽度或直径。各个几何公差的最小区域形状分布与各自几何公差带形状一致，其宽度或直径由被测提取要素自身确定，但其方向和位置需保持与工程图所示基准的关系。

Although minimum zone method is the basis for arbitrating the assessment result, under the premise of meeting the functional requirements, some approximate methods can be used to assess form deviation. For example, to assess the straightness deviation with reference to the straight line connected by the two endpoints. The most often used approximate method is least square method in coordinate measurement. Usually the value from approximate method is greater than the value from minimum zone method.

尽管最小区域法是评定依据，但在满足功能要求的前提下，可采用近似方法评定形状偏差。例如，用两端点连线作为基准评定直线度偏差。坐标测量中最常用的近似方法是最小二乘法，其评定值通常大于最小区域法结果。

4.8.2 Assessment of Form Deviation 形状偏差的评定

Form deviation refers to the variation quantity of the measured extracted feature from its associated feature. Least condition principle is the basic principle to assess the form deviations.

形状偏差指被测提取要素相对于其关联要素的变动量。因形状公差无基准要求，最小条件原则是形状偏差评定的基本原则。

Integral feature（except profile of any line and profile of any surface）整体要素（任意线轮廓度、任意面轮廓度除外）。

For integral feature（except profile of any line and profile of any surface）, least condition is on the condition that the associated feature contacts with the extracted tolerance feature externally（outside the material）, and the variation quantity of extracted tolerance feature from the associated feature is least, shown in Fig. 4-20, in order to assess the straightness error within a certain plane, there are a lot of ideal straight lines which contact with the extracted tolerance feature, such as A_1-B_2, A_2-B_2, A_2-B_3, and the assessment values of straightness error values are h_1, h_2, h_3 respectively. There exists only that one line that meets the least condition. Obviously, ideal straight line A_1-B_2 meets least condition in this example, then $h_1=f$ is the straightness error value, and it should be smaller or equal to the straightness tolerance value.

对于整体要素（任意线轮廓度、任意面轮廓度除外），最小条件是指关联要素在体外与被测提取要素接触，且被测提取要素相对于关联要素的变动量最小。如图4-20所示，评定平面内直线度误差时，存在多条与提取要素接触的理想直线（如A_1-B_2，A_2-B_2，A_2-B_3,），其评定值分别为h_1、h_2、h_3。其中仅有一条直线满足最小条件，本例中理想直线A_1-B_2对应的$h_1=f$即为直线度误差值，应小于或等于直线度公差值。

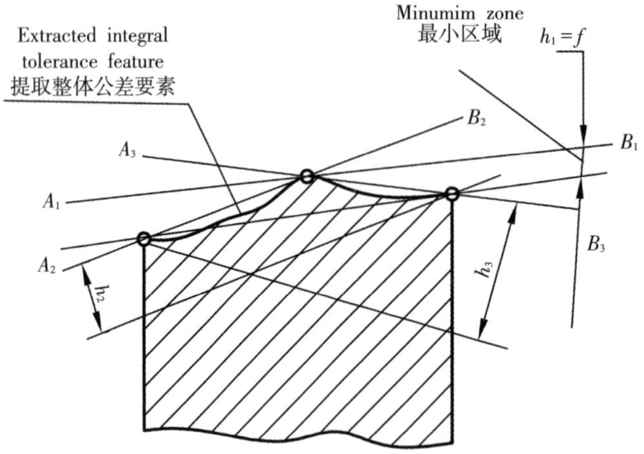

Fig. 4-20　Example of least condition

图4-20　最小条件示例

（1）Discriminate method of minimum zone for straightness deviation直线度偏差最小区域判别法。

The straightness of a single tolerance feature is deemed if the feature is confined between two straight lines, and the distance between these lines is equal to or less than the value of the specified tolerance. The orientation of the straight lines should be chosen to minimize the maximum distance between them.

当单一要素被限制在两条平行直线之间，且两线间距等于或小于公差值时，判定直线度合格。应选择两线方向使最大间距为可能的最小值。

For the straightness tolerance within a certain plane, the separation criterion is used to discriminate the minimum zone. In this case the tolerance zone is within two parallel straight lines, then when assessing the straightness deviation, use two parallel straight lines to envelope the extracted points, and the two straight lines contact with three points with the peak and the nadir separated, shown in Fig. 4-21.

对于平面内直线度公差，采用分离准则判别最小区域。此时公差带为两平行直线之间的区域，评定直线度偏差时，用两条平行直线包容提取点，两直线须与三个分离的点（峰点和谷点）接触，如图4-21所示。

（2）Discriminate method of minimum zone for flatness deviation 平面度偏差最小区域判别法。

The flatness of a single tolerance feature is deemed to be correct when the feature is confined between two planes, and the distance between them is equal to or less than the value of the specified tolerance, The orientation of the planes shall be chosen so that the maximum distance between them is the least possible value.

当单一要素被限制在两平行平面之间，且两平面间距等于或小于公差值时，判定平面度合格。应选择两平面方向使最大间距为可能的最小值。

When assessing the flatness deviation, using two parallel planes to envelope the extracted points, it is confirmed by the minimum zone method if the contact condition is one of the following three criteria.

评定平面度偏差时，若两平行平面与提取点的接触满足以下任一准则，则符合最小区域法：

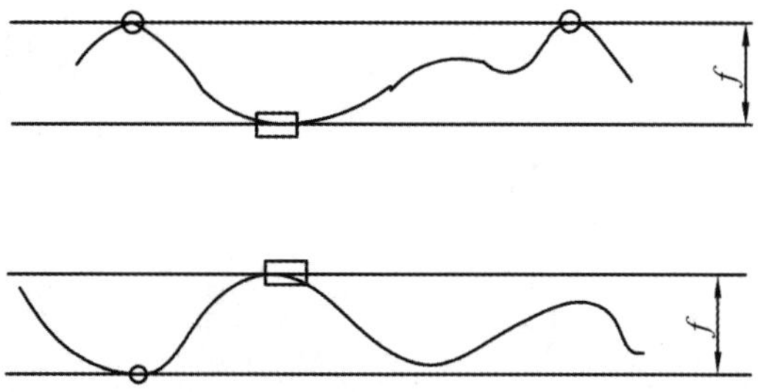

Fig. 4-21 Discriminate method of minimum zone for straightness (separated criterion)

图4-21 直线度最小区域判别法（分离准则）

① Triangle criterion: one plane contacts three peaks (or nadirs) and another plane contacts one nadir (or peak), at the same time, the projection of the one pole is within the triangle made by the three poles, see Fig. 4-22 (a).

三角形准则：一平面与三个峰点（或谷点）接触，另一平面与一个谷点（或峰点）接触，且该点的投影位于三点形成的三角形内，如图4-22（a）所示。

② Separated criterion:two peaks and two nadirs contact the two planes respectively, and the line made by peaks and the line made by two nadirs cross in space, see Fig. 4-22 (b).

分离准则：两峰点和两谷点分别接触两平面，且峰点连线和谷点连线在空间相交，如图4-22（b）所示。

③ Straight line criterion: the projection of one peak (or nadir) lies on the line made by two nadirs (or peaks), see Fig. 4-22 (c).

直线准则：一峰点（或谷点）的投影位于两谷点（或峰点）连线上，如图4-22（c）所示。

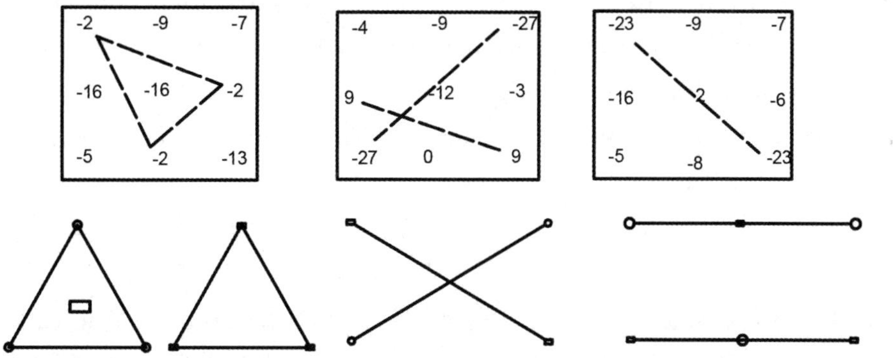

(a) (b) (c)

Fig. 4-22 Discriminate method of minimum zone for flatness

图4-22 平面度最小区域判别法

The circularity of a single tolerance feature is deemed to be correct when the feature is confined between two concentric circles such that the difference in radii is equal to or less than the value of the specified tolerance. The location of the centers of these circles and the value of their radii shall be chosen so that the difference in radii between the two concentric circles is the least possible value.

对于圆度偏差，当单一要素被限制在两同心圆之间，且两圆半径差等于或小于公差值时，判定圆度合格。圆心位置和半径应选择使两同心圆半径差为可能的最小值。

When assessing the circularity deviation, use two concentric circles to envelope the extracted points, it conforms to minimum zone method if there are four poles contacting the two circles, separated inside and outside respectively, as shown in Fig. 4-23.

评定圆度偏差时，若两同心圆包容提取点，且内外侧各有两点接触（即四点接触）则符合最小区域法，如图4–23所示。

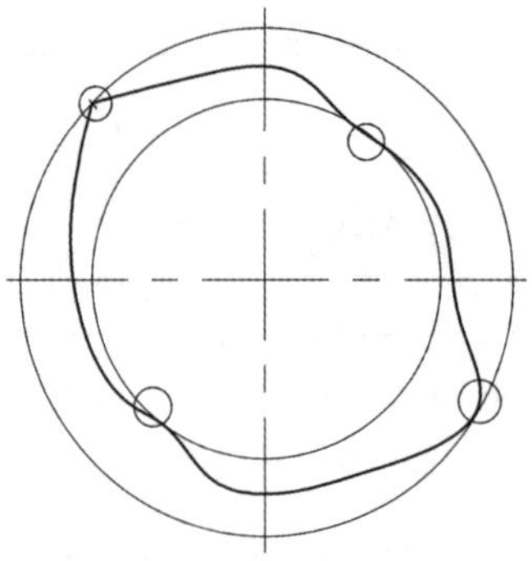

Fig. 4–23 Discriminate method of minimum zone for circularity

图4–23 圆度最小区域判别法

4.8.3 Assessment of Orientation Deviation 方向偏差的评定

Orientation deviation is the variable quantity of extracted measured feature from associated feature which has certain orientation, and the orientation of the associated feature is determined by datum（s）.

方向偏差指被测提取要素相对于具有确定方向的关联要素的变动量，关联要素方向由基准确定。

The orientation deviation value is the width or diameter of orientation minimum zone. Orientation minimum zone refers to a zone whose form and orientation are same with orientation tolerance zone, and it exhibits the minimum width or diameter when enveloping the extracted measured feature in accordance with the specified orientation. For example, see Fig. 4–24（a）, the parallelism deviation of a plane to datum plane is the distance between two parallel planes which envelope the extracted measured plane and are parallel to the datum with the least f. And in Fig. 4–24（b）, the perpendicularity deviation of the axis to the datum plane is the diameter ϕf of the cylindrical surface which envelopes the extracted measured axis and is perpendicular to datum plane and the diameter is least.

方向偏差值为方向最小区域的宽度或直径。方向最小区域是指形状和方向与方向公差带一致，且在按给定方向包容被测提取要素时具有最小宽度或直径的区域。例如，图4–24（a）中，平面相对于基准平面的平行度偏差为包容被测平面且平行于基准的两平行平面间的最小距离f；图4–24（b）中，轴线相对于基准平面的垂直度偏差为包容被测轴线且垂直于基准平面的圆柱面最小直径ϕf。

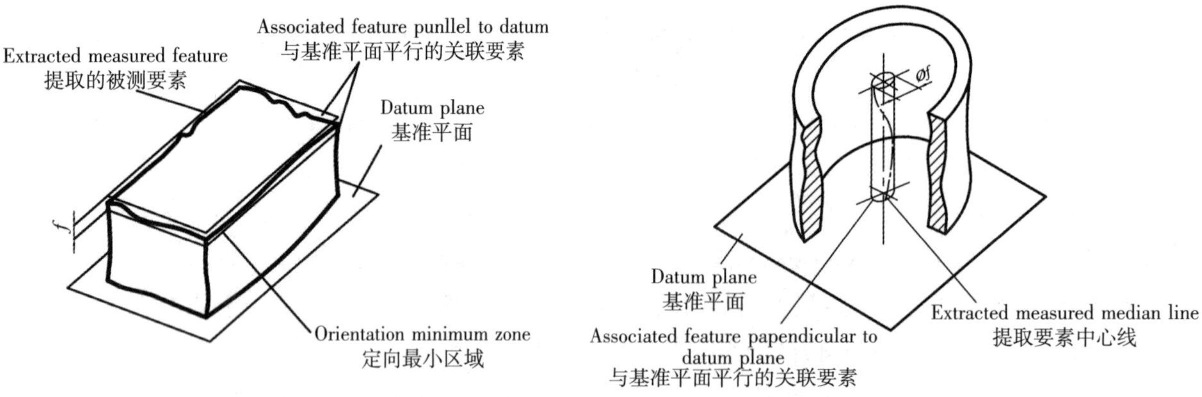

（a）parallelism deviation of a plane

（a）平面平行度偏差

（b）parallel of axis

（b）轴线垂直度

Fig. 4–24 Orientation minimum zone

图4–24 方向偏差示例

4.8.4 Assessment of Location Deviation 位置偏差的评定

Location deviation is the variable quantity of the extracted measured feature from an associated feature which has certain position, and the position of the associated feature is determined by datum (s) and theoretically exact dimension (TED). For coaxiality and symmetry, the TED is zero. Location deviation is the width or diameter of location minimum zone. Location minimum zone is the envelope zone which has the same form as tolerance zone and envelopes the extracted measured feature located by associated feature and has the least width or diameter.

位置偏差指被测提取要素相对于具有确定位置的关联要素的变动量，关联要素位置由基准和理论正确尺寸（TED）确定。对于同轴度和对称度，TED为零。位置偏差值为位置最小区域的宽度或直径。位置最小区域是指形状与位置公差带一致，且包容按关联要素定位的被测提取要素时具有最小宽度或直径的区域。

Some examples are shown in Fig. 4–25. In Fig. 4–25 (a), the coaxiality deviation is the diameter ϕf of a cylinder which is used to envelope the extracted measured median line and the associated axis is coaxial with the datum axis and the diameter is least. In Fig. 4–25 (b), the symmetry deviation of plane to plane is the distance f of the two parallel planes which are used to envelope the extracted measured median plane and the associated median plane is coplane with the datum plane and the distance is least. In Fig. 4–25 (c), the position deviation is the diameter ϕf of the circle which is used to envelope the extracted measured point and the associated center point is at the position limited by datums and the diameter ϕf is least.

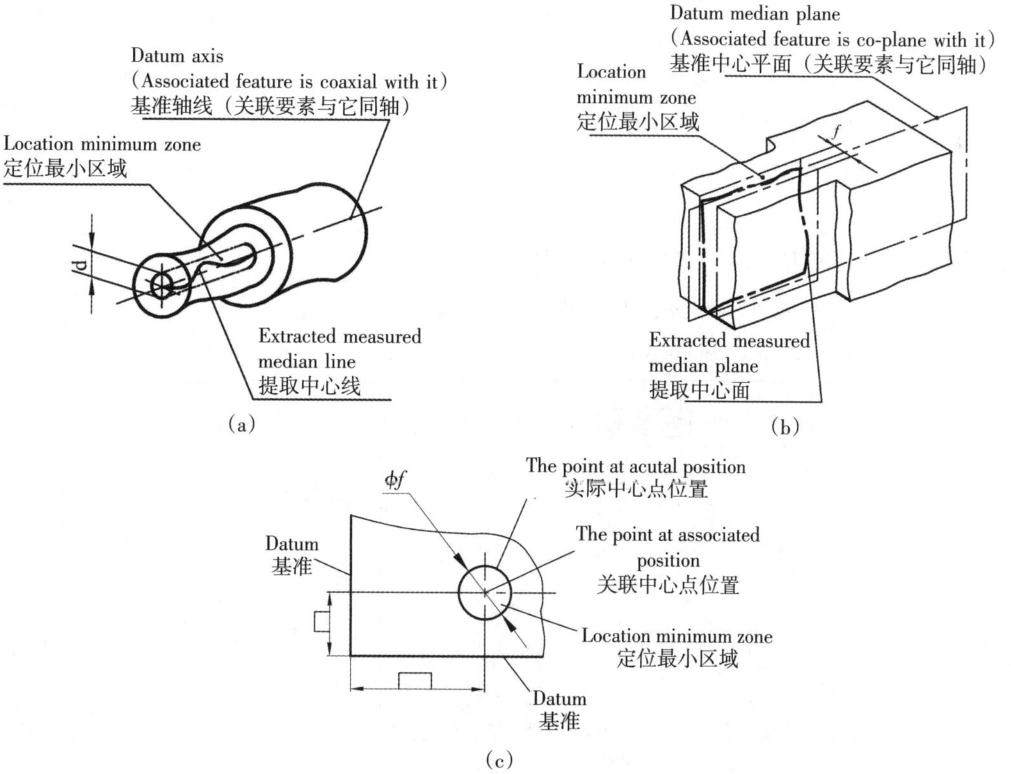

Fig. 4–25 Example location minimum zone

图4–25 位置偏差示例

如图4–25所示：图4–25（a）中，同轴度偏差为包容被测中心线且同轴于基准轴线的最小圆柱面直径ϕf；图4–25（b）中，平面相对于基准平面的对称度偏差为包容被测中心平面且共面于基准平面的两平行平面间的最小距离f；图4-25（c）中，位置度偏差为包容被测点且关联中心点位于基准限定位置的最小圆直径ϕf。

4.8.5 Assessment of Run-out Deviation 跳动偏差的评定

The method of measuring circular run-out and total run-out is to measure the variation quantity of the extracted measured feature from the datum axis in a specified direction during which the extracted measured feature rotates around the datum axis, shown in Fig. 4-26.

圆跳动和全跳动的测量方法是在被测要素绕基准轴线旋转时，测量提取要素在指定方向相对于基准轴线的变动量，如图4-26所示。

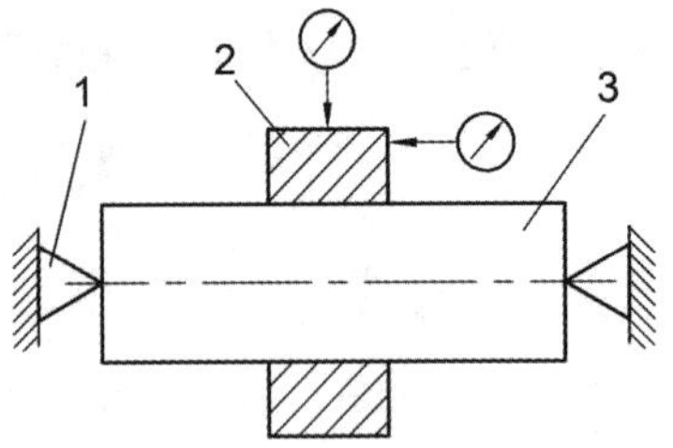

1—locator；2—test piece；3—mandrel

1—定位器；2—被测件；3—心轴

Fig. 4-26 Examples of measuring run-out

图4-26 跳动测量示例

Exercise 4 习题4

4-1 What are the main tolerance principles? What are the differences in their applications?

公差原则有哪几种？其使用情况有何差异？

4-2 What's the difference between Maximum Material Condition（MMC） and Maximum Material Virtual Condition（MMVC)?

最大实体状态和最大实体实效状态的区别是什么？

4-3 Explain the meaning of the annotation shown in Fig. 4-27.

有一标注如图4-27所示，试说明其含义。

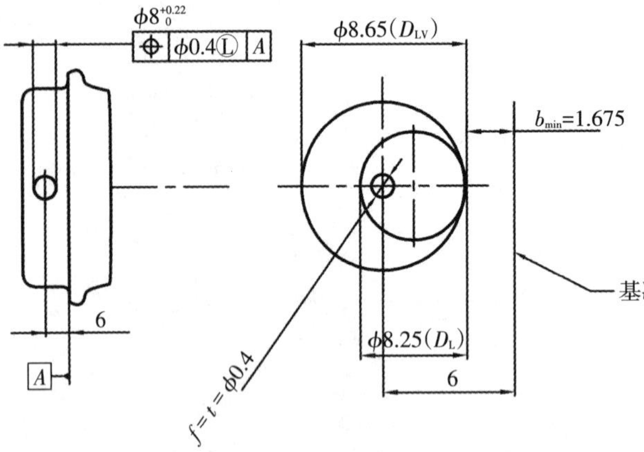

Fig. 4-27 Figure for question 4-3

图4-27 作业题4-3图

4-4 To indicate following geometric tolerance requirements on the engineering drawing shown in Fig. 4-28.

将下列几何公差要求标注在图4–28上。

(1) The hole diameter is ϕ30H7, complying with the envelope requirement;

内孔尺寸为ϕ30H7，遵守包容要求；

(2) The circularity tolerance value of the conical surface is 0.02 mm, and the straightness tolerance value of the generating line is 0.01 mm;

圆锥面的圆度公差为0.02 mm，母线的直线度公差为0.01 mm；

(3) The inclined runout tolerance value is 0.04 mm for conical surface related to the axis of the internal hole;

圆锥面对内孔轴线的斜向圆跳动公差为0.04 mm；

(4) Perpendicularity tolerance value is 0.01 mm for the axis of hole related to right end face;

内孔的轴线对右端面垂直度公差为0.01 mm；

(5) Parallelism tolerance value is 0.02 mm for left end facerelated toright end face;

左端面对右端面的平行度公差为0.02 mm。

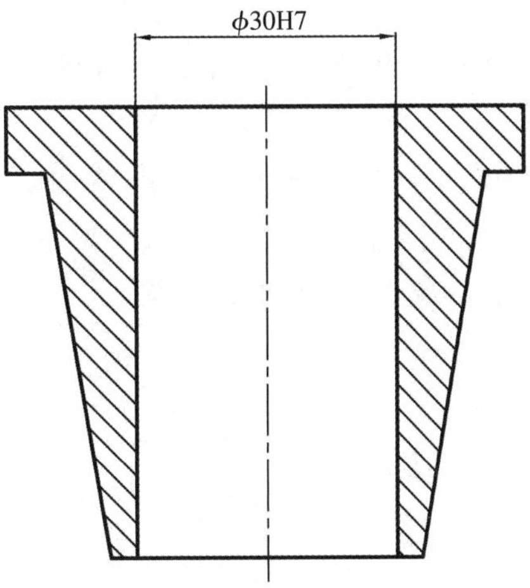

Fig. 4–28 Figure for question 4–4

图4–28 作业题4–4图

4–5 To indicate following geometric tolerance requirements on the engineering drawing shown in Fig. 4–29.

将下列各项几何公差要求标注在题图4–29上。

(1) The flatness tolerance value is 0.01 mm for left end face;

左端面的平面度公差为0.01 mm；

(2) The parallelism tolerance value is 0.01 mm for right end facerelated to left end face;

右端面对左端面的平行度公差为0.01 mm;

(3) The envelope requirement is selected for ϕ70H7; Independency principle is selected for ϕ210b7;

孔ϕ70H7遵守包容要求；ϕ210b7外圆，遵守独立原则；

(4) The perpendicularity tolerance value is 0.02 mm for the axis of hole ϕ70 related to left end face;

ϕ70孔的轴线对左端面的垂直度公差为0.02 mm;

(5) The coaxiality tolerance value is 0.03 mm for ϕ210 mm related to ϕ70 mm;

ϕ210外圆对ϕ70孔轴线的同轴度公差为0.03 mm;

(6) The position tolerance value is 0.15 mm for the axis of 4–ϕ20H8 related to left end plane (first datum) and the axis of ϕ70 mm (second datum), maximum material requirements are selected for them.

4–ϕ20H8对左端面（第一基准）及ϕ70 mm孔轴线（第二基准）的位置度公差为0.15 mm，均采用最大实体要求。

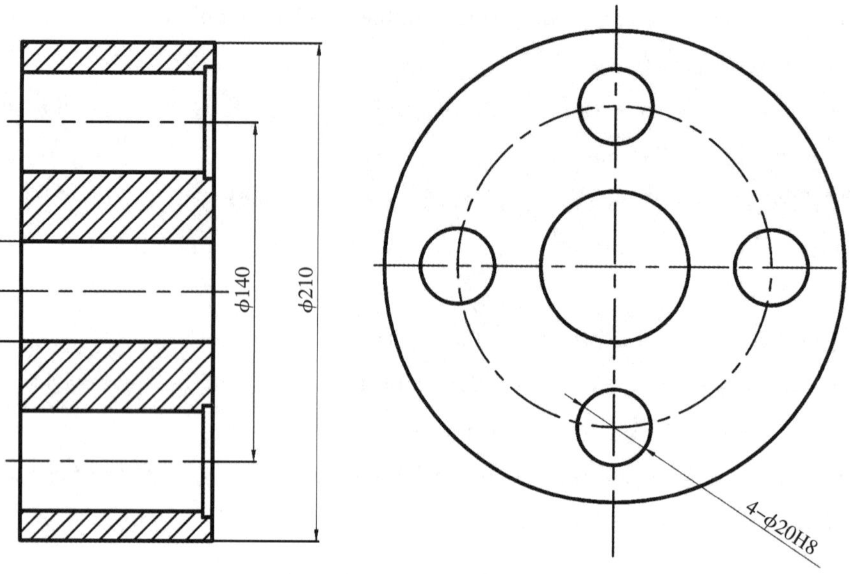

Fig. 4-29 Figure for question 4-5

图4-29 作业题4-5图

4-6 To change the indication mistakes in the legends of Fig. 4-30。

试改正图4-30各图例标注的错误。

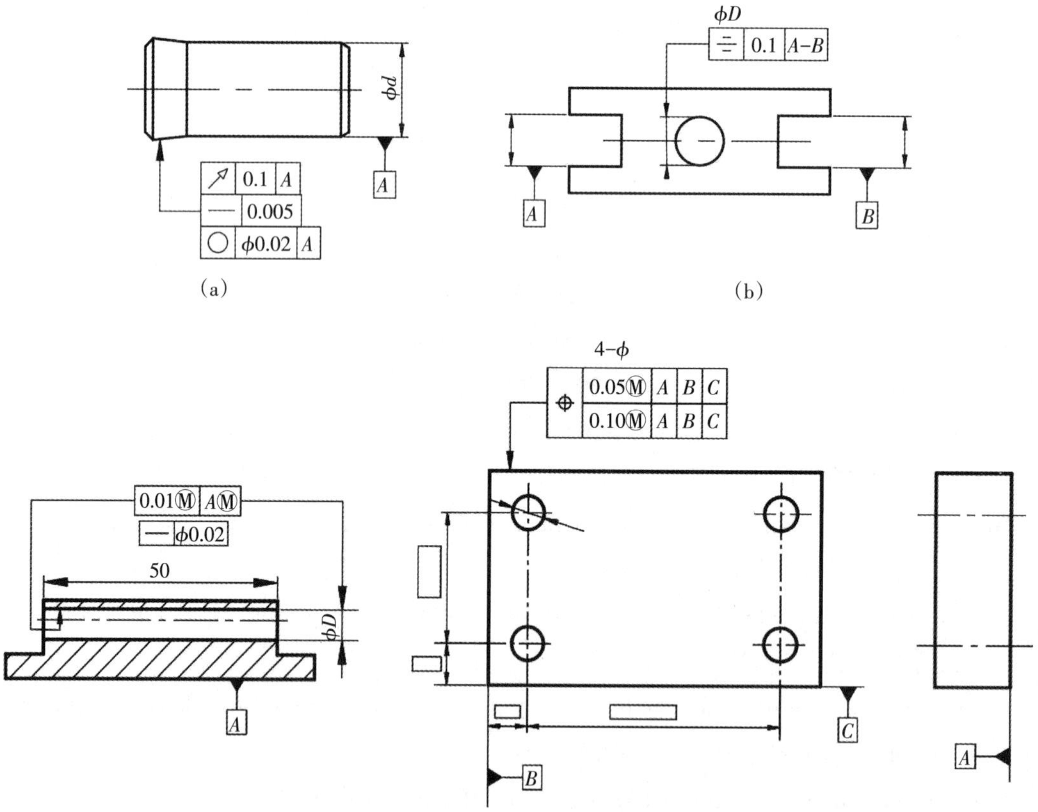

(a) (b) (c) (d)

Fig. 4-30 Figure for question 4-6

图4-30 作业题4-6图

Figure 4-31 illustrates three annotation methods for holes, please complete the following table based on the interpretation of each annotation.

图4-31分别给出了孔的三种图样标注方法，试根据标注的含义填写下表。

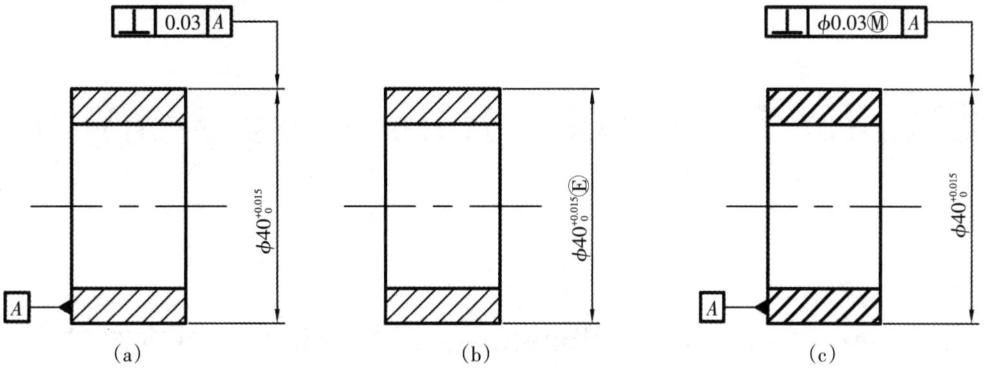

Fig. 4-31 Figure for question 4-7

图4-31 作业题4-7图

Drawing number 图号	Tolerance principle name 采用的公差原则的名称	Boundary name and boundary size 边界名称及边界尺寸/mm	Tolerance value of location under MMC 最大实体状态下的位置公差值/mm	Maximum permissible location deviation 允许的最大位置误差值/mm	Actual size tolerance range 实际尺寸范围/mm
a					
b					
c					

Chapter 5 Surface Roughness 表面粗糙度

5.1 Basic Concept of Surface Roughness 表面粗糙度的基本概念

5.1.1 Definition of Surface Roughness 表面粗糙度的定义

Surface roughness is the micro geometric characteristics composed of small spacing and peaks and valleys on the machined surface, also known as micro roughness. The formation of surface roughness is mainly due to the friction between surfaces of tools and parts, plastic deformation and metal tearing during chip separation, and high-frequency vibration in the processing system. After the part is finished, its profile shape is complex, as shown in Fig. 5-1.

表面粗糙度是指加工表面上具有的较小间距和峰谷所组成的微观几何形状特性，亦称微观不平度。它主要是由在加工过程中刀具和零件表面之间的摩擦、切屑分离时的塑性变形和金属撕裂，以及在工艺系统中存在高频振动等所形成。零件完工后，它的截面轮廓形状是复杂的，如图5-1所示。

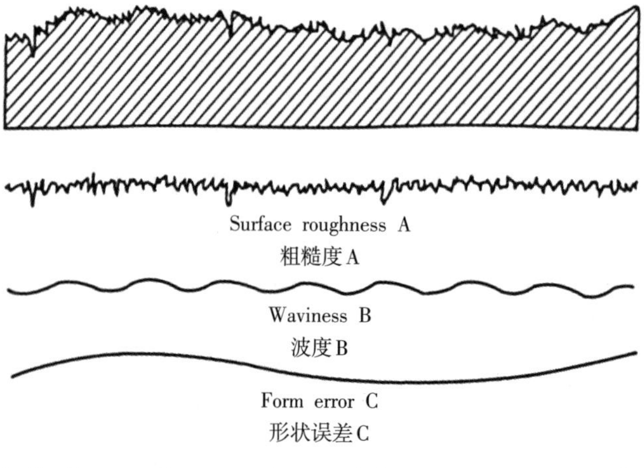

Fig. 5-1 Surface roughness, waviness and form error on part surface

图5-1 零件表面上的表面粗糙度波度和形状误差

Generally, the wave spacing of surface roughness is less than 1 mm, and the wave spacing between 1 and 10 mm belongs to surface waviness; the wave spacing greater than 10 mm belongs to macro-form error.

一般地，表面粗糙度的波距小于1 mm，波距为1 ~ 10 mm的属于表面波度；波距大于10 mm的属于宏观形状误差。

5.1.2 Influence of Surface Roughness on Mechanical Parts Performance 表面粗糙度对机械零件使用性能的影响

The size of surface roughness has a great influence on the service performance and service life of parts, especially the mechanical parts working under high temperature and high speed, and high pressure. In order to select the parameters and allowable values of surface roughness reasonably, it is necessary to understand its influence on performance.

表面粗糙度的大小对零件的使用性能和使用寿命有很大影响，尤其是对高温、高速、高压条件下工作的机械零件影响更大。为了合理地选用表面粗糙度的参数及允许值，首先要了解它对使用性能的影响。

5.1.3 Influence on Friction and Wear of Moving Surface of Parts 对零件运动表面的摩擦和磨损的影响

The two surfaces with micro geometric errors can only contact at the peak of the contour. The actual effective contact area is very small, which leads to the increase of unit pressure. If there is relative movement between the surfaces, the contact between the two peaks will produce friction resistance to the movement and cause wear of the parts. Generally speaking, the rougher the surface is, the greater the resistance is and the faster the part wears. It must be pointed out that a smoother surface does not necessarily create less wear on the surface. In addition to the influence of surface roughness, the wear loss is also related to the scratching effect of worn metal particles, extrusion of lubricating oil and adsorption between molecules, etc. These factors will also increase the frictional resistance.As a result, the wear of particularly smooth surfaces is aggravated.

具有微观几何形状误差的两个表面只能在轮廓的峰顶发生接触。实际有效接触面积很小，导致单位压力增大。若表面间有相对运动，则峰顶间的接触作用就会对运动产生摩擦阻力，同时使零件产生磨损。一般来说，表面越粗糙，则摩擦阻力越大，零件的磨损也越快。必须指出，表面越光滑，磨损量不一定越小。磨损量除受表面粗糙度的影响外，还与磨损下来的金属微粒的刻划作用、润滑油被挤出和分子间的吸附作用等因素有关，这些因素也会使摩擦阻力增大，所以，特别光滑的表面磨损反而加剧。

5.1.4 Influence on Fits Properties 对配合性质的影响

For the parts with fitting requirements, the surface roughness affects the stability of the fit properties of no matter what kind fit. For the clearance fit of sliding bearing, the gap will increase due to the wear of peak tip of surface micro shape in the working process. If the gap caused by the rougher surface increases too much, the original fitting properties will be destroyed. The surface roughness is not related to the nominal size, so the smaller the fit size is, the more serious the effect is. For the transition fit, the rough surface will also enlarge the gap in the process of use and assembly, thus reducing the degree of centering and changing the original fitting properties. For interference fit, due to the uneven surface of the parts, the peak of part surface will be fattened after pressing, so that the actual interference is less than the theoretical interference, thus reducing the connection strength. Therefore, in order to ensure the parts fitting properties, grinding method is usually applied to improve the parts surface roughness.

对于有配合要求的零件表面，无论是哪类配合，表面粗糙度都影响配合性质的稳定性。对于滑动轴承用的间隙配合，会因表面微观形状的峰顶在工作过程中很快被磨损而使间隙增大。如果表面粗糙所引起的间隙增大过多，就会破坏原有的配合性质。由于表面粗糙度与公称尺寸的大小无关，所以配合的尺寸越小，这种影响越严重。对于过渡配合，表面粗糙度会在使用和装配过程中使间隙扩大，从而使定心变差，改变原来的配合性质。对于过盈配合，由于零件表面凸凹不平，配合零件经过压装后，零件表面的峰顶会被挤平，使得实际过盈小于理论的计算过盈量，从而降低了连接强度。因此，为了保证零件的配合性质，通常采用磨削的方法改善其表面粗糙度质量。

5.1.5 Influence on Corrosion Resistance 对抗腐蚀性的影响

Metal corrosion is often caused by chemical or electrochemical action, such as steel rust, copper green. The rougher the surface is, the more serious the corrosion is. Since corrosive gas or liquid is easy to accumulate at the bottom of the pit, corrosion will penetrate into the metal from the valley. The rougher the surface is and the deeper the pit is, the more serious the corrosion is. Therefore, improving the parts surface roughness can enhance its corrosion resistance.

金属腐蚀往往是由化学作用或电化学作用造成的，如钢铁生锈、铜生铜绿，都是腐蚀作用所致。零件表面越粗糙，其腐蚀作用也就越强。由于腐蚀性气体或液体容易在凹坑底部积聚，腐蚀作用便从凹谷深入金属内部。表面越粗糙，凹谷越深，腐蚀作用就越严重。因此，提高零件表面粗糙度质量，可以增强其抗腐蚀能力。

5.1.6 Influence of Anti-Fatigue Strength 对抗疲劳强度的影响

Under the conditions of alternating load, heavy load and high-speed work, the part's fatigue strength is related to the physical and mechanical properties of the material, and also to the surface roughness. Because the rougher the surface is and the deeper the dent is, the smaller the root curvature radius is, and the more sensitive it is to stress concentration. Especially under the action of alternating load, the greater the influence is, the more quickly, the parts will produce fatigue cracks and damage. Therefore, for the parts under alternating load, the fatigue strength can be improved by improving the surface roughness quality, thus the size and weight of the parts can be reduced accordingly.

零件在交变载荷、重载荷及高速工作条件下，其疲劳强度除了与零件材料的物理、力学性能有关外，还与表面粗糙度有很大关系。因为零件表面越粗糙，凹痕越深，其根部曲率半径就越小，对应力集中就越敏感。特别是在交变载荷作用下，其影响更大，零件迅速产生疲劳裂缝而损坏。所以，对于承受交变载荷的零件，可以通过提高其表面粗糙度质量来提高其疲劳强度，从而相应减小零件的尺寸和重量。

5.1.7 Influence on Contact Sealing 对结合密封性的影响

If the two rough joint surfaces only contact at local points, there will be gaps, which will affect the sealing performance. For the static sealing surface without relative sliding between the contact surfaces, if the bottoms of the surface micro unevenness are too deep, and the sealing material cannot completely fill these micro uneven bottoms after the assembly is preloaded, the leakage gap will be left on the sealing surface. Therefore, improving the surface roughness quality of parts can improve its sealing performance. For the relatively sliding dynamic sealing surface, due to the relative movement, there should be a certain thickness of lubricating oil film between the surfaces, so the micro uneven height of the surface should be appropriate, which is generally 4 ~ 5 μm.

若粗糙不平的两个结合表面仅在局部点上接触，必然产生缝隙，这会影响密封性能。对于接触表面之间没有相对滑动的静力密封表面，若表面微观不平度谷底过深，则密封材料在装配过程中受预压后不能完全填满这些微观不平谷底，从而在密封面上留下渗漏间隙。因此，提高零件表面粗糙度质量，可提高其密封性能。对于相对滑动的动力密封表面，由于相对运动，表面间需要有一定厚度的润滑油膜，所以表面微观不平高度应适宜，一般为4 ~ 5 μm。

5.1.8 Other Influences 其他影响

The influence of surface roughness on parts performance is far more than these five aspects, such as the impact on contact stiffness, impact strength, resistance to fluid flow, surface high frequency current, appearance quality and measurement accuracy of machines and instruments.

表面粗糙度对零件性能的影响远不止以上六个方面，还包括对接触刚度的影响,对冲击强度的影响，对流体流动的阻力的影响，对表面高频电流的影响，以及对机器、仪器的外观质量和测量精度等都有很大影响。

In a word, surface roughness is an important parameter in precision design. In order to ensure mechanical parts performance, the requirements of surface roughness must be put forward reasonably.

总之，表面粗糙度是精度设计中的一个重要的参数。为保证机械零件的使用性能，必须合理地提出表面粗糙度要求。

5.2 Evaluation of Surface Roughness 表面粗糙度的评定

For parts with surface roughness requirements, it is necessary to measure and evaluate the qualification of surface roughness after machining.

具有表面粗糙度要求的零件表面，加工后需要测量和评定其表面粗糙度的合格性。

5.2.1 Basic Terms 基本术语

5.2.1.1 Sampling length 取样长度（l_r）

The sampling length refers to the length in the x-axis direction used to distinguish the irregular features of the evaluated profile. It is the length of a datum line specified in the measurement or evaluation of surface roughness, which contains at least five peaks and valleys, as shown in Fig. 5-2. The sampling for evaluating the surface roughness is obtained from the surface profile. The direction of the sampling length $l_r$is consistent with the direction of the profile, that is, the direction of the x-axis is consistent with the direction of the spacing. The purpose of specifying the sampling length is to limit and weaken the influence of other geometric errors, especially the surface waviness, on the measurement. Generally, the rougher the surface is, the larger the sampling length is.

取样长度是在x轴方向上用于区分被评定轮廓的不规则特征的长度，它是测量或评定表面粗糙度时所规定的一段基准线长度，包含至少5个轮廓峰和轮廓谷，如图5-2所示。评定表面粗糙度的取样是从表面轮廓上取得的。取样长度l_r的方向与轮廓走向一致，即x轴方向与间距方向一致。规定取样长度的目的在于限制和减弱其他几何形状误差，特别是表面波度对测量的影响。一般来说，表面越粗糙，取样长度就越大。

5.2.1.2 Evaluation length 评定长度（l_n）

The evaluation length is used to determine the length in the x-axis direction of the evaluated contour. Due to the uneven surface roughness of parts, in order to reflect the characteristics of surface roughness reasonably, the minimum length specified in the measurement and evaluation of surface roughness is called evaluation length（l_n）. The evaluation length may include one or more sampling lengths, as shown in Fig. 5-2. Generally, $l_n = 5l_r$; if the measured surface is relatively uniform, select $l_n < 5l_r$; if the uniformity is poor, choose $l_n > 5l_r$.

评定长度用于判别被评定轮廓的x轴方向上的长度。由于零件表面粗糙度不均匀，为了合理地反映表面粗糙度特征，在测量和评定表面粗糙度时所规定的一段最小长度称为评定长度（l_n）。评定长度可包含一个或几个取样长度，如图5-2所示。一般情况下，取$l_n = 5l_r$；若被测表面比较均匀，可选$l_n < 5l_r$；若均匀性差，可选$l_n > 5l_r$。

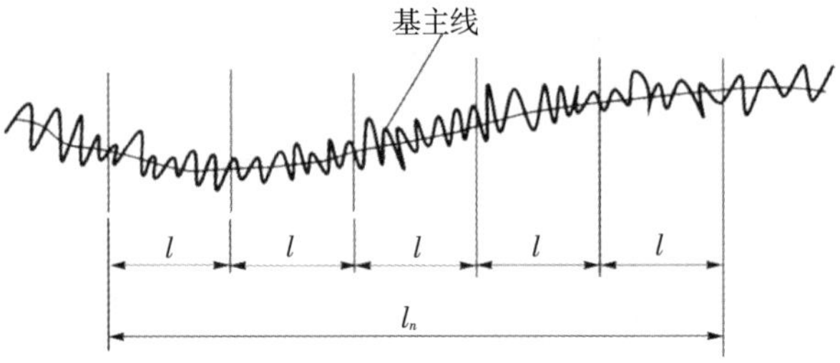

Fig. 5-2 Sampling length and evaluation length

图5-2 取样长度和评定长度

5.2.1.3 Median line 中线

The median line is the datum line which has the geometric contour shape and divides the contour. There are two kinds of datum lines: the least square median line of the contour and the arithmetic mean median line of the contour. The least square median line of contour is the line that makes the offset distance of each point on the contour line the minimum, that is, the sum of squares of ordinate values y_i within the sampling length, and the calculation formula is shown in Eq.（5-1）.

$$\sum_{i=1}^{n} F_i = \sum_{i=1}^{n} F_i' \tag{5-1}$$

中线是具有几何轮廓形状并划分轮廓的基准线。基准线有轮廓最小二乘中线和轮廓算术平均中线两种。轮廓最小二乘中线是指在取样长度内，使得轮廓线上各点到该线的偏移距离之和最小的那条线，即在采样长度内，各点纵坐标值Y_i的平方和为最小的线，计算公式如式（5-1）所示。

The arithmetic-mean line of the contour is a line that divides the actual contour into upper and lower parts within the sampling length and makes the area of the upper and lower parts equal,as shown in Fig. 5-3. It is difficult to determine the least-square median line on the contour graph. So the arithmetic mean median line is often used by visual estimation. As shown in Fig. 5-4, the arithmetic mean median line is usually determined by visual estimation.

轮廓算术平均中线是指在取样长度内，将实际轮廓划分为上、下两部分，且使上、下两部分面积相等的线，如图5-3所示。在轮廓图形上确定最小二乘中线的位置比较困难，可使用轮廓算术平均中线，如图5-4所示，通常用目测估计确定算术平均中线。

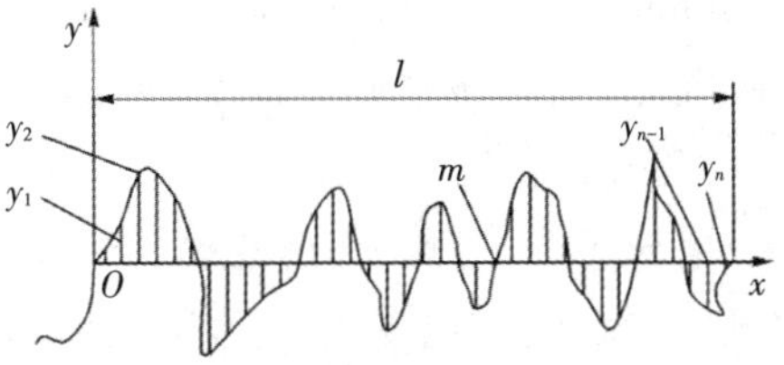

Fig. 5-3 The least square median line of the contour

图5-3 轮廓最小二乘中线示意图

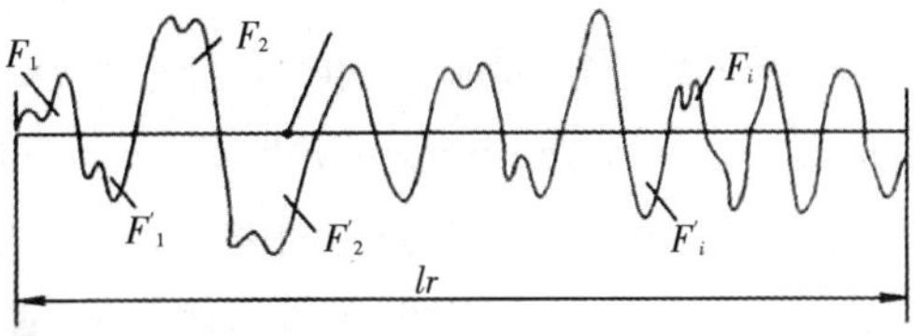

Fig. 5-4 The arithmetic mean median line of the contour

图5-4 粗糙度轮廓算术平均中线

5.2.2 Evaluation Parameters 评定参数

In order to meet the different functional requirements for the surface of parts, *Geometrical Product Specifications* (*GPS*) *-Surface Texture: Profile Method-Terms, Definitions and Surface Texture Parameters* (GB/T3505-2009) specifies the corresponding evaluation parameters from three aspects of surface micro geometry amplitude, spacing and shape.

为了满足对零件表面不同的功能要求,《产品几何技术规范（GPS）表面结构 轮廓法术语、定义及表面结构参数》(GB/T3505—2009 ）从表面微观几何形状幅度、间距和形状三个方面的特征，规定了相应的评定参数。

5.2.3 Amplitude Parameter 幅度参数

5.2.3.1 The arithmetic mean deviation (*Ra*) 轮廓的算术平均偏差 (*Ra*)

The arithmetic mean deviation *Ra* of the contour is the arithmetic mean of the absolute value of the ordinate value within a sampling length. It is shown as Fig. 5-5, and the calculation formula is shown in Eq. (5-2).

轮廓的算术平均偏差*Ra*是在一个取样长度内纵坐标值绝对值的算术平均值，如图5-5所示。

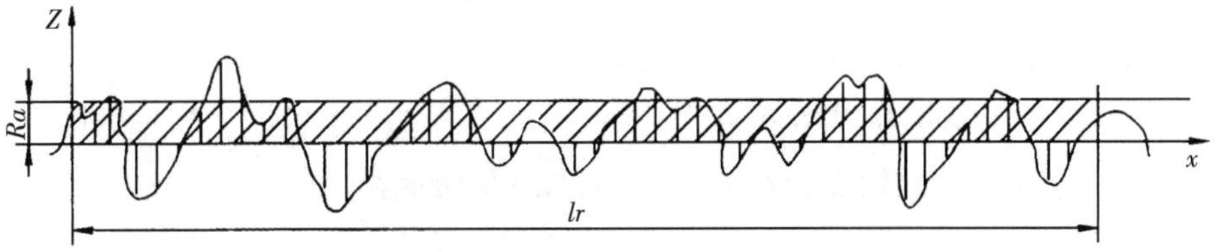

Fig. 5-5 Arithmetic mean deviation of the profile *Ra*

图5-5 轮廓算术平均偏差*Ra*的确定

计算公式如式5-2所示。

$$Ra=\frac{1}{n}\sum_{i=1}^{n}|Z_i| \quad \text{or} \quad Ra=\frac{1}{l_r}\int_0^{l_r}|z(x)|\mathrm{d}x \tag{5-2}$$

Where, z—the contour offset distance (The distances from each point on the profile to the reference line);

Z_i—the profile deviation at the ith point ($i=1, 2, \cdots, n$).

式中，z——轮廓偏距（轮廓上各点至基准线的距离）；

Z_i——第i点的轮廓偏距（$i=1, 2, \cdots, n$）。

The value of *Ra* can objectively reflect the micro geometric characteristics of the measured surface. The smaller the *Ra* value is, the smaller the amplitude of the micro peaks and valleys on the measured surface is, the smoother the surface is; conversely, the larger the *Ra* value is, the rougher the surface is. The *Ra* value is measured using a stylus-type inductive profile meter. Due to the limitation of the stylus radius and the instrument's measuring principle, the *Ra* value should not be used as the evaluation parameter of the too rough or too smooth surface, but only suitable for the surface with *Ra* value of 0.025 ~ 6.3 μm.

*Ra*值的大小能客观地反映被测表面微观几何特性，*Ra*值越小，被测表面微小峰谷的幅度越小，表面越光滑；反之，*Ra*越大，被测表面越粗糙。*Ra*值使用触针式电感轮廓仪测得，受触针半径和仪器测量原理的限制，*Ra*值不宜用作过于粗糙或过于光滑表面的评定参数，仅适用于*Ra*值为0.025 ~ 6.3 μm的表面。

5.2.3.2 The maximum height（*Rz*）轮廓的最大高度（*Rz*）

The maximum height *Rz* of the contour is the height of the sum of the maximum contour peak height Zp_i（$Zp_i = Zp$）and the maximum contour valley depth Zv_i（$Zv_i = Zv$）within a sampling length. The calculation formula is shown in Eq.（5−3）. The peak of surface roug hness profile refers to the part of surface roughness profile connecting two adjacent intersection points outward; the surface roughness profile valley refers to the part of roughness profile connecting two adjacent intersection points inward.

轮廓的最大高度*Rz*是在一个取样长度内，最大轮廓峰高Zp_i（$Zp_i = Zp$）和最大轮廓谷深Zv_i（$Zv_i = Zv$）之和，计算公式如式5−3所示。表面粗糙度轮廓的峰是指连接两个相邻交点并向外凸出的轮廓部分；表面粗糙度轮廓的谷则是指连接两个相邻交点并向内凹陷的轮廓部分。

$$Rz = |Zp_{max} + Zv_{max}| \tag{5-3}$$

Where, Zp —the maximum peak height of profile;

Zv —the largest valley depth of profile.

式中，Zp ——最大轮廓峰高；

Zv ——最大轮廓谷深。

Amplitude parameters（*Ra*, *Rz*）are the parameters that must be marked in the standard, so they are also called basic parameters. Within the selected sampling length, there should generally be more than five（at least six）contour peaks and valleys, shown as in Fig. 5−6.

幅度参数（*Ra*，*Rz*）是标准规定必须标注的参数，故又称基本参数。在所选取的取样长度内，一般应包含5个以上（至少6个）的轮廓峰和轮廓谷，如图5−6所示。

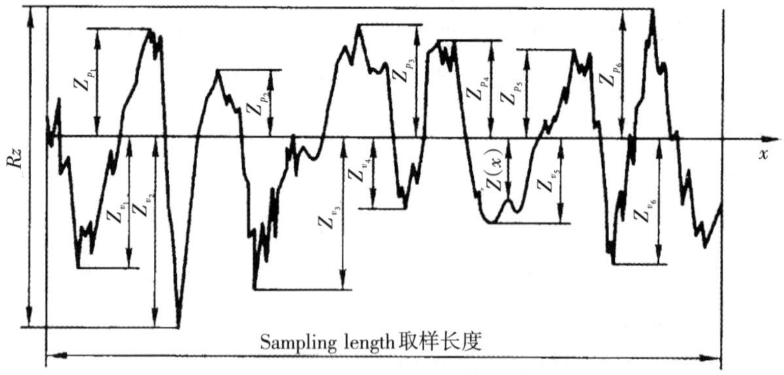

Fig. 5−6 Maximum height of profile *Rz*

图5−6 轮廓最大高度*Rz*的确定

5.2.4 Spacing Parameter 间距参数

The spacing parameter is expressed by the average width *Rsm* of contour units, which is defined as the average of contour units within a sampling length(l_r), as shown in Fig. 5-7. The calculation formula is shown in Eq.(5-4).

间距参数用轮廓单元的平均宽度*Rsm*来表示，*Rsm*是在一个取样长度（l_r）内轮廓单元宽度的平均值，如图5-7所示。计算公式如式（5-4）所示。

$$Rsm = \frac{1}{m}\sum_{i=1}^{m} Xs_i \tag{5-4}$$

Where， Xs_i—the profile width;

m—the spacing number of the profile irregularities within the sampling length.

式中， Xs_i——轮廓宽度；

m——取样长度内轮廓不规则部分的间距数量。

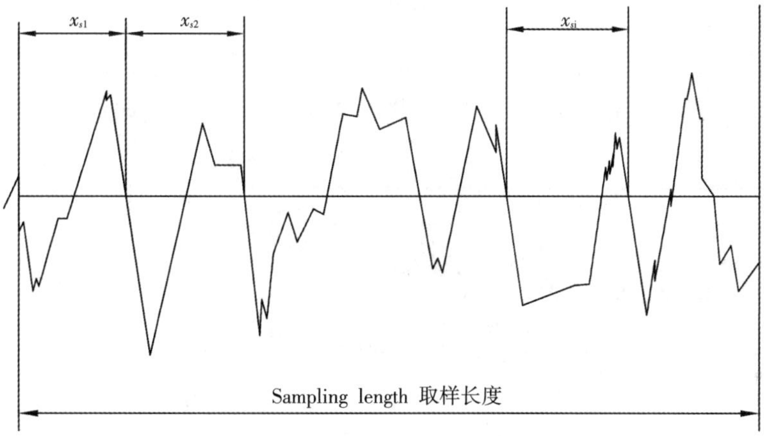

Fig. 5-7 The average width of contour units *Rsm*

图5-7 轮廓单元的平均宽度 *Rsm*的确定

The outward and inward parts of the evaluation profile at the beginning or the end of the sampling length are regarded as a profile peak or valley of surface roughness. When several surface roughness profile units are determined over several continuous sampling lengths, the peaks and valleys evaluated at the beginning or end of each sampling length are included only once at the beginning of each sampling length.

在取样长度始端或末端的评定轮廓的向外部分和向内部分可看作一个表面粗糙度轮廓峰或轮廓谷。当在几个连续的取样长度上确定几个表面粗糙度轮廓单元时，在每一个取样长度的始端或末端评定的峰和谷仅在每个取样长度的始端计入一次。

Rsm is used when the workpiece has requirements for paint performance, crack resistance and corrosion resistance. *Rsm* reflects the density of peaks and valleys on the contour surface. The greater the *Rsm*, the thinner the peaks and valleys, and the worse the tightness. The surface shown in Fig. 5-8（b）is better sealed than the surface shown in Fig. 5-8（a).

*Rsm*用于零件对涂漆性能、抗裂纹和抗腐蚀等有要求的情况下。它反映了轮廓表面峰谷的疏密程度，*Rsm*越大，峰谷越稀，密封性越差。图5-8（b）比图5-8（a）的密封性好。

(a) (b)

Fig. 5-8 The height parameters are similar, the density is different, and the sealing properties are different

图5-8 高度参数相近，疏密度不同，密封性不同

5.2.5 Mixing Parameters 混合参数

The mixing parameter is expressed as the ratio of the solid material length of the contour $Rmr(c)$. The ratio of the solid material length of the contour Rmr （c） is defined as defined as the ratio of the solid material length of the contour to the evaluation length at a given horizontal position c. as shown in Fig. 5-9. The calculation formula is shown in Eq. (5-5).

$$Rmr(c) = \frac{Ml(c)}{l_n},\ Ml(c) = Ml_1 + Ml_2 + \cdots\cdots + Ml_n \tag{5-5}$$

混合参数用轮廓的支承长度率$Rmr(c)$来表示。轮廓的支承长度率Rmr（c）定义为在给定水平位置c上轮廓的实体材料长度与评定长度的比率，如图5-9所示。计算公式如式5-5所示。

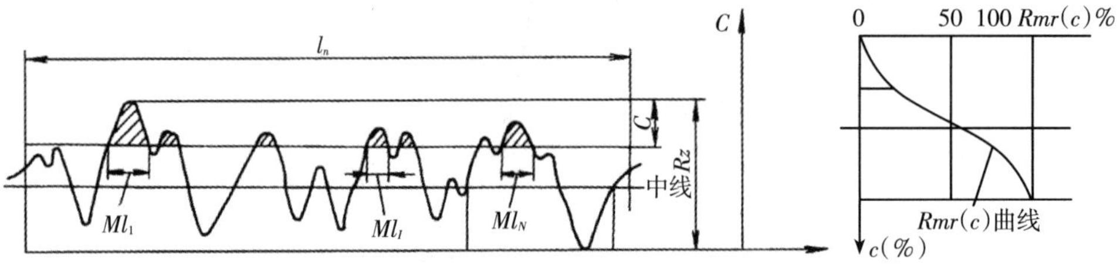

Fig. 5-9 Ratio of the solid material length of the contour

图5-9 支承长度率曲线

The ratio of the solid material length of the contour serves as a highly effective indicator of surface wear resistance. At any given horizontal position, an increase in the $Rmr(c)$ value corresponds to a longer stretch of solid material on the surface, which in turn enhances both contact stiffness and wear resistance. As shown in Fig. 5-10, despite having similar amplitude parameters, variations in contour support length ratios lead to differing levels of contact stiffness. Specifically, Fig. 5-10 (a) demonstrates superior contact stiffness and surface abrasion resistance compared to Fig. 5-10 (b).

支撑长度率曲线对于反映表面耐磨性具有显著的功效。同样水平位置下，Rmr（c）值越大，表面实体材料越长，接触刚度和耐磨性越好。如图5-10所示，幅度参数相同，轮廓支撑长度率不同，接触刚度不同，图5-10（a）比图5-10（b）的接触刚度和表面耐磨性好。

(a) (b)

Fig. 5-10 The amplitude parameter is the same, the contour support length rate is different, and the contact stiffness is different

图5-10 幅度参数相同，轮廓支撑长度率不同，接触刚度不同

The spacing parameter（*Rsm*）and the mixing parameter *Rmr*（*c*）are relative to the basic parameters, which are called additional parameters. Only when the important surface of a few parts has special use requirements can the two additional evaluation parameters can be selected. The additional parameters cannot be separately indicated on the drawing, but can only be used as auxiliary parameters of amplitude parameters.

间距参数（*Rsm*）与混合参数*Rmr*(*c*) 是相对于基本参数而言的，它们被称为附加参数。只有在少数零件的重要表面有特殊使用要求时，才可以选择这两个附加评定参数。附加参数不能单独在图样上标出，只能作为幅度参数的辅助参数使用。

5.3 Parameter Value of Surface Roughness and Its Selection 表面粗糙度的参数值及其选用

5.3.1 Parameter Values of Surface Roughness 表面粗糙度的参数值

The parameter values of surface roughness have been standardized, so the design should be selected according to the national standard *Geometrical Product Specifications*（*GPS*）—*Surface Texture: Profile Method-Surface Roughness Parameters and Their Values*（GB/T 1031-2009). The amplitude parameter values are shown in Tab. 5-1 and Tab. 5-2, the interval parameter values and the mixing parameter values are shown in Tab. 5-3.

表面粗糙度的参数值已经标准化，设计时应按国家标准《产品几何技术规范（GPS)—表面结构、轮廓法、表面粗糙度参数及其数值》(GB/T 1031—2009 ）规定的参数值系列选取。幅度参数值见表5-1和表5-2，间距参数值和混合参数值见表5-3。

Tab. 5-1 Parameter values of *Ra*（From GB/T1031-2009）

表5-1 *Ra*的数值（摘自GB/T 1031—2009）

Ra/μm（Priority series优先系列）	0.012	0.025		0.05	0.1		0.2	0.4		0.8
	1.6	3.2		6.3	12.5		25	50		100
Ra/μm（Supplement series补充系列）	0.008	0.010	0.016	0.020	0.032		0.040	0.063	0.080	0.125
	0.160	0.25	0.32	0.50	0.63		1.00	1.25	2.0	2.5
	4.0	5.0	8.0	10.0	16.0	20	32	40	63	80

Tab. 5-2 Parameter values of *Rz*（From GB/T 1031-2009）

表5-2 *Rz*的数值（摘自GB/T 1031—2009）

Parameter values of *Rz* *Rz*的数值/μm	0.025	0.050	0.100	0.20	0.40
	0.8	1.6	3.2	6.3	12.5
	25	50	100	200	400
	800	1600	—	—	—

Tab. 5-3 Parameter values of *Rsm*，*Rmr*(*c*)（From GB/T 1031-2009）

表5-3 *Rsm*，*Rmr*(*c*）的数值（摘自GB/T 1031—2009）

Parameter values of *Rsm* *Rsm*的数值/μm	0.006	0.0125	0.025	0.05	0.1
	0.2	0.4	0.8	1.6	3.2
	6.3	12.5	—	—	—
Parameter values of *Rmr*(*c*) *Rmr*(*c*）的数值	10%	15%	20%	25%	30%
	40%	50%	60%	70%	80%
	90%	—	—	—	—

In general, when measuring *Ra* and *Rz*, it is recommended to select the corresponding sampling length and evaluation length value according to Tab. 5-4. At this time, the sampling length value can be omitted in the drawing. When the values in Tab. 5-5 cannot be selected for special requirements, the sampling length value shall be marked on the drawing.

在一般情况下，测量*Ra*和*Rz*时，推荐按表5-4选用对应的取样长度及评定长度值。此时，在图样上可省略标注取样长度值。当有特殊要求不能选用表5-4中数值时，应在图样上标注出取样长度值。

Tab. 5-4 Parameter values of l_r and l_n

表5-4 l_r和l_n的参数值

Ra/μm	*Rz*/μm	l_r/mm	l_n/mm
≥0.008 ~ 0.02	≥0.025 ~ 0.10	0.08	0.40
> 0.02 ~ 0.10	> 0.10 ~ 0.50	0.25	1.25
> 0.10 ~ 2.0	> 0.50 ~ 10.0	0.8	4.0
> 2.0 ~ 10.0	> 10.0 ~ 50.0	2.5	12.5
> 10.0 ~ 80.0	> 50.0 ~ 320	8.0	40.0

5.3.2 Selection of Surface Roughness 表面粗糙度的选用

5.3.2.1 Selection of evaluation parameters 评定参数的选用

Amplitude parameters are the basic parameters specified in the standard and can be selected independently. For the surface with surface roughness requirements, an amplitude parameter must be selected. For the surface roughness parameter value of 0.025 ~ 6.3 μm in amplitude direction, *Ra* is recommended to be preferred, because *Ra* can reflect the micro peak valley characteristics of the measured surface. When the surface roughness requirements are particularly high or particularly low, or $Ra < 0.025$ μm or $Ra > 6.3$ μm, *Rz* should be selected.

幅度参数是标准规定的基本参数，可以独立选用。对于有表面粗糙度要求的表面，必须选用一个幅度参数。幅度方向的表面粗糙度参数值为0.025 ~ 6.3 μm时，推荐优先选用*Ra*，因为*Ra*能够比较全面地反映被测表面的微小峰谷特征。在表面粗糙度要求特别高或特别低，$Ra < 0.025$ μm或$Ra > 6.3$ μm时，选用*Rz*。

A few additional parameters cannot be used as additional parameters for general evaluation. They are only selected as additional parameters for the important surfaces of a few parts that have special usage requirements. *Rsm* is mainly used when there are requirements for painting performance, such as uniform spraying, excellent adhesion and smoothness of coating. In addition, when it is required to resist crack, vibration, corrosion and reduce the friction resistance of fluid flow, it is also selected.

附加评定参数一般不能作为独立参数选用。只有少数零件的重要表面，有特殊使用要求时才附加选用。*Rsm*主要在对涂漆性能有要求时使用，如喷涂均匀、涂层有极好的附着性和光洁性等。另外，在要求抗裂纹、抗震、抗腐蚀以及减少流体流动摩擦阻力时，也会选择使用*Rsm*。

For the selection of mixed parameters, the support length ratio *Rmr*(*c*) of the contour is mainly used in the case of high requirements of wear resistance and contact stiffness.

对于混合参数的选用，轮廓的支承长度率*Rmr*(*c*)主要在耐磨性接触刚度要求较高等场合附加选用。

5.3.2.2 Selection of parameter values 参数值的选用

The reasonable choice of surface roughness parameters not only has a great impact on the performance of products, but also has a direct impact on the quality and manufacturing cost of products. Generally speaking, the smaller the surface roughness value (evaluation parameter value) is, the better the working performance and the longer the service life of the parts is. The selection of surface roughness parameters should consider not only the functional requirements of parts, but also the manufacturing cost. The selection principle of surface roughness parameters is to meet the functional requirements of parts, and the second is to consider the possibility of economy and technology.

表面粗糙度参数值选择的合理与否，不仅对产品的使用性能有很大的影响，而且直接关系到产品的质量和制造成本。一般来说，表面粗糙度值（评定参数值）越小，零件的工作性能越好，使用寿命也越长。选择表面粗糙度参数值，不仅要考虑零件的功能要求，还要考虑其制造成本。表面粗糙度参数值的选用原则是首先满足零件的功能要求，其次考虑经济性及工艺的可能性。

The following principles should be followed when selecting parameter values:

选择表面粗糙度参数值时应遵循如下原则：

(1) On the same part, the *Ra* or *Rz* value of the working surface is smaller than that of the non-working surface. The value of *Ra* or *Rz* of the friction surface is smaller than that of the non-friction surface; The roughness parameter of the rolling friction surface is smaller than that of the sliding friction surface.

在同一零件上，工作表面的*Ra*或*Rz*值应小于非工作表面的值；摩擦表面的*Ra*或*Rz*值应小于非摩擦表面的值；滚动摩擦表面的粗糙度参数应小于滑动摩擦表面的值。

(2) The surface roughness of the fillet groove of important parts with high velocity, high pressure per unit area and alternating stress should be smaller.

运动速度高、单位面积压力大，以及受交变应力作用的重要零件，其圆角、沟槽的表面粗糙度值都应较小。

(3) The surface roughness values of the fitting surfaces with high fitting properties (such as those with small clearance fit) and interference fit surfaces with reliable connection and heavy load should be smaller.

配合性能要求高（如小间隙配合）的配合表面以及连接可靠、承受重载的过盈配合表面的粗糙度值应更小。

(4) The surface roughness of the clearance fit should be smaller than that of the interference fit. For the same fit property, the smaller the part size is, the smaller the surface roughness parameter value is; for the same tolerance grade, the surface roughness parameter value of small size is smaller than that of large size, and that of the shaft is smaller than that of the hole. When determining the parameter value of surface roughness, it should be noted that it is compatible with dimensional tolerance and geometric tolerance. The smaller the dimensional tolerance and geometric tolerance are, the smaller the *Ra* or *Rz* value of surface roughness is.

间隙配合的表面粗糙度应小于过盈配合的表面粗糙度。配合性质相同时，零件尺寸越小，则表面粗糙度参数值越小；同一公差等级，小尺寸比大尺寸、轴比孔的表面粗糙度参数值要小。在确定表面粗糙度参数值时，应注意其与尺寸公差和几何公差的协调。尺寸公差和几何公差越小，表面粗糙度的*Ra*或*Rz*值也应越小。

The surface roughness value required for corrosion resistance and good sealing performance or beautiful appearance should be small. If the relevant standards have specified the surface roughness requirements, the surface roughness parameters should be determined according to the standards.

对于要求耐腐蚀、密封性能好或外观美观的表面，其粗糙度值应较小。如果相关标准已规定了表面粗糙度要求，则应根据标准确定表面粗糙度参数。

In engineering practice, because the relationship between surface roughness and function is very complex, it is difficult to accurately determine the allowable value of parameters. In the specific design, except for the surface with special requirements, the empirical statistical data are generally used and the class comparison method is used. After the surface roughness is preliminarily determined according to the analogy method, the appropriate adjustment is made according to the working conditions.

在工程实践中，由于表面粗糙度和功能的关系十分复杂，因而很难准确地确定参数的允许值。在具体设计时，除有特殊要求的表面外，一般采用经验统计资料，用类比法来选用。用类比法初步确定表面粗糙度后，再对比工作条件做适当调整。

5.4 Surface Roughness Symbol and Its Marking 表面粗糙度符号及其标注

The surface roughness symbols and codes marked on the drawings are the requirements after the surface is completed. The marking of surface roughness shall comply with the national standard *Geometric Product Specifications*（*GPS*）*—Indication of Surface Texture in Technical Product Documentation*（GB/T 131-2006).

图样上所标注的表面粗糙度符号和代号是该表面完工后的要求。表面粗糙度的标注应符合国家标准《产品几何技术规范（GPS）技术产品文件中表面结构的表示法》(GB/T 131—2006）的规定。

5.4.1 Basic Symbols of Surface Roughness 表面粗糙度的基本符号

Tab. 5-5 shows the basic symbols and descriptions of the surface roughness of parts on the drawing. When only machining is required either removing material or without removing material, but there is no requirement for other provisions on surface roughness, it is allowed to note only the surface roughness symbol.

图样上表示零件表面粗糙度的基本符号及其说明见表5-5。当仅需要采用去除材料的方法或不去除材料的方法加工，但对表面粗糙度的其他规定没有要求时，允许只注表面粗糙度符号。

Tab. 5-5 Symbols of surface roughness

表5-5 表面粗糙度符号

Symbol 符号	Significance and explanation 意义及说明
	Basic symbol indicating that the surface can be obtained by any method 基本符号，表示表面可用任何方法获得
	The basic symbol plus a dash indicates that the surface is obtained by removing the material.For example: turning,milling,drilling,grinding,etc. 基本符号加一段短划线，表示表面是用去除材料的方法获得。例如：车削、铣削、钻削、磨削等
	The basic symbol plus a small circle indicates that the surface is obtained with-out removing the material. For example:casting,forging,stamping,etc. 基本符号加一个小圆，表示表面是用不去除材料的方法获得。例如：铸造，锻造，冲压成型等
	A horizontal line can be added on the long side of the above three symbols to mark the relevant parameters and instructions 在前述3个符号的长边上均可加一段横线，用于标注有关参数和说明
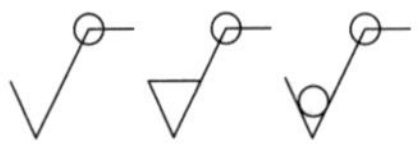	All surfaces with the same surface roughness can be indicated by adding one small circle to the above symbols 在前述3个符号的上均可加一个小圆，表示所有表面具有相同的表面粗糙度要求

5.4.2 Marking of Surface Roughness Symbols 表面粗糙度符号的标注

5.4.2.1 Symbols of surface roughness 表面粗糙度的符号

Around the surface roughness symbol, it is required to write some values and relevant regulations. The positions of these values and relevant regulations are shown in Fig. 5-11. The surface roughness symbol, these values and various relevant regulations constitute the surface roughness code. It is stipulated in the standard that when the number of allowable measured values of surface roughness parameters is less than 16% of the total, the upper or lower limit value of surface roughness parameters shall be marked on the drawing, that is, "16% Rule". When it is required that all measured values of surface roughness parameters shall not exceed the specified values, the maximum value of surface roughness parameters, i.e. "Maximum Rule", shall be marked on the drawing.

在表面粗糙度符号周围，要求注写若干数值以及相关规定，这些数值和有关规定注写位置，如图5-11所示。表面粗糙度符号和这些数值以及各种有关规定共同组成表面粗糙度代号。国家标准规定，当表面粗糙度参数允许测量值的数量占总量的比例小于16%时，应在图样上标注表面粗糙度参数的上限值或下限值，即"16%规则"。当要求表面粗糙度参数的所有实测值不得超过规定值时，应在图样上标注表面粗糙度参数的最大值，即"最大规则"。

In Fig. 5-11, the letter "a" denotes the first surface roughness requirement (unit: μm); the letter "b" is the second surface roughness requirement (unit: μm); The letter "c" is the processing method (turning, milling, etc.); The letter "d" is the processing texture direction symbol; The letter "e" is the machining allowance (unit: mm).

图5-11中，a为第一表面粗糙度要求（单位：μm）；b为第二表面粗糙度要求（单位：μm）；c为加工方法车、铣等；d为加工纹理方向符号；e为加工余量（单位：mm）。

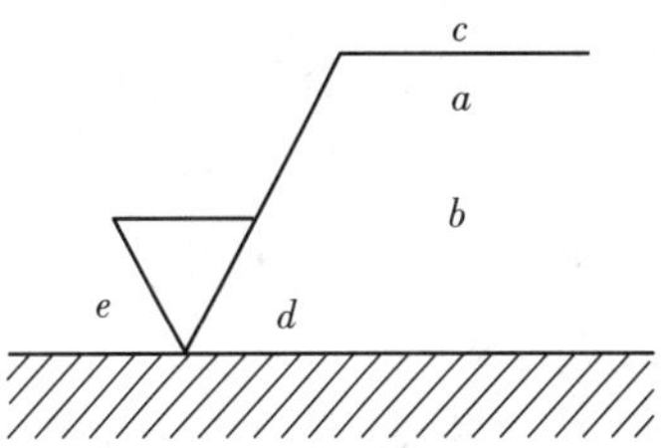

Fig. 5-11 Marking method of surface roughness code
图5-11 表面粗糙度代号的注法

5.4.2.2 Marking of Surface Roughness Amplitude Parameters 表面粗糙度幅度参数的标注

The surface roughness amplitude parameters are marked at the positions of codes a and b. The complete graphic symbols are shown in Fig. 5-12. The figure includes marking of upper or lower limit: upper limit symbol "U" and lower limit symbol "L" shall be marked when indicating a bidirectional limit. If the same parameter has two-way limit requirement, the labels of "U" and "L" can be omitted without ambiguity. If it is a one-way lower limit value, it is necessary to add "L".

表面粗糙度幅度参数标注在代号a和b位置。完整图形符号如图5-12所示，图中包含上限或下限的标注：表示双向极限时应标注上限符号"U"和下限符号"L"。如果同一参数具有双向极限要求，在不引起歧义时，可省略"U"和"L"的标注。若为单向下限值，则必须加注"L"。

Marking of transmission band and sampling length.Transmission band refers to the wavelength range (unit: mm) between cut-off wavelength values of two filters. The cut-off wavelength of the long wave filter is the sampling length l_n. When marking the transmission belt, the short wave is in the front and the long wave is in the back, separated by a hyphen " – ", as "0.0025-0.8" shown in Fig. 5-12. In some cases, only one filter should be marked in the label of transmission band, and the hyphen " – " should be reserved to distinguish short waves from long waves. Namely, 0.0025−denotes a short-wave filter, and −0.8 denotes a long-wave filter.

传输带和取样长度的标注。传输带是指两个滤波器的截止波长值之间的波长范围（单位为mm）。长波滤波器的截止波长值就是取样长度l_n。标注传输带时，短波在前，长波在后，并用连字号“-”隔开，如图5-12中“0.0025-0.8”。在某些情况下，标注传输带，只标一个滤波器，并应保留连字号“-”，以此区别是短波还是长波。即0.0025-为短波滤波器，-0.8为长波滤波器。

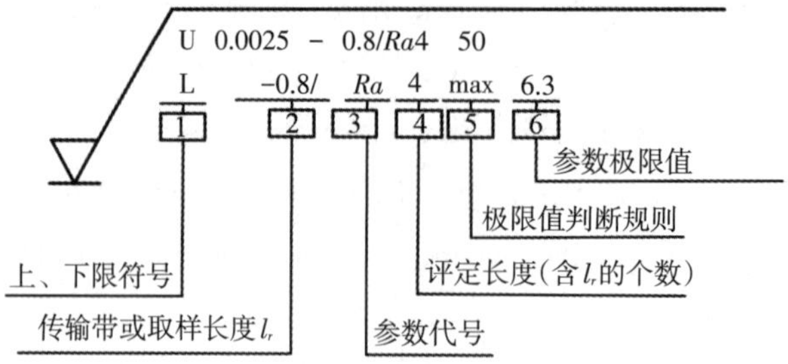

1—Upper and lower limit symbols 2—Conveyor belt or sampling length l_r

3—Parameter code 4—Evaluation length(including number of l_r)

5—Limit value judgment rule 6—Parameter limit value

Fig. 5-12 Example of marking surface roughness

图5-12 表面粗糙度标注示例

Marking of parameter code: the parameter code is marked after the conveyor belt or sampling length, which is separated by “/”, and is marked with *Ra* or *Rz*.

参数代号的标注：参数代号标注在传输带或取样长度后，它们之间用“/”隔开，并注明*Ra*或*Rz*。

Marking of evaluation length: if the default evaluation length is selected, the annotation can be omitted. If it is not equal to $5l_r$, the number of sampling lengths l_r should be noted.

评定长度的标注：如果默认评定长度，可省略标注。如果它不等于$5l_r$，则应注明取样长度l_r的个数。

Limit value judgment rule and limit value annotation: the annotation of limit value judgment rule is shown in Fig. 5-12; the upper limit is “16% Rule”, and the lower limit is “Maximum Rule”. To avoid misunderstanding, insert a space between the parameter code and the limit value.

极限值判断规则和极限值的标注：极限值判断规则的标注如图5-12所示，上限为“16%规则”,下限为“最大规则”。为了避免误解，在参数代号和极限值之间插入一个空格。

The marking method and significance of surface roughness amplitude parameters are shown in Tab. 5-6.

表面粗糙度幅度参数的标注方法及意义见表5-6。

Tab. 5-6 Marking method and significance of surface roughness amplitude parameters

表5-6 表面粗糙度幅度参数的标注方法及意义

Symbol 符号	Significance and explanation 意义及说明
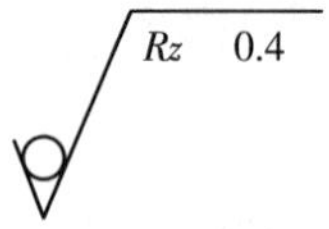	Material removal is not allowed, one-way upper limit value,default conveyor belt, maximum height of contour is 0.4 μm, evaluation length is 5 sampling lengths (default), 16% rule (default) 不允许去除材料，单向上限值，默认传输带，轮廓的最大高度为0.4 μm，评定长度为5个取样长度（默认），16%规则（默认）
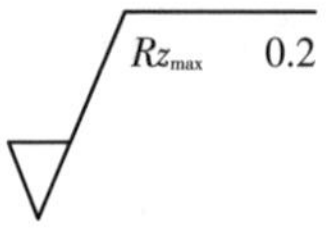	Material removal, one-way upperlimit, default conveyor belt, maximum height of contour is 0.2 μm, evaluation length is 5 sampling lengths (default), maximum rule 去除材料，单向上限值，默认传输带，轮廓的最大高度为0.2 μm，评定长度为5个取样长度（默认），最大规则

Tab. 5-6（Continued）表5-6（续）

Symbol 符号	Significance and explanation 意义及说明
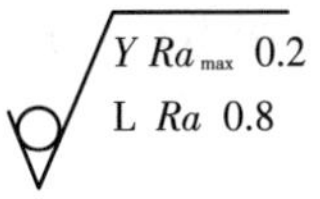	Material removal is not allowed. The two limit values are the default conveyor belt.The upper limit value: arithmetic mean deviation is 3.2 μm, evaluation length is 5 sampling lengths (default), maximum rule; the lower limit value: arithmetic mean deviation is 0.8 μm,evaluation length is 5sampling lengths (default), 16%Rule (default) 不允许去除材料，双向极限值，两个极限值均为默认传输带，上限值：算术平均偏差为3.2 μm，评定长度为5个取样长度（默认），最大规则；下限值：算术平均偏差为0.8 μm，评定长度为5个取样长度（默认），16%规则（默认）
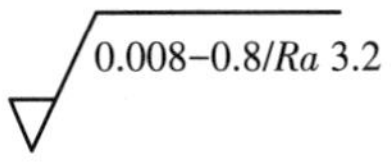	Remove the material, one-way upper limit value,conveyor belt 0.008 ~ 0.8 mm, arithmetic mean deviation is 3.2 μm, evaluation length is 5 sampling length (default), 16% Rule (default) 去除材料，单向上限值，传输带0.008 ~ 0.8 mm，算术平均偏差为3.2 μm，评定长度 为5个取样长度（默认），16%规则（默认）
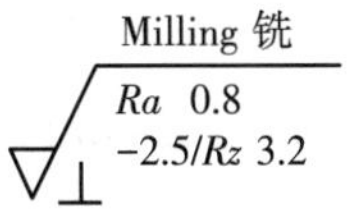	Remove the material, two one-way upper limit values. The former:default conveyorbelt and defaultevaluation length,arithmetic mean deviation is 0.8 μm, 16%rule (default); the latter: conveyor belt is −2.5 mm,default evaluation length, maximum height of contour is 3.2 μm,16%rule (default). The surface texture is perpendicular to the projection plane of the view. The processing method is milling 去除材料，两个单向上限值。前者：默认传输带和默认评定长度，算术平均偏差为0.8 um，16%规则（默认）；后者：传输带为−2.5 mm，默认评定长度，轮廓的最大高度为3.2 μm，16%规则（默认）。表面纹理垂直于视图所在的投影面。加工方法为铣削
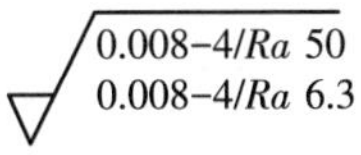	Remove the material, bidirectional limit value.The arithmetic mean deviation of the upper limit value is 50 um, the arithmetic mean deviation of the lower limit value is 6.3 μm, the upper and lower limit of the conveyor belt is 0.008 ~ 4 mm, the default evaluation length,16% rule (default). The machining allowance is 3 mm 去除材料，双向极限值。上限值的算术平均偏差为50 um，下限值的算术平均偏差 为6.3 μm，上下极限的传输带均为0.008 ~ 4 mm，默认评定长度，16%规则（默认）。加工余量为3 mm

5.4.2.3 Marking of other items of surface roughness 表面粗糙度其他项目的标注

The letter "*c*" position in surface roughness code is the location for marking the processing method. If the surface roughness of a surface is required to be obtained by specified processing methods (such as polishing, milling, plating, etc.), it can be marked with text.

表面粗糙度代号的字母*c*位置是标注加工方法的位置。如果某表面的表面粗糙度要求由指定的加工方法（如抛光、铣削、镀覆等）获得，则可以用文字标注。

The letter "*d*" position in surface roughness code is the position for marking the processing texture direction. If it is necessary to control the processing texture direction of the part surface, it can be marked according to the symbol of the processing texture direction.

表面粗糙度代号的字母*d*位置是标注加工纹理方向的位置。如果需要控制零件表面的加工纹理方向，则可按加工纹理方向的符号进行标注。

The letter "*e*" position in surface roughness code is the position for marking the machining allowance, which refers to the total allowance of the part surface before achieving the surface roughness requirement.

表面粗糙度代号的字母*e*位置是标注加工余量的位置。加工余量是指获得本表面粗糙度要求前零件表面的总余量。

5.4.3 Marking Position and Direction of Surface Roughness Symbol 表面粗糙度符号的标注位置与方向

Surface roughness requirements should be marked only once, and should be marked on the same view of corresponding size and tolerance as far as possible. Unless specified, the marked surface roughness requirements apply to finished workpiece surfaces.

表面粗糙度要求对每个表面一般只标注一次，并尽可能标注在相应的尺寸及其公差的同一视图上。除另有说明外，所标注的表面粗糙度要求是针对已经加工完的工件表面。

According to the standard, the marking and reading direction of surface roughness is consistent with that of size, as shown in Fig. 5-13.

标准规定，表面粗糙度的标注和读取方向与尺寸的标注和读取方向一致，如图5-13所示。

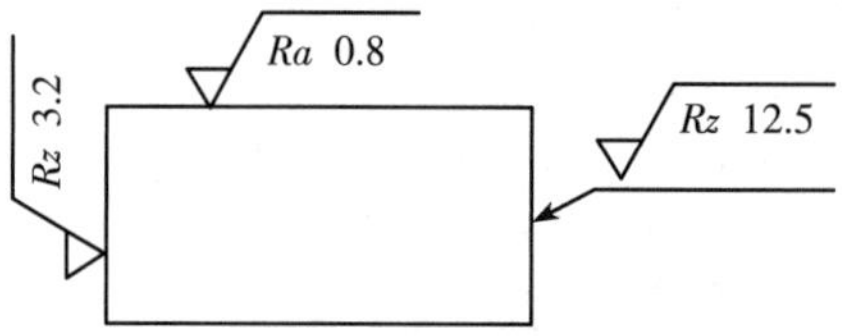

Fig. 5-13 Marking direction of surface roughness requirements

图5-13 表面粗糙度要求的标注方向

The surface roughness requirements can be marked on the contour line , and the symbol should point from the outside of the material and contact the surface. If necessary, the surface roughness symbol can also be marked using a leader with an arrow or black dot, as shown in Fig. 5-14 and Fig. 5-15.

表面粗糙度要求可标注在轮廓线上，其符号应从材料外指向并与表面接触。必要时，表面粗糙度符号也可用带有箭头或黑点的指引线引出标注，如图5-14和图5-15所示。

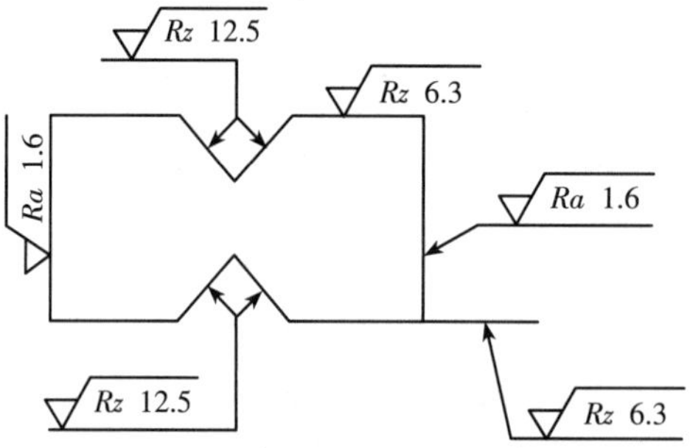

Fig. 5-14 Marking of surface roughness on contour line

图5-14 表面粗糙度在轮廓线上的标注

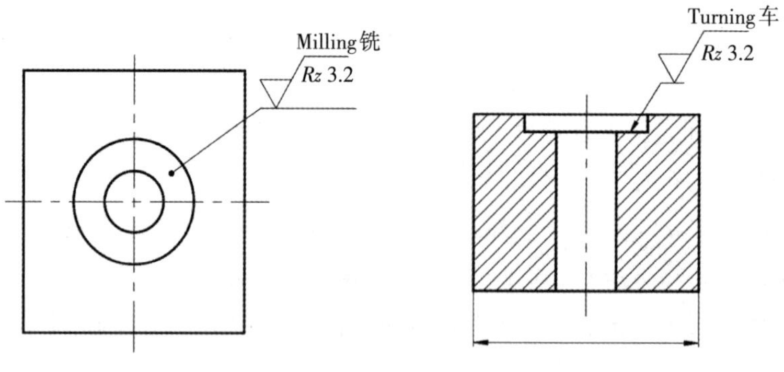

Fig. 5-15 Marking surface roughness with leader

图5-15 用指引线引出标注表面粗糙度

Surface roughness can be marked on a given size without causing misunderstanding, as shown in Fig. 5-16.

图5-16所示，表面粗糙度可以标注在给定尺寸上，而不会引起误解。

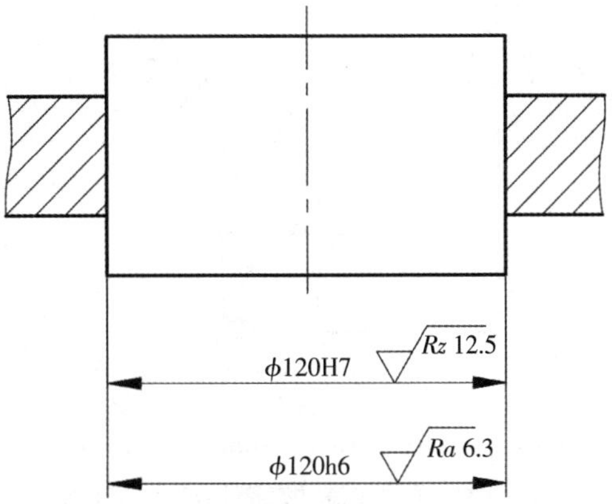

Fig. 5-16　Marking of surface roughness

图5-16　表面粗糙度标注在尺寸线上

The surface roughness requirements can be marked above the geometric tolerance frame, as shown in Fig. 5-17.

表面粗糙度要求可标注在几何公差框格的上方，如图5-17所示。

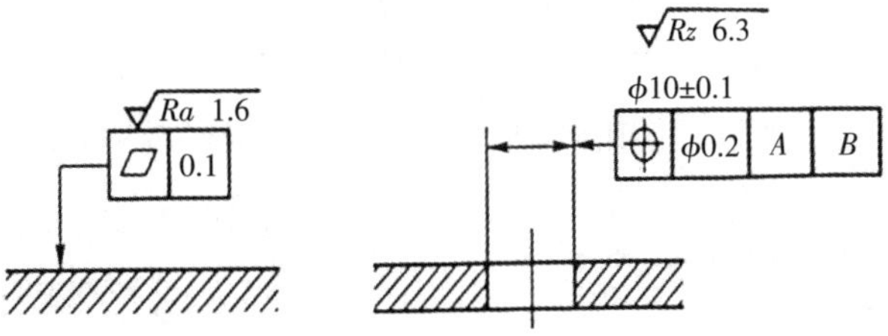

Fig. 5-17　Marking of surface roughness above geometric tolerance frame on size line

图5-17　表面粗糙度标注在几何公差框格的上方

The surface roughness requirements can be directly marked on the extension line of the size line, or can be led out by the leader with an arrow. The surface roughness of cylinders and prisms is required to be marked only once, as shown in Fig. 5-18. For each prism surface with different surface roughness requirements, it should be marked separately, as shown in Fig. 5-19.

表面粗糙度要求可以直接标注在尺寸线的延长线上，或用带箭头的指引线引出标注。圆柱和棱柱表面的粗糙度要求只标注一次，如图5-18所示。对于每一个有不同表面粗糙度要求的棱柱面，则应分别单独标注，如图5-19所示。

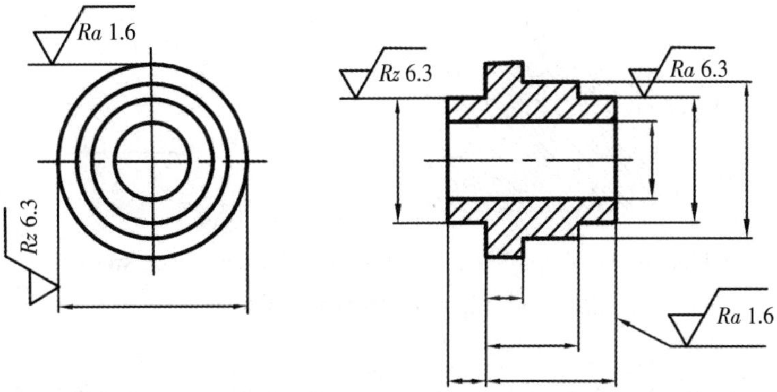

Fig. 5-18　Marking of surface roughness on cylinder feature extension line

图5-18　表面粗糙度标注在圆柱特征的延长线上

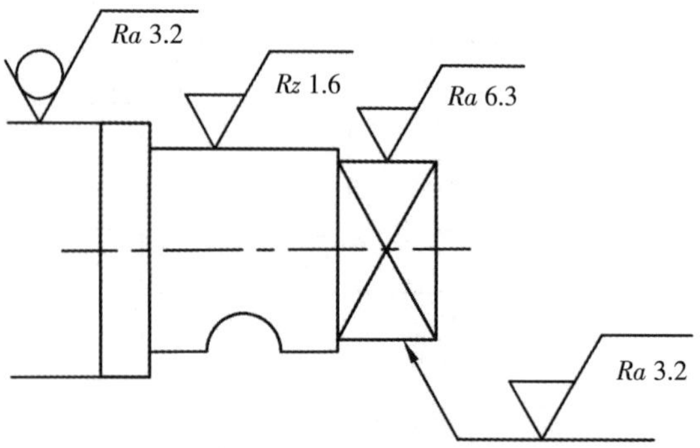

Fig. 5-19 Marking of surface roughness on cylinder and prism

图5-19 圆柱和棱柱表面粗糙度的注法

A simplified method for surface roughness. If most (or all) surfaces of the workpiece have the same surface roughness requirements, the surface roughness requirements can be uniformly marked near the Title Block of the drawing, as shown in Fig. 5-20 (a) and (b). The surface with the same surface roughness requirements can be simplified in the form of an equation near the drawing or Title Block with complete symbols with letters, as shown in Fig. 5-21.

表面粗糙度的简化标注方法。如果工件的大部分（或全部）表面具有相同的表面粗糙度要求，可以在图纸的标注栏附近统一标注表面粗糙度要求，如图5-20（a）和（b）所示。对于具有相同表面粗糙度要求的表面，可以在图形或标题栏附近以带有字母的完整符号形式进行简化标注，如图5-21所示。

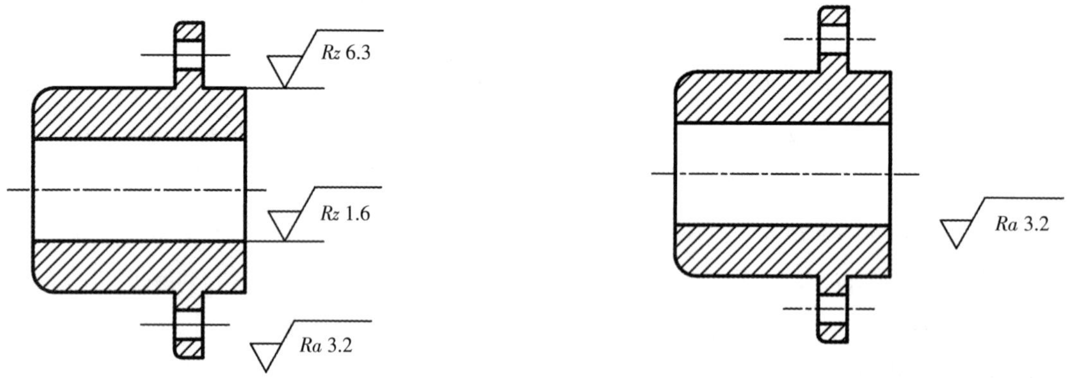

(a) Most surfaces roughness require the same

(a) 大多数表面要求相同

(b) All surfaces require the same surface roughness

(b) 全部表面要求相同

Fig. 5-20 Marking of most or all surfaces with the same surface roughness requirements

图5-20 大多数或全部表面有相同的表面粗糙度要求的标注

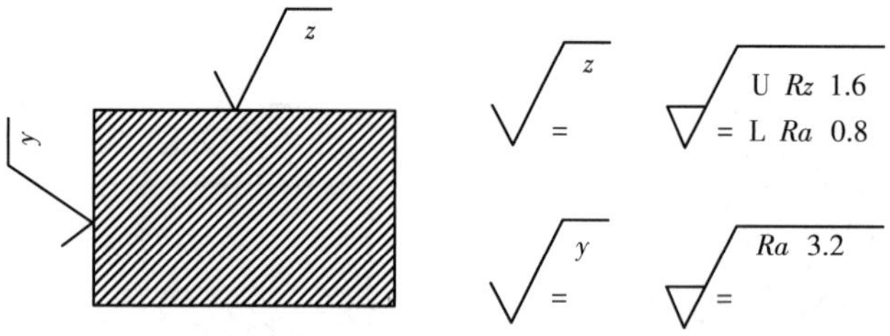

Fig. 5-21 Simplified annotation for limited paper space

图5-21 在图纸空间有限时的简化标注

5.5 Measurement of Surface Roughness 表面粗糙度的检测

For surface roughness, if the measurement section direction f is unspecified, the measurement should be carried out in the direction of the maximum amplitude parameter, generally speaking, in the direction perpendicular to the surface processing texture. The measuring methods of surface roughness include the comparison method, the light cutting method, the interference method, the needle tracing method, the impression method and the three-dimensional geometric surface measurement method.

对于表面粗糙度来说，如果未指定测量截面的方向，则测量应在最大幅度参数的方向上进行，一般而言，即在垂直于表面处理纹理的方向上进行标注。表面粗糙度的测量方法包括比较法、光切法、干涉法、针描法、印模法和三维几何表面测量法。

Exercise 5 习题5

5-1 Comparison of roughness profile amplitude parameters for each group of holes and explanations.

（1）ϕ60H6 hole and ϕ30H6 hole.

（2）The two holes in ϕ50H7/k6 and ϕ50H7/g6.

（3）Two ϕ30H7 holes with cylindricity tolerances of 0.01 mm and 0.02 mm respectively.

比较下列每组中两孔的粗糙度轮廓幅度参数值的大小，并说明原因。

（1）ϕ60H6孔与ϕ30H6孔。

（2）ϕ50H7/k6与ϕ50H7/g6中的两孔。

（3）圆柱度公差分别为0.01mm和0.02mm的两个ϕ30H7孔。

5-2 Please mark the following surface roughness technical requirements on Fig.5-22 (use default standardized values for unspecified items).

（1）The maximum value of surface roughness parameter *Ra* for the ϕ10 hole is 3.2 μm.

（2）The left end face of the ϕ40 shaft is machined by milling. The upper limit of the surface roughness parameter *Rz* is 12.5μm, and the machining texture is approximately radial.

（3）For the remaining surfaces, the maximum value of the surface roughness parameter *Ra* is 25 μm.

试将下列表面粗糙度技术要求标注在题图5-22上（未指明者皆采用默认的标准化值）。

（1）ϕ10孔的表面粗糙度参数Ra的最大值为3.2 μm。

（2）ϕ40轴的左端面采用铣削加工，表面粗糙度参数*Rz*的上限值为12.5 μm，加工纹理呈近似放射形。

（3）其余表面的表面粗糙度参数*Ra*的最大值为25 μm。

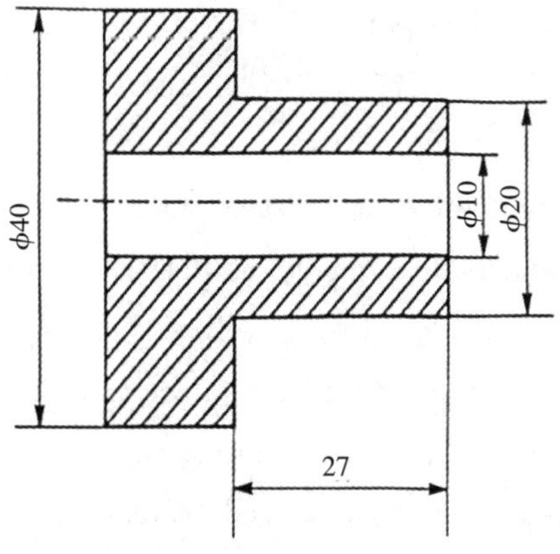

Fig. 5-22 Assiqnment question 2 Figure

图5-22 作业题2图

Chapter 6 Inspection of Plain Workpiece Sizes 光滑工件尺寸的检测

Plain limit gauge (hereinafter referred to as gauge) is a fixed-value testing tool without scale. When using the gauge to inspect the parts, we can only determine whether the parts are qualified, but we cannot accurately measure the actual size of the parts. However, for mass-produced parts, they can be considered qualified as long as the size of the parts is within the specified tolerance range. Therefore, the gauge inspection is very convenient, and because of its simple structure and high inspection efficiency, it is widely used in production.

光滑极限量规（以下简称量规）是一种没有刻度的定值用检验工具。在使用量规检验零件时，只能确定零件是否合格，而不能准确测量出零件的实际尺寸数值。但是对于大量生产的零件来说，只要能够确认零件的尺寸在规定的公差范围内，即可认为零件合格。因此，量规检验非常方便，而且由于结构简单、检验效率高，其在生产中得到了广泛应用。

The relevant standards in GB are GB/T 1957-2006 (*Tolerances and General Features for Plain Limit Gauges*) and GB/T 3177-2009 (*Geometrical Product Specifications (GPS)-Inspection of Plain Workpiece Sizes*). These two standards give the methods to inspect sizes and geometrical errors of plain workpiece.

相关国家标准包括GB/T 1957—2006（《光滑极限量规的公差与一般特性》）和GB/T 3177—2009（《产品几何技术规范（GPS)—光滑工件尺寸的检验》)。这些标准提供了检验光滑工件尺寸和几何误差的方法。

6.1 Misacceptance and Misabandonment 误收与误废

A workpiece that is unqualified but judged as qualified is called "misacceptance", while a qualified workpiece rejected as waste is "misabandonment".

工件不合格却被判定为合格的情况被称为“误收”;而工件合格却被误判为废品的情况则被称为“误废”。

Because of the measurement error, there are misacceptance and misabandonment of the workpiece, as shown in Fig. 6-1. The former impairs the quality of the parts, while the latter makes the production uneconomical. In other words, the measurement error will in fact change the prescribed tolerance zone of workpiece, making it shrink or expand. Considering the effect of measurement error, the possible minimum manufacturing tolerances of workpiece are called production tolerances; and the possible maximum manufacturing tolerances of workpiece are called guaranteed tolerance.

由于测量误差的存在，工件存在误收和误废的情况，如图6-1所示。误收会损害零件的质量，而误废则会使生产变得不经济。也就是说，测量误差实际上会改变工件规定的公差带，使其缩小或扩大。

考虑到测量误差的影响，工件可能的最小制造公差被称为生产公差；而工件可能的最大制造公差则被称为保证公差。

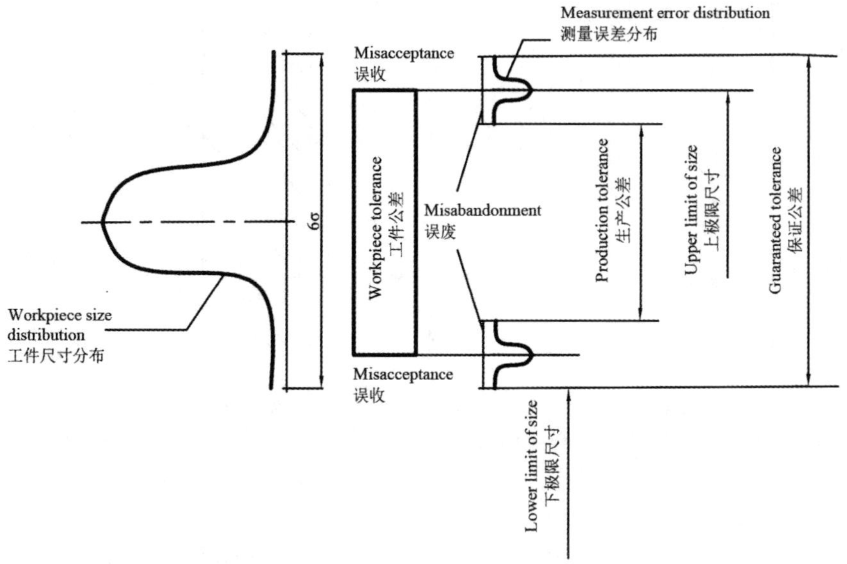

Fig. 6-1 The effect of measurement error

图6-1 测量误差的影响

6.1.1 Workpiece Acceptance Principles 验收原则

When the purpose is to detect whether the discrimination of measured parameters meets the requirements of tolerance, the detection method shall comply with the provisions of the principle of tolerance, according to the acceptance method the workpiece should be accepted only falling within the size limits, which only allows misabandonment and not misacceptance.

当测量参数用于判别是否符合公差要求时，检测方法应遵循公差原则的规定。因此，按照验收方法，只有尺寸在公差范围内的工件才应被接受，只允许误废而不允许误收。

6.1.2 Acceptance Limit 接收极限

According to the requirements of the engineering drawings, the actual size should be between the maximum and minimum limits of size. Considering the workshop actual conditions, usually the form errors of workpiece are determined by the accuracy of the machining devices and processing devices, and the workpiece is pass or fail that is judged by a single measuring. The effect of temperature and pressing-shrink and the symmetrical errors from measurement tools and standard gauges are not amended. Therefore, the measurement sizes are not the true size, i.e. actual size equals true size plus or minus measurement error. Acceptance limits are determined by one of the following:

根据工程图纸的要求，实际尺寸应在最大和最小尺寸极限之间。考虑到车间的实际情况，工件的形状误差取决于加工设备和工艺装备的精度，工件的合格与否，只按尺寸测量结果来判断，不考虑温度和压陷效应的影响，以及测量工具和标准量具带来的对称误差。因此，测量得到的尺寸并非真实尺寸，即实际尺寸 = 真实尺寸±测量误差。接收极限可以通过以下某一种方式确定：

（1）Acceptance limits are determined by moving a safety margin into the tolerance zone from the specified maximum material size（MMS）and least material size（LMS）respectively, as shown inFig. 6-2.

验收极限是通过将安全裕度分别从规定的最大实体尺寸（MMS）和最小实体尺寸（LMS）向公差带内移动来确定的，如图6-2所示。

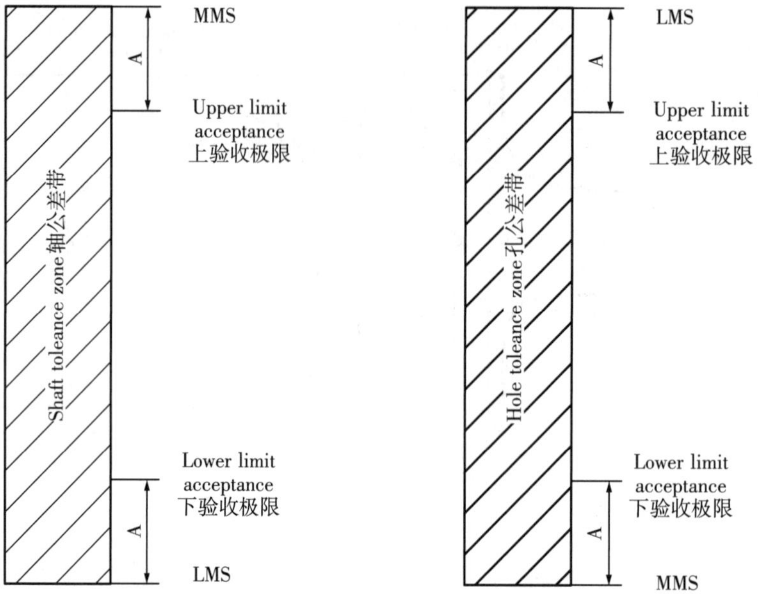

Fig. 6-2 Acceptance limits and safety margin

图6-2 验收极限与安全裕度

The limits of acceptance for hole:

孔尺寸的验收极限：

$$\text{Upper acceptance limit} = D_L - A$$
$$\text{Lower acceptance limit} = D_M + A \qquad (6\text{-}1)$$
$$\text{上验收极限} = D_L - A$$
$$\text{下验收极限} = D_M + A$$

The limits of acceptance for shaft:

轴尺寸的验收极限：

$$\text{Upper acceptance limit} = d_M - A$$
$$\text{Lower acceptance limit} = d_L + A \qquad (6\text{-}2)$$
$$\text{上验收极限} = d_M - A$$
$$\text{下验收极限} = d_L + A$$

Where, A —is safety margin. Usually $A = T/10$;

T —is the tolerance of shafts or holes. The values are shown in Tab. 6-1.

式中，A ——安全裕度，通常$A = T/10$；

T ——零件公差，其数值见表6-1。

(2) Acceptance limits are equal to the maximum material size and the least material size respectively, i.e. the safety margin equal to zero.

验收极限分别等于最大实体尺寸和最小实体尺寸，即安全裕度为0。

(3) One of acceptance limits is determined by moving a safety margin into the tolerance zone from the specified maximum material size, and the other acceptance limit is equal to the least material size.

其中一个验收极限是通过将安全裕度从规定的最大实体尺寸移入公差带内来确定的，而另一个验收极限则等于最小实体尺寸。

Tab. 6-1 The safety margin *A* and the permissible measurement uncertainty of the measuring instrument

表6-1 安全裕度*A*与计量器具的测量不确定度允许值 单位：μm

Tolerance grade 公差等级		IT6					IT7					IT8					IT9				
Nominal size 公称尺寸/mm		T	A	u_1			T	A	u_1			T	A	u_1			T	A	u_1		
>大于	End至			Ⅰ	Ⅱ	Ⅲ			Ⅰ	Ⅱ	Ⅲ			Ⅰ	Ⅱ	Ⅲ			Ⅰ	Ⅱ	Ⅲ
—	3	6	0.6	0.54	0.9	1.4	10	1.0	0.9	1.5	2.3	14	1.4	1.3	2.1	3.2	25	2.5	2.3	3.8	5.6
3	6	8	0.8	0.72	1.2	1.8	12	1.2	1.1	1.8	2.7	18	1.8	1.6	2.7	4.1	30	3.0	2.7	4.5	6.8
6	10	9	0.9	0.81	1.4	2.0	15	1.5	1.4	2.3	3.4	22	2.2	2.0	3.3	5.0	36	3.6	3.3	5.4	8.1
10	18	11	1.1	1.0	1.7	2.5	18	1.8	1.7	2.7	4.1	27	2.7	2.4	4.1	6.1	43	4.3	3.9	6.5	9.7
18	30	13	1.3	1.2	2.0	2.9	21	2.1	1.9	3.2	4.7	33	3.3	3.0	5.0	7.4	52	5.2	4.7	7.8	12
30	50	16	1.6	1.4	2.4	3.6	25	2.5	2.3	3.8	5.6	39	3.9	3.5	5.9	8.8	62	6.2	5.6	9.3	14
50	80	19	1.9	1.7	2.9	4.3	30	3.0	2.7	4.5	6.8	46	4.6	4.1	6.9	10	74	7.4	6.7	11	17
80	120	22	2.2	2.0	3.3	5.0	35	3.5	3.2	5.3	7.9	54	5.4	4.9	8.1	12	87	8.7	7.8	13	20
120	180	25	2.5	2.3	3.8	5.6	40	4.0	3.6	6.0	9.0	63	6.3	5.7	9.5	14	100	10	9.0	15	23
180	250	29	2.9	2.6	4.4	6.5	46	4.6	4.1	6.9	10	72	7.2	6.5	11	16	115	12	10	17	26
250	315	32	3.2	2.9	4.8	7.2	52	5.2	4.7	7.8	12	81	8.1	7.3	12	18	130	13	12	19	29
315	400	36	3.6	3.2	5.4	8.1	57	5.7	5.1	8.4	13	89	8.9	8.0	13	20	140	14	13	21	32
400	500	40	4.0	3.6	6.0	9.0	63	6.3	5.7	9.5	14	97	9.7	8.7	15	22	155	16	14	23	35

Tolerance grade 公差等级		IT10					IT11					IT12				IT13			
Nominal size 公称尺寸/mm		T	A	u_1			T	A	u_1			T	A	u_1		T	A	u_1	
>大于	End至			Ⅰ	Ⅱ	Ⅲ			Ⅰ	Ⅱ	Ⅲ			Ⅰ	Ⅱ			Ⅰ	Ⅱ
—	3	40	4.0	3.6	6.0	9.0	60	6.0	5.4	9.0	14	100	10	9.0	15	140	14	13	21
3	6	48	4.8	4.3	7.2	11	75	7.5	6.8	11	17	120	12	11	18	180	18	16	27
6	10	58	5.8	5.2	8.7	13	90	9.0	8.1	14	20	150	15	14	23	220	22	20	33
10	18	70	7.0	6.3	11	16	110	11	10	17	25	180	18	16	27	270	27	24	41
18	30	84	8.4	7.6	13	19	130	13	12	20	29	210	21	19	32	330	33	30	50
30	50	100	10	9.0	15	23	160	16	14	24	36	250	25	23	38	390	39	35	59
50	80	120	12	11	18	27	190	19	17	29	43	300	30	27	45	460	46	41	69
80	120	140	14	13	21	32	220	22	20	33	50	350	35	32	53	540	54	49	81
120	180	160	16	15	24	36	250	25	23	38	56	400	40	36	60	630	63	57	95
180	250	185	18	17	28	42	290	29	26	44	65	460	46	41	69	720	72	65	110
250	315	210	21	19	32	47	320	32	29	48	72	520	52	47	78	810	81	73	120
315	400	230	23	21	35	52	360	36	32	54	81	570	57	51	86	890	89	80	130
400	500	250	25	23	38	56	400	40	36	60	90	630	63	57	95	970	97	87	150

Note: u_1 is divided into Grades Ⅰ, Ⅱ, and Ⅲ. In general, Grade I should be prioritized, followed by Grades Ⅱ and Ⅲ.

注：u_1分Ⅰ，Ⅱ，Ⅲ档，一般情况下应优先选用Ⅰ档，其次选用Ⅱ档、Ⅲ档。

The limits of acceptance for hole:

孔尺寸的验收极限：

Upper acceptance limit = D_L

上验收极限 = D_L (6–3)

Lower acceptance limit = $D_M + A$

下验收极限 = $D_M + A$

The limits of acceptance for shaft:

轴尺寸的验收极限：

Upper acceptance limit = $d_M - A$ (6–4)

上验收极限 = $d_M - A$

Lower acceptance limit = d_L

下验收极限 = d_L

6.1.3 Selection of the Way of the Acceptance Limits 验收极限方式的选择

Selection of acceptance limits should consider the functional requirements of size, size tolerance grades, the uncertainty of measurement and the process capacity. Based on 6.1.2 Acceptance Limit, the specific methods are as follows.

验收极限的选择应考虑尺寸的功能要求、尺寸公差等级、测量的不确定性以及工艺能力等。根据6.1.2接收极限，具体如下：

Case ①is used when an envelope requirement adopted or for higher size tolerance grades.

情况①适用于采用包容要求或较高尺寸公差等级的情况。

Case ② is used for the process capacity CP≥1, non-fit size and general tolerance.

情况②适用于工艺能力指数CP≥1、非配合尺寸以及一般公差的情况。

Case ③is used for the process capacity CP≥1 and when the envelope requirement is adopted or for the partial size distribution.

情况③适用于工艺能力指数CP≥1且采用包容要求或者尺寸分布符合的情况。

6.1.4 Selection of Measurement Tools 计量器具的选择

The selection of measuring instruments in machinery manufacturing mainly depends on the technical and economic indicators when considering these indicators, the following two requirements should be considered:

在机械制造中，测量仪器的选择主要取决于测量仪器的技术指标和经济指标。考虑这些指标时，主要需满足以下两个要求：

（1）When selecting measuring instrament according to the measured workpiece shape, position and size, the measuring range of the measuring instruments selected should be able to meet the requirements. Such as a projector or tool microscope, usually only used for measuring small workpieces.

根据被测工件的形状、位置和尺寸来选择测量仪器，所选测量仪器的测量范围应能满足要求。例如，投影仪或工具显微镜通常仅用于测量小型工件。

（2）When selecting measuring instruments, according to the size tolerances of measured workpiece, considering the errors of the measuring instruments will cause reducing workpiece manufacturing tolerances, in the choice of measuring instruments, not only is it necessary to strictly limit the actual size of the workpiece to be measured so as not to exceed the specified limits of size; but also we should avoid that the production cost is increased because the production tolerances of the workpiece are decreased too much. Smaller and stricker measurment error limits of instruments lead to higher costs and requirement for use the environment and the operator. Therefore, using high - precision measuring instruments to detect low precision workpieces is not economical. According to the workpiece accuracy requirements, reasonable choice of measuring instruments is an important task. The permissible values for the uncertainty of the measuring instruments are shown as Tab. 6-2 to Tab. 6-4.

根据被测工件的尺寸公差来选择测量仪器。考虑到测量仪器的误差会引起工件的制造公差减小，因此，在选择测量仪器时，不仅需要严格限制被测工件的实际尺寸，以确保其不超过规定的尺寸范围，同时应避免由工件生产公差过度减小而增加生产成本。测量仪器的测量误差极限越小，仪器的成本就越高，对使用环境和操作人员的要求也就越高。因此，使用高精度测量仪器来检测低精度工件是不经济的。根据工件的精度要求，合理选择测量仪器是一项重要工作。表6-2至表6-4列出了有关计量器具不确定度的允许值。

Tab. 6-2 The uncertainty of micrometer and vernier caliper

表6-2 千分尺和游标卡尺的不确定度

单位：mm

<table>
<tr><th rowspan="3">Range of sizes
尺寸范围</th><th colspan="4">Type of measuring instrument 计量器具类型</th></tr>
<tr><th>Outer diameter micrometer with a division value of 0.01
分度值为0.01的外径千分尺</th><th>Inner diameter micrometer with a division value of 0.01
分度值为0.01的内径千分尺</th><th>Vernier caliper with a division value of 0.02
分度值为0.02的游标卡尺</th><th>Vernier caliper with a division value of 0.05
分度值为0.05的游标卡尺</th></tr>
<tr><th colspan="4">Uncertainty 不确定度</th></tr>
<tr><td>0 ~ 50</td><td>0.004</td><td rowspan="3">0.008</td><td rowspan="6">0.020</td><td rowspan="4">0.05</td></tr>
<tr><td>50 ~ 100</td><td>0.005</td></tr>
<tr><td>100 ~ 150</td><td>0.006</td></tr>
<tr><td>150 ~ 200</td><td>0.007</td><td rowspan="3">0.013</td></tr>
<tr><td>200 ~ 250</td><td>0.008</td><td rowspan="6">0.100</td></tr>
<tr><td>250 ~ 300</td><td>0.009</td></tr>
<tr><td>300 ~ 350</td><td>0.010</td><td rowspan="3">0.020</td><td rowspan="4"></td></tr>
<tr><td>350 ~ 400</td><td>0.011</td></tr>
<tr><td>400 ~ 450</td><td>0.012</td></tr>
<tr><td>450 ~ 500</td><td>0.013</td><td>0.025</td></tr>
</table>

Note: When comparative measurement is used, the uncertainty of the micrometer can be less than the value specified in this table, typically reduced by 40%.

注：当采用比较测量时，千分尺的不确定度可小于本表规定的数值，一般可减小40%。

Tab. 6-3 The uncertainty of comparator

表6-3 比较仪的不确定度

单位：mm

<table>
<tr><th colspan="2" rowspan="2">Size Range
尺寸范围</th><th colspan="4">Measuring Instrument 所使用的计量器具</th></tr>
<tr><th>Comparator with a division value of 0.0005 (equivalent to a magnification of 2000 times)
分度值为0.0005(相当于放大倍数2000倍)的比较仪</th><th>Comparator with a division value of 0.001 (equivalent to a magnification of 1000 times)
分度值为0.001(相当于放大倍数1000倍)的比较仪</th><th>Comparator with a division value of 0.002 (equivalent to a magnification of 500 times)
分度值为0.002(相当于放大倍数500倍)的比较仪</th><th>Comparator with a division value of 0.005 (equivalent to a magnification of 200 times)
分度值为0.005(相当于放大倍数200倍)的比较仪</th></tr>
<tr><th>>大于</th><th>End 至</th><th colspan="4">Uncertainty 不确定度</th></tr>
<tr><td>—</td><td>25</td><td>0.0006</td><td>0.0010</td><td>0.0017</td><td>0.0030</td></tr>
</table>

Tab. 6-3（Continued）表6-3（续）

<table>
<tr><th colspan="2" rowspan="2">Size Range
尺寸范围</th><th colspan="4">Measuring Instrument 所使用的计量器具</th></tr>
<tr><th>Comparator with a division value of 0.0005（equivalent to a magnification of 2000 times）
分度值为0.0005(相当于放大倍数2000倍)的比较仪</th><th>Comparator with a division value of 0.001（equivalent to a magnification of 1000 times）
分度值为0.001(相当于放大倍数1000倍)的比较仪</th><th>Comparator with a division value of 0.002（equivalent to a magnification of 500 times）
分度值为0.002(相当于放大倍数500倍)的比较仪</th><th>Comparator with a division value of 0.005（equivalent to a magnification of 200 times）
分度值为0.005(相当于放大倍数200倍)的比较仪</th></tr>
<tr><td>25</td><td>40</td><td>0.0007</td><td>0.0010</td><td rowspan="3">0.0018</td><td rowspan="4">0.0030</td></tr>
<tr><td>40</td><td>65</td><td>0.0008</td><td rowspan="2">0.0011</td></tr>
<tr><td>65</td><td>90</td><td>0.0008</td></tr>
<tr><td>90</td><td>115</td><td>0.0009</td><td>0.0012</td><td rowspan="2">0.0019</td></tr>
<tr><td>115</td><td>165</td><td>0.0010</td><td>0.0013</td><td rowspan="4">0.0035</td></tr>
<tr><td>165</td><td>215</td><td>0.0012</td><td>0.0014</td><td>0.0020</td></tr>
<tr><td>215</td><td>265</td><td>0.0014</td><td>0.0016</td><td>0.0021</td></tr>
<tr><td>265</td><td>315</td><td>0.0016</td><td>0.0017</td><td>0.0022</td></tr>
</table>

Note：During measurement, the standard used consists of four Grade 1（or Class 4）gauge blocks.

注：测量时，使用的标准器由4块1级（或4等）量块组成。

Tab. 6-4 The uncertainty of indicator gauge

表6-4 指示表的不确定度

单位：mm

<table>
<tr><th colspan="2" rowspan="3">Size Range
尺寸范围</th><th colspan="4">The measuring instrument 所使用的计量器具</th></tr>
<tr><th>A dial indicator with a division value of 0.001（Class 0 over the entire range,Class 1 is within 0.2 mm）
分度值为0.001的千分表（0级在全程范围内，1级在0.2 mm内）</th><th>A dial indicator with a division value of 0.001,0.002 and 0.005（Class 1 over the entire range）
分度值为0.001、0.002、0.005的千分表（1级在全程范围内）</th><th rowspan="2">A dial indicator with a division value of 0.01（Class 0 over the entire range, Class 1 is within 1 mm）
分度值为0.01的百分表（0级在全程范围内，1级在任意1 mm内）</th><th rowspan="2">A dial indicator with a division value of 0.01（Class 1 over the entire range）
分度值为0.01的百分表（1级在全程范围内）</th></tr>
<tr><th>A dial indicator with a division value of 0.002（Within one revolution）
分度值为0.002的千分表（在1转范围内）</th><th>A dial indicator with a division value of 0.01（Class 0 within any 1 mm）
分度值为0.01的百分表（0级在任意1 mm内）</th></tr>
<tr><th>>大于</th><th>End至</th><th colspan="4">Uncertainty 不确定度</th></tr>
<tr><td>—</td><td>25</td><td rowspan="5">0.005</td><td rowspan="9">0.010</td><td rowspan="9">0.018</td><td rowspan="9">0.030</td></tr>
<tr><td>25</td><td>40</td></tr>
<tr><td>40</td><td>65</td></tr>
<tr><td>65</td><td>90</td></tr>
<tr><td>90</td><td>115</td></tr>
<tr><td>115</td><td>165</td><td rowspan="4">0.006</td></tr>
<tr><td>165</td><td>215</td></tr>
<tr><td>215</td><td>265</td></tr>
<tr><td>265</td><td>315</td></tr>
</table>

Example 6-1: The measured workpiece is ϕ140H11 Ⓔ, try to determine limits and select appropriate measuring device.

例6-1：试确定被测件ϕ1140H11 Ⓔ 的验收极限，并选择计量器具。

Solution:（1）Determine safety margin *A* and acceptance limit.

Refer to Tab. 2-5（Chapter 2）, then determine the upper and lower limit deviations of the tolerance zone for ϕ140H11 mm are $\phi 140_{0}^{+0.25}$ mm, look up Tab. 6-1: A = 0. 025 mm, u_1 = 0.023 mm.

The workpiece dimensions adopt envelope requirement, then

Upper acceptance limit $= D_{max} - A = (140 + 0.25 - 0.025)$ mm = 140.225 mm

Lower acceptance limit $= D_{min} + A = (140 + 0.025)$ mm = 140.025 mm

(2) Selection of measurement tools.

Look up the permissible values for the uncertainty of the measuring instruments value of the workshop gauge in Tab. 6-2, then determine vernier caliper, the uncertainty is 0.020 mm$\leqslant u_1$ = 0.023 mm, and the value is the closest, which can meet the requirements.

解：(1) 确定安全裕度A和验收极限。

查表2-5（第二章）确定ϕ140H11 mm公差带的上、下极限偏差为$\phi 140_{0}^{+0.25}$ mm，再根据表6-1，查得A = 0.025 mm，u_1 = 0.023 mm。

工件尺寸采用包容要求，则

上验收极限尺寸 $= D_{max} - A = (140 + 0.25 - 0.025)$ mm = 140.225 mm

下验收极限尺寸 $= D_{min} + A = (140 + 0.025)$ mm = 140.025 mm

(2) 选择计量器具。

由表6-2查得分度值为0.02 mm的游标卡尺，它的测量不确定度为0.020 mm $< u_1$ = 0.023 mm，且数值最为接近，可以满足要求。

6.2 Function and Type of the Plain Limit Gauge 光滑极限量规的功用和种类

Plain limit gauge is generally divided into two kinds: smooth limit gauge for hole and smooth limit gauge for axis.

光滑极限量规一般分为两种：孔用光滑极限量规和轴用光滑极限量规。

6.2.1 Classify the Workpieces According to Their Shape 按工件形状不同分类

The gauge for holes is called a "plug gauge", and the gauge for shaft diameter is called a "ring gauge". Both plug gauges and ring gauges have go gauges (ends) and not-go gauges (ends). The go gauge simulates the maximum material boundary of the hole or shaft, with its nominal size equal to the maximum material (MMS) size of the hole or shaft ($D_{min} = D_M$, $d_{max} = d_M$). It checks whether the actual contour of the hole or shaft (the combination of actual size and shape error) exceeds its maximum material boundary, i.e. whether the external function size of the hole or shaft exceeds its MMS. The not-go gauge checks whether the actual size of the hole or shaft exceeds its least material size (LMS). The not-go gauge has its nominal size equal to the LMS of the hole or shaft, that is $D_{min} = D_L$ or $d_{max} = d_L$.

检验孔径的量规称为"塞规",检验轴径的量规称为"卡规"。塞规和卡规均有通规（通端）和止规（止端）之分。通规用来模拟体现被测孔或轴的最大实体边界，量规的公称尺寸等于孔和轴的最大实体尺寸（$D_{min} = D_M$，$d_{max} = d_M$）。检验孔或轴的实际轮廓（实际尺寸和形状误差的综合结果）是否超出其最大实体边界，即检验孔或轴的体外作用尺寸是否超出其最大实体尺寸。止规用来检验被测孔或轴的实际尺寸是否超出其最小实体尺寸。止规的公称尺寸等于孔和轴的最小实体尺寸，即：$D_{min} = D_L$，$d_{max} = d_L$。

6.2.2 Different Classification According to the Use of the Specifications 按量规的用途不同分类

According to the different usage occasions of the gauge, the gauge can be divided into the following three categories:

根据不同的使用场合，可将量规分为以下三类:

（1）Workshop gauges. In the part manufacturing process, the gauge used by the operator to inspect the part is called the workshop gauge, the go gauge is represented by "T", and the not-go gauge is represented by "Z". To ensure the accuracy of the processed parts, the operator should use new or less worn go gauges.

工作量规。在零件制造过程中，操作者对零件进行检验所使用的量规称为工作量规,，通规用"T"表示，止规用"Z"表示。为了保证加工零件的精度，操作者应该使用新的或者磨损较小的通规。

（2）Inspection garages. The gauge used by the inspection department or the user representative in the acceptance of the product is called the acceptance gauge. Generally, the inspectors or the user representatives use the go gauges which are old workshop gauges, these old workshop gauges have been worn out quite seriously but not exceeding the wear limit. Hence, the operator passes the self-inspection, the inspector or user representative will also find it qualified, so as to ensure the qualified rate of the parts.

验收量规。检验部门或用户代表在验收产品时所使用的量规称为验收量规。一般来说，检验员或用户代表使用的是车间中的旧通规，尽管这些旧量规已经磨损得相当严重，但尚未达到磨损极限。这样，由操作者自检合格，检验人员或用户代表验收时也一定合格，从而保证了零件的合格率。

（3）Reference gauges. The measuring gauge that checks whether the shaft gauges（cylindrical ring gauge or gap gauge）conform to the manufacturing tolerance during manufacturing and whether they have reached the wear limit in use is called the proof-reading gauge. Since the shaft（with external dimensions）gauges have the inner dimensions（hard to measure), we use the reference gauge. The reference gauge is the outer size and can be detected with a general meter. The hole gauge forhole（plug gauge）itself is an external size, which can be easily detected by a general measuring meter, so there is no reference gauge for it. The reference gauges are further divided into three types:

校对量规。检验轴用量规（环规或卡规）在制造时是否符合制造公差以及在使用中是否已达到磨损极限的量规称为校对量规。由于轴用量规是内尺寸，不易检验，所以我们使用校对量规。校对量规是外尺寸，可以用通用量仪检测。孔用量规（塞规）本身是外尺寸，可以较方便地用通用量仪检测，所以不设校对量规。校对量规又可分为以下三类:

① "Verify Go-Go" "jiao tong-tong" gauges（codenamed "TT"), uesd to check out the go gauges for the shaftgauge.

"验证通过"——"校通－通"量规（代号"TT"）是检验轴的校对量规。

② "Verify No Go-Go" "jiao zhi-tong" gauges（codenamed "ZT"), uesd for checking out the not-go gauges for the shaftgauge.

"验证不通过"——"校止－通"量规（代号"ZT"），是用于校验轴用量规的止规。

③ "Verify Go-Wear" "jiao tong-sun" gauges（codenamed "TS"), uesd for checking out the wear limits of the go gauges of the shaft gauge.

"校验通规磨损（情况)"——"校通－损"量规（代号"TS"），是检验轴用工作量规是否达到磨损极限的校对量规。

Exercise 6 习题6

6-1 What are the characteristics of the smooth limit gauge? How to judge the qualification of the workpiece?

光滑极限量规的特点是什么？如何判断工件的合格性？

6-2 What categories can the plain limit gauge be divided into? What principles should be followed in the gauge design? What is the meaning of the principle?

普通极限量规可以分为哪些类别？量规设计应遵循哪些原则？这些原则的意义是什么？

6-3 What is the relationship between the size tolerance zone of the gauge and the dimension tolerance zone of the workpiece? Draw its tolerance zone diagram.

量规的尺寸公差带与工件的尺寸公差带之间有什么关系？画出其公差带图。

6-4 The dimensions of the following,try to determine the limits of acceptance and to select the measurement tools following the shaft or hole:

（1）ϕ40 g 6 （2）ϕ50H8 （3）ϕ125T9

测量如下工件，试确定验收极限并选择合适的计量器具。

6-5 What are the structural forms of the gauge? How should it be considered in its practical application?

量规的结构形式有哪些？在实际应用中应如何考虑？

Chapter 7 Tolerance and Fit of Rolling Bearings 滚动轴承的公差与配合

7.1 Overview 概述

A bearing that performs rolling motion between parts supporting a load and moving relative to each other is called a rolling bearing. Rolling bearings are widely used as standard components on machines, which can reduce the friction of moving pairs and improve efficiency. Rolling bearings have many structural changes due to their different uses and working conditions, but their basic structure consists of inner rings, outer rings, rolling elements (steel balls or rollers), and cages (also known as retainers or separation rings), as shown in Fig. 7-1. In addition, various types of bearings with different structures are matched with parts such as dust covers, sealing rings, stop washers and locking sleeves. The basic structural functions of rolling bearings are shown in Tab. 7-1.

支撑载荷和彼此相对运动的零件间做滚动运动的轴承称为滚动轴承。滚动轴承是机器上广泛应用的标准部件，可以减小运动副的摩擦，提高效率。滚动轴承由于用途和工作条件不同，其结构变化很多。基本结构由内圈、外圈、滚动体（钢球或滚子）和保持架（又称保持器或隔离圈）组成，如图7-1所示。除此之外，一些结构不同的轴承还配有防尘盖、密封圈、止动垫圈及紧定套等零件。滚动轴承的基本结构作用见表7-1。

The inner ring of a rolling bearing is assembled with the journal, while the outer ring is assembled with the housing bore. The rolling element is a component that carries and forms rolling friction in the bearing. Its size, shape and quantity are determined by factors such as bearing capacity and load direction. The cage is a set of isolation elements that evenly separate a group of rolling elements inside the bearing, allowing each rolling element to take turns bearing equal loads, and maintaining normal rolling of the rolling element between the inner and outer raceways of the bearing.

滚动轴承的内圈与轴颈装配，外圈与外壳孔装配。滚动体是轴承中承载并使轴承形成滚动摩擦的元件。它们的尺寸大小、形状和数量由承载能力和载荷方向等因素决定。保持架是一组隔离元件，其作用是将轴承内一组滚动体均匀分开，使每个滚动体均匀地轮流承受相等的载荷，并保持滚动体在轴承内、外滚道间正常滚动。

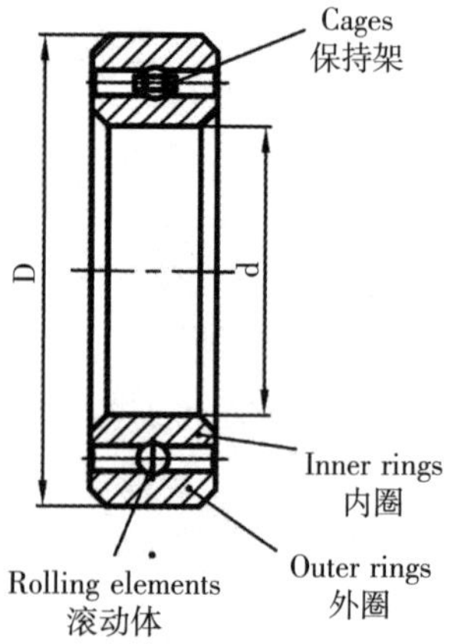

Fig. 7-1 Inner ring
图7-1 滚动轴承

Rolling bearings are a standard parts with two types of interchangeability. The fit between the rolling bearing inner ring the journal, as well as the fit between the outer ring and the hole seat, belongs to external interchangeability (complete interchangeability); the fit between the rolling element and the inner uses group selection assembly for internal interchangeability based on manufacturing economy.

滚动轴承是具有两种互换性的标准零件。滚动轴承内圈与轴颈的配合，以及外圈与孔座的配合，属于部件的外互换，采用完全互换；而滚动体与轴承内外圈的配合，从制造经济性角度出发，采用分组选择装配，为内互换。

Rolling bearings are divided into radial bearings that mainly bear radial loads, radial thrust bearings that simultaneously bear radial and axial loads, and thrust bearings that only bear axial loads according to the direction in which they bear the load. According to their rolling body shape, they are divided into ball bearings and roller bearings. The working performance of rolling bearings depends on the manufacturing accuracy of the rolling bearings itself, the fitting properties of the rolling bearings with the shaft and housing holes, as well as the dimensional accuracy, geometric tolerances and surface roughness of the shaft and housing holes. In the design process, reasonable selection should be made based on the above factors.

滚动轴承按其承受负荷的方向，分为主要承受径向负荷的向心轴承、同时承受径向和轴向负荷的向心推力轴承和仅承受轴向负荷的推力轴承。按其滚动体形状，分为球轴承和滚子轴承。滚动轴承的工作性能取决于滚动轴承本身的制造精度、滚动轴承与轴和轴承座孔的配合性质，以及轴和孔的尺寸精度、几何公差和表面粗糙度等因素。设计时，应根据以上因素合理选用。

7.2 Precision Level and Application of Rolling Bearings 滚动轴承精度等级及应用

7.2.1 Accuracy Level of Rolling Bearings 滚动轴承的精度等级

Rolling bearings are classified by dimensional accuracy and rotational accuracy. The dimensional accuracy of bearings refers to the manufacturing accuracy of the inner diameter (d), outer diameter (D), inner width (B), outer width (C), and assembly width (T) of the bearing. The deviation or variation of these dimensions represents the level of accuracy, as shown in Fig. 7-2.

滚动轴承按照尺寸精度和旋转精度分级。轴承的尺寸精度是指轴承内圈内径（d）、外圈外径（D）、内圈宽度（B）、外圈宽度（C）和装配宽度（T）的制造精度。这些尺寸的偏差或变动量表示精度的水平，如图7-2所示。

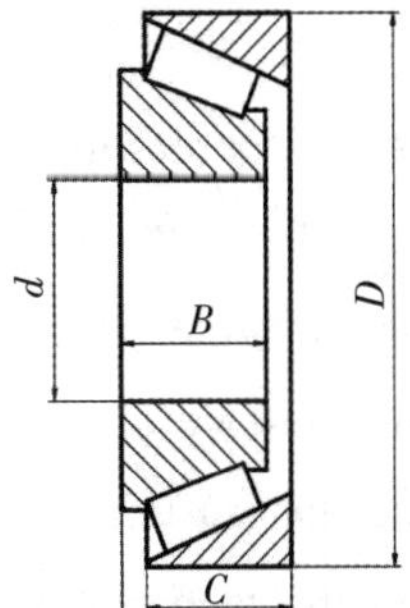

Fig. 7-2 Basic size of rolling bearings

图7-2 滚动轴承的基本尺寸

Excessive interference fit between the shaft neck and housing bore can cause significant contact stress between the bearing rolling element and the raceway of the ring, and increase the friction and heat generation of the bearing, there by reducing its lifespan.

轴颈与外壳孔之间的过盈配合会使轴承滚动体与滚道之间产生较大的接触应力，增加轴承摩擦和发热，以致降低轴承寿命。

7.2.2 Application of Rolling Bearings 滚动轴承的应用

Level 0 bearings belong to ordinary bearings, which are most widely used in the mechanical manufacturing industry. They are often used in general mechanisms with medium load, medium speed, and low rotational accuracy requirements, such as bearings used in variable speed mechanisms in ordinary machine tools, ordinary motors, water pumps, compressors, internal combustion engines, and other rotating mechanisms.

0级轴承属于普通轴承，在机械制造业中应用最广。其经常用在中等负荷、中等转速、旋转精度要求不高的普通机构中，如普通机床中的变速机构、普通电动机、水泵、压缩机、内燃机等旋转机构中所用的轴承。

Level 6, 6X, and 5 bearings are used in mechanisms with high rotational accuracy and speed, such as the spindle bearings of ordinary machine tools （usually the rear spindle bearings） and bearings used in precision machine tool transmission shafts.

6级、6X级及5级轴承用于旋转精度和转速较高的机构中，如普通机床的主轴轴承（一般为主轴后轴承）、精密机床传动轴所使用的轴承。

Level 4 bearings are used in rotating mechanisms with high rotational accuracy and speed, such as spindle bearings in precision machine tools, precision instruments, and mechanical rotating mechanisms.

4级轴承用于旋转精度高和转速高的旋转机械中，如精密机床、精密仪器和机械的旋转机构中的主轴轴承等。

Level 2 bearings are used in high-precision and high-speed special precision parts, such as the spindle bearings of precision coordinate boring machines and the main supports of high-precision gear grinders.

2级轴承用于高精度、高转速的特殊精密部件上，如精密坐标镗床的主轴轴承、高精度齿轮磨床的主要轴承。

7.3 Inner and Outer Diameter Tolerance Zones of Rolling Bearings 滚动轴承内、外径公差带

Rolling bearings are standard components: the outer ring-housing bore fit uses a base shaft system, and the inner ring journal adopts a base hole system. In most cases, the inner ring of the bearing rotates together with the shaft. In order to prevent wear caused by relative sliding of the mating surface between the inner ring and the shaft neck, a certain degree of interference is required for the fit. However, since the inner ring is a thin-walled part, the interference should not be too large. The outer ring of the bearing is installed in the housing hole and usually does not rotate. When the temperature rises during operation, it will cause the shaft to expand. One end of the bearings at both ends should be a movable support. Therefore, the fit between the outer diameter of the bearing and the housing hole can be slightly loosened to compensate for the thermal expansion and elongation of the shaft. The inner and outer rings of bearings are thin-walled parts that are prone to deformation during manufacturing and free state, and corrected after assembly. Fig. 7-3 is the diagram of inner and outer diameter tolerance zones for bearings. According to the national standard for rolling bearings, the tolerance zone of the average inner diameter of a single plane of each level of bearing is distributed below the zero line, that is, the upper limit deviation is zero, the lower limit deviation is negative, and it is consistent with the tolerance zone of the reference hole with the basic deviation code H.

滚动轴承是标准件，其外圈与壳体孔的配合采用基轴制，内圈与轴颈的配合采用基孔制。多数情况下，轴承内圈与轴一起旋转。为了防止内圈和轴颈的配合面发生相对滑动而产生磨损，要求配合具有一定的过盈量。但由于内圈是薄壁零件，过盈量不能太大。轴承外圈安装在壳体孔中，通常不旋转。当工作时温度升高，会使轴膨胀。两端轴承中有一端是游动支承。因此，轴承外径与壳体孔的配合稍微松一些，使之能补偿轴的热胀伸长。轴承的内、外圈都是薄壁零件，在制造和自由状态下都易变形，在装配后可以得到校正。轴承的内、外径公差带图如图7-3所示。关于滚动轴承的国家标准规定，各级轴承的单一平面平均内径的公差带都分布在零线下侧，即上极限偏差为零，下极限偏差为负值，与基本偏差代号为H的基准孔公差带一致。

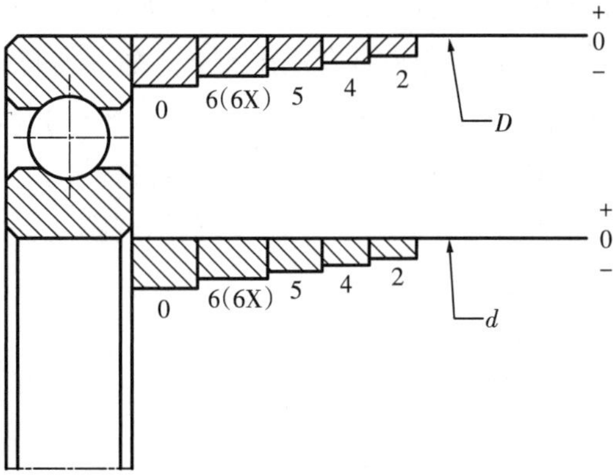

Fig. 7-3 Size tolerance zones for inner and outer diameters of rolling bearings

图7-3 滚动轴承内、外径公差带

Based on the above characteristics of bearings, the national standard for rolling bearing tolerances has made two provisions for the inner and outer diameter dimensions of bearings: First, it specifies the allowable deviation of the maximum and minimum values of the inner and outer diameter dimensions, i.e., the deviation of a single inner and outer diameter. Its main purpose is to limit the amount of deformation. The second is to specify the average deviation between the maximum and minimum actual dimensions of the inner and outer diameters, i.e., the average inner and outer diameter deviation of a single plane. The purpose is for bearing fitting. All qualified rolling bearings should meet the requirements of both specified tolerances simultaneously.

根据轴承的以上特点，关于滚动轴承公差的国家标准对轴承内径和外径尺寸做了两种规定：一是规定了内、外径尺寸的最大值和最小值所允许的偏差，即单一内、外径偏差。其主要目的是限制变形量。二是规定了内、外径实际尺寸的最大值和最小值的平均偏差，即单一平面平均内、外径偏差。其目的是满足轴承的配合需求。所有合格的滚动轴承，应同时满足所规定的这两种公差的要求。

7.4 Tolerance Zones for Journals and Casings 轴颈和外壳的公差带

The national standard GB/T 307.1−2017 *Tolerances for Radial Bearings Rolling Bearings* specifies the dimensional tolerance zones, geometric tolerances, and basic principles and requirements for fitting selection of the shaft and housing holes that fit with the inner and outer rings of the bearing. Some commonly used tolerance zones are shown in Fig. 7−4 and Fig. 7−5, which are selected from the tolerance zones for shafts and holes in GB/T 1800.1−2020, respectively. Due to the fact that rolling bearings are standard parts, the fit between the inner ring of the bearing and the journal belongs to the base hole system, while the fit between the outer ring of the bearing and the housing hole belongs to the base shaft system. The tolerance zones for the journal and housing holes are selected from the national standards for smooth cylinders, and they are combined with the inner and outer rings of the bearing to obtain various fits with different degrees of tightness. It should be pointed out that the fit between the inner ring of the bearing and the journal belongs to the base hole system, and the bearing tolerance zone adopts a unidirectional distribution with zero upper limit deviation and negative lower limit deviation. Therefore, the fit between the inner ring of the bearing and the journal corresponds to a smoother cylinder formed by the base hole system, which is tighter.

国家标准GB/T 307.1—2017《滚动轴承、向心轴承产品几何技术规范（GPS）和公差值》规定了与轴承内、外圈相配合的轴和壳体孔的尺寸公差带、几何公差，以及配合选择的基本原则和要求。其中一些常用公差带如图7-4和图7-5所示，它们分别选自GB/T 1800.1—2020中的轴、孔公差带。滚动轴承是标准零件，所以轴承内圈与轴颈的配合为基孔制的配合，轴承外圈与壳体孔的配合为基轴制的配合。轴颈和壳体孔的公差带均在光滑圆柱体的国家标准中选择，它们分别与轴承内、外圈结合，可以得到松紧程度不同的各种配合。需要指出的是，轴承内圈与轴颈的配合属基孔制，且轴承公差带均采用上极限偏差为零、下极限偏差为负的单向分布。因此，轴承内圈与轴颈得到的配合比相应光滑圆柱体按基孔制所形成的配合紧一些。

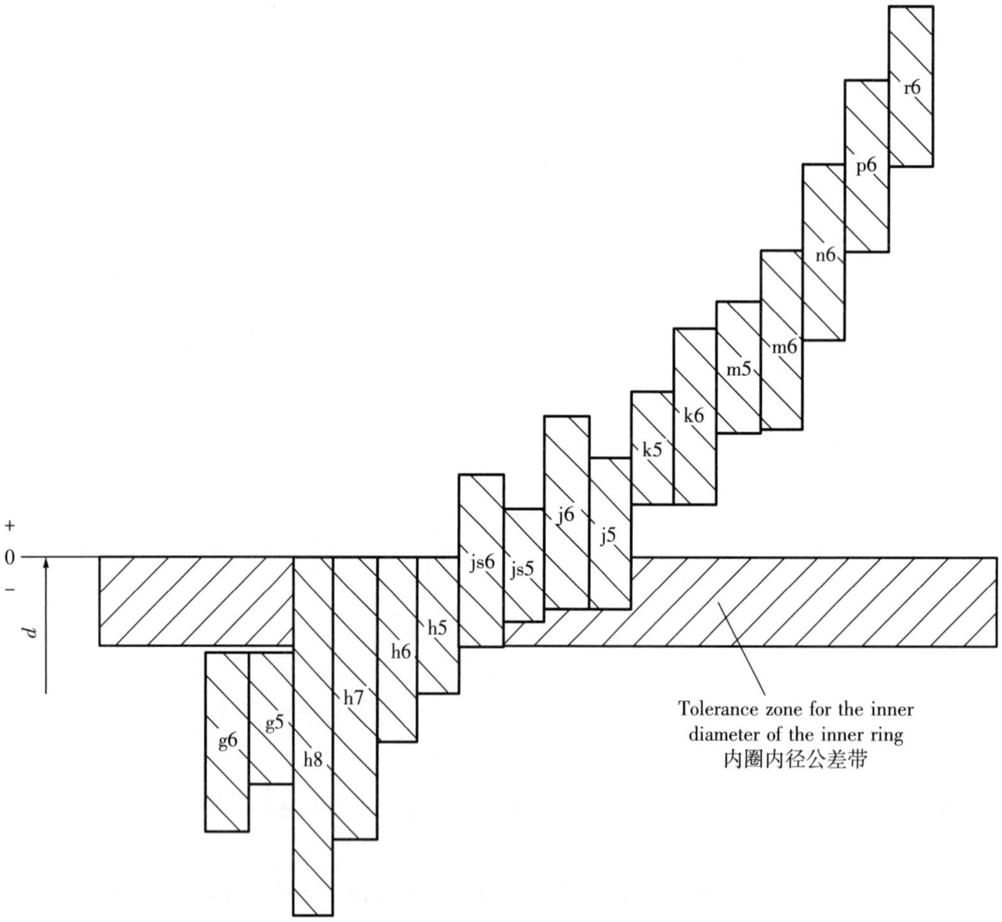

Fig. 7−4　Size tolerance zones of shaft fitting with rolling bearings

图7−4　与滚动轴承配合的轴的公差带

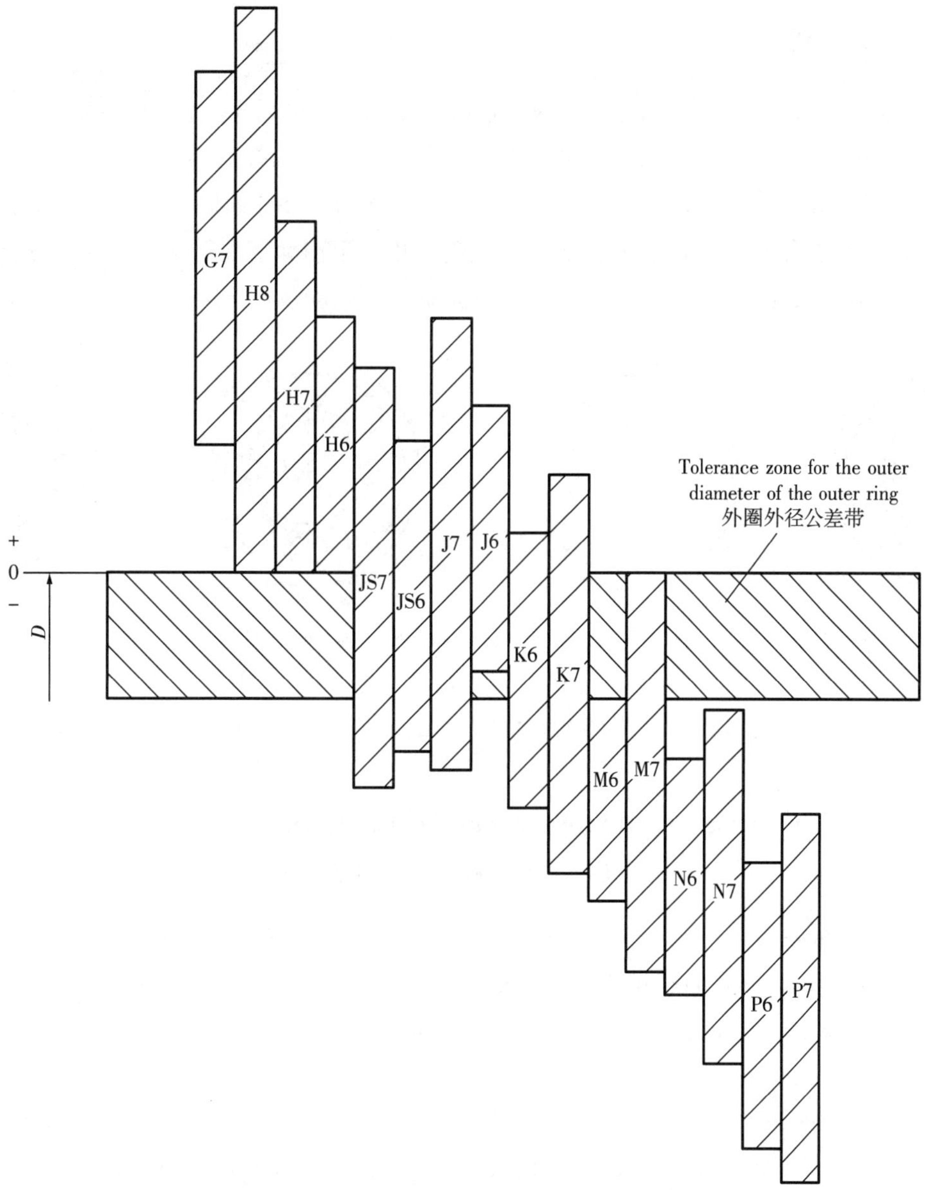

Fig. 7–5 Size tolerance zones of hole fitting with rolling bearings

图7–5 与滚动轴承配合的外壳孔的公差带

7.5 Selection of Rolling Bearing Fit 滚动轴承配合的选择

When selecting a bearing fit, the following factors should be comprehensively considered: the working conditions of the bearing, the magnitude, direction, and nature of the load acting on the bearing, the working temperature, bearing type and size, rotational accuracy and speed, and a series of other factors. The main factors are analyzed as follows.

选择轴承配合时，应综合考虑以下方面的因素：轴承的工作条件，作用在轴承上的负荷的大小、方向和性质，工作温度，轴承类型和尺寸，旋转精度和速度等一系列其他因素。现对主要因素的分析如下。

7.5.1 Load Type 负荷类型

When the bearing rotates, according to the combined radial load acting on the bearing relative to the rotation of the bearing ring, the load can be classified into three types: directional load (also known as local load), rotating load (also known as cyclic load), and swing load (oscillating load), as shown in Fig. 7-6.

轴承转动时，根据作用于轴承上的合成径向载负荷相对于轴承套圈的旋转情况，可将所受负荷分为定向负荷（也称局部负荷）、旋转负荷（也称循环负荷）和摆动负荷三类，如图7-6所示。

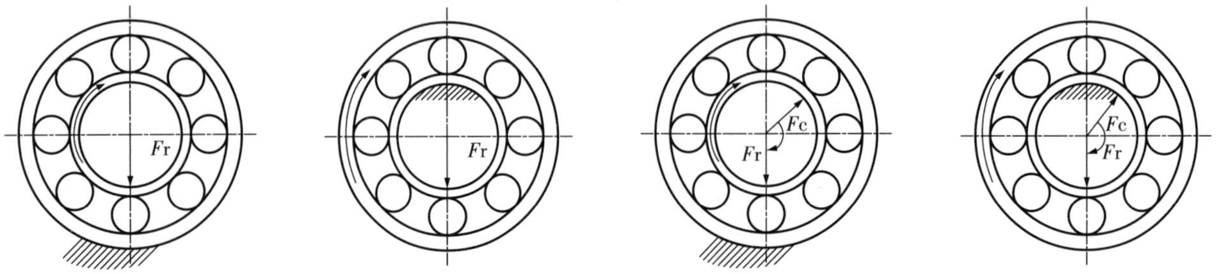

(a) Rotating innerring load and directional outer ring load
旋转的内圈负荷和固定的外圈负荷

(b) Directional innerring load and rotating outerring load
固定的内圈负荷和旋转的外圈负荷

(c) Rotating innerring load and swing outer ring load
旋转的内圈负荷和外圈承受摆动负荷

(d) Swing innerring load and rotating outerring load
内圈承受摆动负荷和旋转的外圈负荷

Fig. 7-6 The types of loads sustained by the bearing

图7-6 轴承承受的负荷类型

(1) Directional load: the composite radial load acting on the bearing is relatively stationary relative to the bearing ring, i.e., the load direction always remains constant and is concentrated on the local area of the ring raceway. The load borne by the ring is called directional load, as shown in Fig. 7-6 (a) for the outer ring and (b) fot the inner ring. In Fig. 7-6, the fit between the ring and the housing hole or shaft that bears this type of load is generally selected as a loose transition fit or smaller clearance fit, in order to allow the friction torque between the ring and the raceway to drive the rotation and extend bearing service life.

定向负荷：作用于轴承上的合成径向负荷与轴承套圈相对静止，即负荷方向始终不变地作用在套圈滚道的局部区域上。该套圈所承受的这种负荷称为定向负荷，如图7-6（a）中的轴承外圈和图7-6（b）中的轴承内圈所示。图7-6中承受这类负荷的套圈与壳体孔或轴的配合，一般选较松的过渡配合或较小的间隙配合，以便让套圈与滚道间的摩擦力矩能驱动旋转，从而延长轴承的使用寿命。

(2) Rotating load: the composite radial load acting on the bearing rotates relative to the ring, i.e., the composite radial load sequentially acts on the entire circumference of the ring raceway. The nature of the load borne by the ring is called rotating load, as shown in Fig. 7-6 (a) and (c) for inner ring and Fig. 7-6 (b) and (d) for outer ring. The fit between the ring and the shaft (or housing hole) that usually bears rotating loads should be an interference fit or a tighter transition fit, and the magnitude of the interference should be based on the principle of not a causing a crawling phenomenon between the ring and the surface of the shaft or housing hole.

旋转负荷：作用于轴承上的合成径向负荷与套圈相对旋转，即合成径向负荷顺次作用在套圈滚道的整个圆周上。该套圈所承受的这种负荷性质称为旋转负荷，如图7-6（a）、（c）所示轴承内圈和图7-6（b）、（d）所示轴承外圈。通常承受旋转负荷的套圈与轴（或壳体孔）相配，应选过盈配合或较紧的过渡配合，其过盈量的大小以不使套圈与轴或壳体孔配合表面间产生爬行现象为原则。

（3）Oscillating load: the composite radial load acting on the bearing and the bearing ring swing relative to each other within a certain area, i.e., its load vector frequently changes and acts on the local circumference of the ring raceway. The nature of the load borne by the ring is called oscillating load, as shown in Fig. 7-6（c）outer ring and Fig. 7-6（d）inner ring. The fitting requirements for the ring that bears the oscillating load are the same or slightly loose as those for the cyclic load.

摆动负荷：作用于轴承上的合成径向负荷与所承受的套圈在一定区域内相对摆动，即其负荷向量经常变动地作用在套圈滚道的局部圆周上。该套圈所承受的负荷性质称为摆动负荷，如图7-6（c）所示轴承外圈和图7-6（d）所示轴承内圈。承受摆动负荷的套圈，其配合要求与循环负荷相同或略松一些。

7.5.2 Load Size 负荷大小

The minimum interference fit between the rolling bearing ring and the shaft or housing hole depends on the magnitude of the load. Generally, the ratio of the radial equivalent dynamic load Pr to the radial rated dynamic load Cr of the bearing: $Pr/Cr \leqslant 0.07$ is called light load, $0.07 < Pr/Cr < 0.12$ is called normal load, those with $Pr/Cr > 0.12$ are called heavy loads.

滚动轴承套圈与轴或壳体孔配合的最小过盈量取决于负荷的大小。一般把轴承的径向当量动负荷Pr与轴承的径向额定动负荷Cr的比值：$Pr/Cr \leqslant 0.07$称为轻负荷，$0.07 < Pr/Cr < 0.12$称为正常负荷，$Pr/Cr > 0.12$称为重负荷。

When bearing heavy loads or impact loads, the bearing will undergo significant deformation, resulting in a decrease in actual interference between the mating surfaces and an increase in actual clearance inside the bearing. In order to ensure normal operation of the bearing, a larger interference fit should be selected. Similarly, for lighter loads, a smaller interference fit can be chosen.

承受较重的负荷或冲击负荷时，轴承会发生较大的变形，使结合面间实际过盈减小和轴承内部的实际间隙增大。为了使轴承运转正常，应选较大的过盈配合。同理，承受较轻的负荷，可选用较小的过盈配合。

When subjected to cyclic loads, the raceway will not experience localized wear. At this time, it is necessary to prevent the sleeve from causing wear and heat on the mating surface relative to the shaft neck or housing hole. The fit between the collar and the shaft neck should be tight, and a transition fit or interference fit is generally selected.

承受循环负荷时，滚道不会出现局部磨损。此时，需要防止套圈相对于轴颈或壳体孔的配合表面产生磨损和发热。套圈与轴颈的配合应紧密，一般选择过渡配合或过盈配合。

When subjected to directional loads, the fit should be relatively loose and there may be a small gap, so that under the frictional force of the rolling element, there is a possibility of occasional periodic movement of the ring relative to the shaft neck or shell hole, thereby eliminating local raceway wear and facilitating assembly and disassembly. A transition fit or clearance fit can generally be selected.

受定向载荷时，配合应较松，可以有小的间隙，以便在滚动体摩擦力作用下，使套圈相对于轴颈或外壳孔有偶尔周期游动的可能，从而消除滚道局部磨损，并便于装配和拆卸。一般可选用过渡配合或间隙配合。

The selection of matching with rolling bearings is generally analogical; Tab. 7-1 to Tab. 7-4 can be used as a reference.

与滚动轴承配合的选择常用类比法，表7-1～表7-4可作参考。

Tab. 7-1　Tolerance zone code for the fit between radial bearings and shafts

表7-1　向心轴承和轴的配合轴公差带代号

Operational status 运转状态		Load status 负荷状态	Deep groove ball bearings, self-aligning ball bearings, and angular contact ball bearings 深沟球轴承、调心球轴承和角接触球轴承	Cylindrical roller bearings and conical roller bearings 圆柱滚子轴承和圆锥滚子轴承	Self-aligning roller bearings 调心滚子轴承	Tolerance zone 公差带
Description 说明	Example 举例		Nominal inside diameter of bearing 轴承公称内径/mm			
Rotating inner ring load and oscillating load 旋转的内圈负荷及摆动负荷	General machinery, electric motors, machine tool spindles, pumps, internal combustion engines, spur gear transmissions, crushers, etc. 一般通用机械，电动机、机床主轴、泵、内燃机、正齿轮传动装置、破碎机等	Light load 轻负荷	≤18	—	—	h5
			>18～100	≤40	≤40	j6①
			>100～200	>40～140	>40～140	k6①
			—	>140～200	140～200	m6①
		Normal load 正常负荷	≤18	—	—	j5，js5
			>18～100	≤40	≤40	k5②
			>100～140	>40～100	>40～65	m5②
			>140～200	>100～140	>65～100	m6
			>200～280	>140～200	100～140	n6
			—	>200～400	140～280	p6
			—	—	280～500	r6
	General machinery, electric motors, machine tool spindles, pumps, internal combustion engines, spur gear transmissions, crushers, etc. 一般通用机械、电动机、机床主轴、泵、内燃机、正齿轮传动装置、破碎机等	Heavy load 重负荷	—	>50～140	>50～100	n6③
				>140～200	>100～140	p6③
				>200	>140～200	r6③
				—	>200	r7③
Fixed inner ring load 固定的内圈负荷	Various wheels 静止轴上的各种轮子	All loads 所有负荷	All sizes 所有尺寸			f6，g6
	Tensioning wheels, rope wheels on stationary axes, vibrating screens, inertial vibrators 张紧轮、绳轮，振动筛，惯性振动器					h6，j6
Only axial load 仅有轴向负荷		All sizes 所有尺寸				j6，js6

Notes：① In situations where high precision is required, j5, k5, and m5 should be selected to replace j6, k6, andm6 respectively.

② The fit of conical roller bearings and angular contact ball bearings has little impact on clearance, so k6 and m6 can be selected to replace k5 and m5 respectively.

③ For bearings under heavy loads, a clearance greater than Group N should be selected.

注：① 对精度有较高要求的场合，应该选用j5，k5，m5 以分别代替j6，k6，m6。

② 圆锥滚子轴承、角接触球轴承配合对游隙的影响不大，可以选用k6，m6 分别代替k5，m5。

③ 重负荷下轴承游隙应选用大于N组的游隙。

Tab. 7–2 Tolerance zone code for the fit between radial bearings and housing holes

表 7–2 向心轴承和外壳孔的配合孔公差带代号

Operational status 运转状态		Load status 负荷状态	Other conditions 其他状况	Tolerance zone 公差带	
Description 说明	Example 举例			Ball bearings 球轴承	Roller bearings 滚子轴承
Fixed outer ring load 固定的外圈负荷	Generalmachinery, electric motors, pumps, crankshaft main bearings 一般机械、电动机、泵、曲轴主轴承	Light, normal, heavy load 轻、正常、重	Easy axial movement, split-type housing can be used 轴向易移动，可采用剖分式外壳	H7、G7	
		Impact 冲击	Axial movement is possible, integral or split-type housing can be used 轴向能移动，可采用整体或剖分式外壳	J7、JS7	
Oscillating load 摆动负荷		Light, normal load 轻、正常			
		Normal, heavy load 正常、重	No axial movement, integral housing should be used 轴向不移动，采用整体式外壳	K7	
		Impact 冲击		M7	
Rotating outer ring load 旋转的外圈负荷	Tension pulleys, hub bearings 张紧滑轮、轮毂轴承	Light load 轻		J7	K7
		Normal load 正常		K7、M7	M7、N7
		Heavy load 重			N7、P7

Notes: ① The parallel tolerance zones are selected from left to right as the size increases. When a higher degree of rotational accuracy is necessary, the tolerance grade can be elevated by one level accordingly.

② Not applicable to split-type housings.

注：① 并列公差带随尺寸的增大从左至右选择，对旋转精度有较高要求时，可相应提高一个公差等级。

② 不适用于剖分式外壳。

Tab. 7–3 Tolerance zone code for the fit between thrust bearings and shafts

表 7–3 推力轴承和轴的配合轴公差带代号

Operational status 运转状态	Load status 负荷状态	Thrust ball and thrust roller bearings 推力球和推力滚子轴承	Thrust self-aligning roller bearings 推力调心滚子轴承	Tolerance zone 公差带
		Nominal inside diameter of bearing 轴承公称内径/mm		
Axial load only 仅有轴向负荷		All size 所有尺寸		j6、js6
Fixed race load 固定的座圈负荷	Combined radial and axial load 径向和轴向联合负荷	— —	≤250 ＞250	j6 js6
Rotating shaft load or oscillating load 旋转的轴圈负荷或摆动负荷		— — —	≤200 ＞200~400 ＞400	k6 m6 n6

Notes: ① When a smaller interference fit is required, j6, k6, and m6 can be used to replace k6, m6, and n6 respectively.

② This also includes thrust tapered roller bearings and thrust angular contact ball bearings.

注：① 当要求较小过盈时，可分别用j6，k6，m6代替k6，m6，n6。

② 也包括推力圆锥滚子轴承、推力角接触球轴承。

Tab. 7-4 Tolerance zone code for the fit between thrust bearings and housing holes

表7-4 推力轴承和外壳孔的配合孔公差带代号

Operational status 运转状态	Load status 负荷状态	Bearing type 轴承类型	Tolerance zone 公差带	Remarks 备注
Only axial load 仅有轴向负荷		Thrust ball bearing 推力球轴承	H8	
		Thrust cylindrical and conical roller bearing 推力圆柱、圆锥滚子轴承	H7	
		Thrust self-aligning roller bearing 推力调心滚子轴承		The clearance between the housing bore and the race is 0.001D （D is the nominal outer diameter of the bearing） 外壳孔与座圈间间隙为0.001D（D为轴承公称外径）
Fixed race load 固定的座圈负荷	Combined radial and axial load 径向和轴向联合负荷	Thrust angular contact ball bearing, thrust self - aligning roller bearing, thrust conical roller bearing 推力角接触球轴承、推力调心滚子轴承、推力圆锥滚 子轴承	H7	
Rotating race load旋转的座圈负荷			K7	Normal operating conditions 普通使用条件
			M7	Subjected to significant radial loads 有较大径向负荷时

7.5.3 Working Conditions of Bearings 轴承的工作条件

7.5.3.1 The influence of working temperature 工作温度的影响

During bearing operation, due to frictional heating and the influence of other heat sources, the temperature of the ring will be higher than that of the matching parts. The thermal expansion of the inner ring will cause it to loosen its fit with the shaft neck, while the thermal expansion of the outer ring will cause it to tighten its fit with the outer shell hole. Therefore, the working temperature of bearings should generally be below 100℃. For bearings operating above this temperature, the selected fit should be appropriately adjusted.

轴承工作时，由于摩擦生热和其他热源的影响，套圈的温度会高于配合零件的温度。内圈的热膨胀会使其与轴颈配合松动，而外圈的热膨胀则会使其与外壳孔的配合变紧。因此，轴承工作温度一般应低于100 ℃，在高于此温度中工作的轴承，应适当调整所选的配合。

7.5.3.2 The influence of bearing rotation accuracy and speed 轴承的旋转精度和速度的影响

For bearings with high loads and high rotational accuracy requirements, clearance fit should be avoided to eliminate the effects of elastic deformation and vibration. For light weight bearings of precision machine tools, small clearance fits are often used to avoid the influence of shape errors between holes and shafts on bearing accuracy. For bearings with high rotational speeds and working under impact vibration loads, it is best to choose an interference fit with the shaft neck and housing hole, and the higher the rotational speed, the tighter the fit should be.

对于承受高负荷且对旋转精度要求较高的轴承，为了消除弹性变形和振动的影响，应避免采用间隙配合。对于精密机床的轻负荷轴承，为避免孔与轴的形状误差对轴承精度的影响，常采用较小的间隙配合。对于旋转速度较高，又在冲击振动负荷下工作的轴承，它与轴颈和壳体孔的配合最好选择选用过盈配合，且转速越高，配合应越紧。

7.5.4 Structure and Materials of Shaft and Bearing Seat Holes 轴和轴承座孔的结构与材料

For the convenience of installation and disassembly, split bearing seat holes can be selected. If the fit between the split bearing seat holes and the outer ring is tight, it will cause elliptical deformation of the outer ring. Therefore, it is advisable to use a looser fit. When bearings are installed on thin-walled bearing holes, light alloy bearing holes, or thin-walled hollow shafts, in order to ensure sufficient support stiffness and strength for bearing operation, the fit used should be tighter than that installed on thick walled bearing holes, cast iron bearing holes, or solid shafts.

为了安装和拆卸方便，可以选用剖分式轴承座孔。如果剖分式轴承座孔与外圈采用的配合较紧，会使外圈产生椭圆变形。因此，宜采用较松配合。当轴承安装在薄壁轴承座孔，轻合金轴承座孔或薄壁的空心轴上时，为了保证轴承工作有足够的支撑刚度和强度，所采用的配合应比装在厚壁轴承座孔、铸铁轴承座孔或实心轴上的要紧些。

7.5.5 Conditions for Installing and Disassembling Bearings 安装和拆卸轴承的条件

Considering the convenience of bearing installation and disassembly, it is advisable to use a looser fit, which is particularly important for large and extra large bearings used in heavy-duty machinery. If convenient loading and unloading are required, but tight fitting is required, separable bearings or bearings with tapered holes in the inner ring, locking sleeves, and release sleeves can be used.

考虑到轴承安装与拆卸方便，宜采用较松的配合，这对重型机械中使用的大型和特大型轴承尤为重要。若要求装卸方便，而又需紧配合，可采用分离型轴承或内圈带锥孔及锁紧套和退卸套的轴承。

In addition to the above conditions, it should also be considered that when the inner or outer ring of the bearing is required to move axially, a looser fit should be selected between the inner ring and the shaft or the outer ring and the hole. The larger the size of the rolling bearing, the tighter the selected fit should be. If the working temperature of rolling bearings is higher than 100 ℃, and appropriate adjustments should be made to the selected fit.

除上述条件外，还应考虑到，当要求轴承的内圈或外圈能沿轴向移动时，应在内圈与轴或外圈与孔之间选择较松的配合。滚动轴承的尺寸越大，选取的配合应越紧。滚动轴承的工作温度高于100 ℃，应对所选的配合进行适当调整。

7.5.6 Geometric Tolerance Requirements for Shafts and Holes Fitted with Rolling Bearings 滚动轴承的轴和孔配合的几何公差要求

The geometric tolerance requirements for shafts and housings are the cylindricity andrun-out of the shaft and housing hole. The requirements and the values are shown in Tab. 7-5.

轴和轴承座的几何公差要求主要包括轴和轴承座孔的圆柱度和跳动。具体的要求和数值见表7-5。

Tab. 7-5 Geometric tolerance values of shaft and housing hole (From GB/T 275-2015)

表7-5 轴和轴承座孔的几何公差（摘自GB/T 275—2015）

公称尺寸/mm		圆柱度t/μm				轴向圆跳动t_1/μm			
		轴颈		轴承座孔		轴肩		轴承座孔肩	
		轴承公差等级							
>	≤	普通级	6(6X)	普通级	6(6X)	普通级	6(6X)	普通级	6(6X)
—	6	2.5	1.5	4	2.5	5	3	8	5
6	10	2.5	1.5	4	2.5	6	4	10	6
10	18	3	2	5	3	8	5	12	8
18	30	4	2.5	6	4	10	6	15	10
30	50	4	2.5	7	4	12	8	20	12
50	80	5	3	8	5	15	10	25	15
80	120	6	4	10	6	15	10	25	15
12	180	8	5	12	8	20	12	30	20
18	250	10	7	14	10	10	12	30	20

7.5.7 Surface Roughness Requirements of Shafts and Holes Fitted with Rolling Bearings 滚动轴承的轴和孔配合的表面粗糙度要求

The surface roughness of the shafts and holes fitted with rolling bearings is shown in Tab. 7-6.

滚动轴承的轴和孔滚动轴承的轴和孔的表面粗糙度见表7-6所示。

Tab. 7-6 Surface roughness of the fitting surfaces and end faces (From GB/T 275-2015)

表7-6 配合表面及端面的表面粗糙度（摘自GB/T 275—2015）

轴或轴承座孔直径/mm		轴或轴承座孔配合表面直径公差等级					
		IT7		IT6		IT5	
		表面粗糙度Ra/μm					
>	≤	磨	车	磨	车	磨	车
—	80	1.6	3.2	0.8	1.6	0.4	0.8
80	500	1.6	3.2	1.6	3.2	0.9	1.6
500	1250	3.25	6.3	1.6	3.2	1.6	3.2
端面		3.2	6.3	6.3	6.3	6.3	3.2

Example 7-1: Given: deep grove ball bearing type 6206 ($d=30$ mm, $D=62$ mm, $B=16$ mm, Grade 0), $Pr=1250$ N, $Cr=25$ kN. The inner ring rotates and outer ring rests, to design the precision requirements of the shaft and housing fitted with the rolling bearings.

例7-1：深沟球轴承，型号6206（$d=30$ mm，$D=62$ mm，$B=16$ mm，0级），$Pr=1250$ N，$Cr=25$ kN，轴承内圈旋转、外圈固定。试确定轴、轴承座与滚动轴承配合的精度要求。

Solution: $Pr/Cr=1250/25000=0.05<0.07$

It belongs to light load.

According to Tab. 7-1 , the shaft tolerance zone symbol is ϕ30j6, according to Tab. 7-2 the housing bore tolerance zone symbol is ϕ62H7.

According to Tab. 7-5, the cylindricity value is 0.004 and 0.008 for shaft bearing and housing bore respectively.

According to Tab. 7-5, the end-face run-out is 0.010 and 0.025 for shaft bearing and housing bore respectively.

According to Tab. 7-6, the roughness is *Ra* 1.6 and *Ra* 6.3 for shaft bearing and shoulder,and is *Ra* 3.2 and *Ra* 6.3 for housing bore hole and shoulder.

The results are indicated in the engineer drawing as shown in Fig. 7 -7.

解：$Pr/Cr = 1250/25000 = 0.05 < 0.07$

属于轻负荷。

根据表7-1，轴的公差为ϕ30j6；根据表7-2，轴承座孔的公差为ϕ62H7。

根据表7-5，轴承配合的轴颈和轴承座孔的圆柱度值分别为0.004和0.008。

同样地，根据表7-5，轴承配合的轴和轴承座孔的端面跳动值分别为0.010和0.025。

根据表7-6，轴承配合的轴颈和轴肩的表面粗糙度分别为*Ra* 1.6和*Ra* 6.3，而轴承座孔和孔肩的表面粗糙度则为*Ra* 3.2和*Ra* 6.3。

将这些选择的各项公差标注在图样上，如图7-7所示。

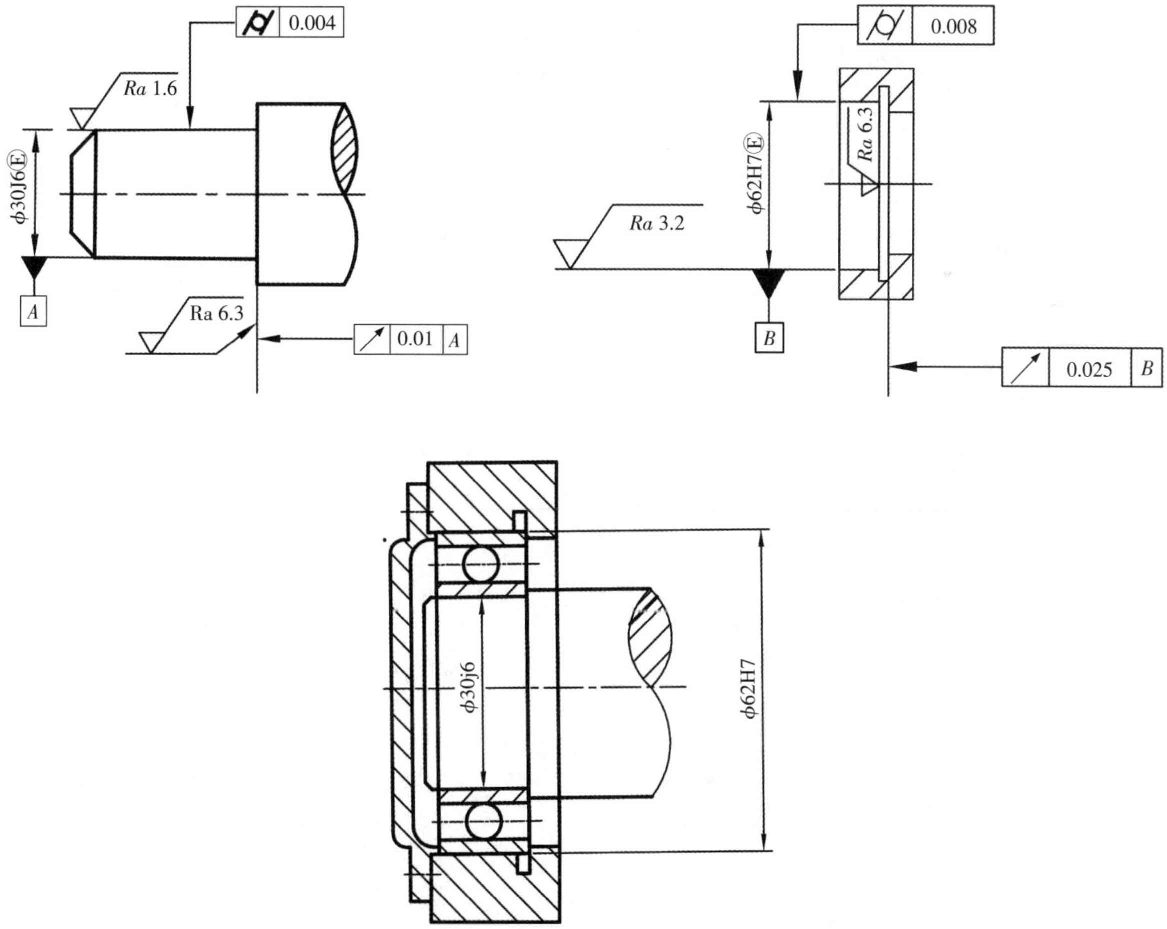

Fig. 7-7 Indication tolerance requirement in engineer drawings

图7-7 公差标标注图样

Exercise 7 习题7

7-1 Please mark symbols of tolerance and fitting in Fig. 7-8 drawing correctly.
请在图7-8上正确标注公差和配合符号。

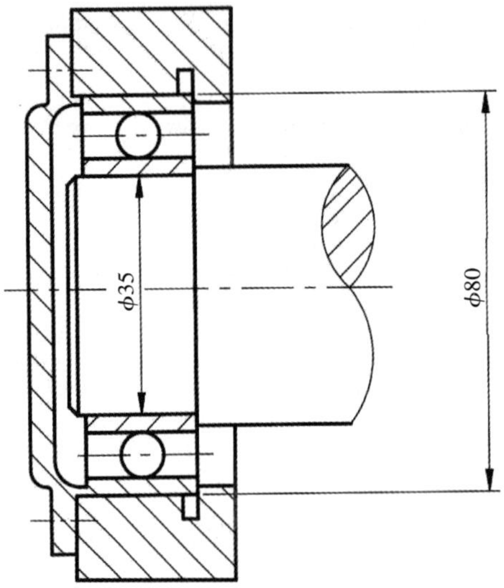

Fig. 7-8 Assignment figure
图7-8 作业题图

Chapter 8 Interchangeability of Thread Joint 螺纹结合的互换性

8.1 Overview 概述

8.1.1 Types of Threads and Their Usage Requirements 螺纹的种类及其使用要求

In practical production, the application of threads is very extensive, and they are the most typical connection structure with interchangeability. Although the production of most components has been standardized and regulated, it is necessary to understand their relevant tolerance regulations and testing methods when using them in order to effectively grasp their performance.

在实际生产中，螺纹的应用非常广泛，它是一种最具有互换性的典型连接结构。虽然现在大多数零件的生产已经标准化和规范化，但在使用时仍需了解其相关的公差规定和检测方法，以便有效地掌握其性能。

In order to meet the requirements for the use of threads and ensure their interchangeability, China has issued a series of national standards for ordinary threads, mainly including GB/T 14791- 2013 *Thread Vocabulary*, GB/T 192- 2003 *Basic Profile, General Threads*, GB/T 197- 2018 *General Purpose Metric Screw Threads - Tolerances*. Threads can be divided into three categories based on their bonding properties and usage requirements:

为满足螺纹的使用要求，保证其互换性，我国发布了一系列普通螺纹国家标准，主要有GB/T 14791—2013《螺纹 术语》、GB/T 192—2003《普通螺纹基本牙型》、GB/T0 197—2018《普通螺纹 公差》。螺纹按照结合性质和使用要求可分成三类：

(1) Tightening threads are mainly used for connecting and fastening components, which are also the most widely used type.

紧固螺纹主要用于连接和紧固零部件，这也是使用最广泛的一种。

(2) Transmission threads are mainly used to transmit displacement and power. For example, the lifting screw of a jack.

传动螺纹主要用于传递位移和动力。例如，千斤顶的起重螺杆。

(3) Sealing threads are mainly used under conditions that require airtightness or watertightness. For example, pipe threaded connections must not leak air, oil or water in the pipeline. The main requirement is to have good screw-in and sealing properties.

紧密螺纹主要用于对气密性或水密性有要求的条件下。例如，管螺纹连接必须确保管道中不得漏气、漏油或漏水。主要要求是有良好的旋合性和密封性。

8.1.2 Basic Tooth Shape and Geometric Parameters of Threads 螺纹的基本牙型和几何参数

The tooth profile of a cylindrical thread refers to the shape of the thread contour on the profile passing through the thread axis, formed by the original triangle, with the bottom edge of the triangle parallel to the thread axis. The basic tooth shape of a cylindrical thread refers to the shape formed by flattening the top and bottom of the original triangle at a specified height, as shown in Fig. 8-1.

圆柱螺纹的牙型是指在通过螺纹轴线的剖面上螺纹轮廓的形状，该形状由原始三角形形成，且该三角形的底边平行于螺纹轴线。圆柱螺纹的基本牙型是指将原始三角形的顶部和底部按规定高度削平后所形成的形状，如图8-1所示。

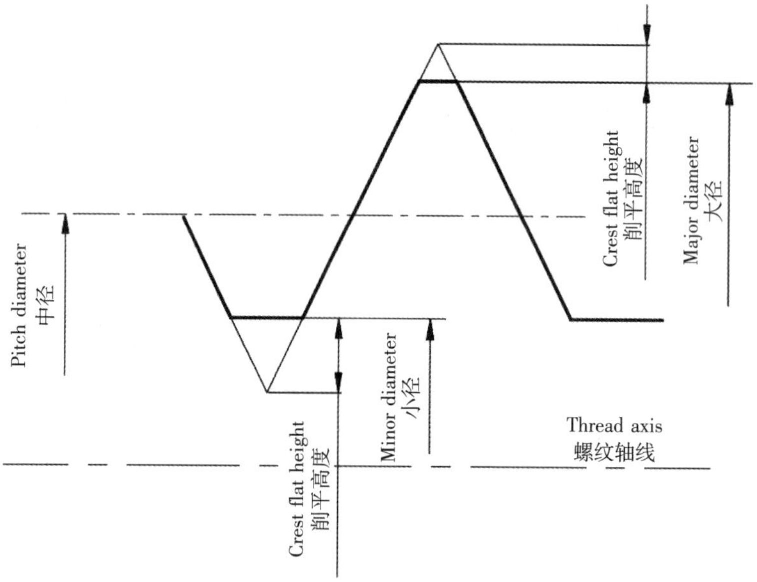

Fig. 8-1 The basic tooth shape of a cylindrical thread

图8-1 圆柱螺纹的基本牙型

8.2 Formation of Cylindrical Thread Profiles 圆柱螺纹牙型形成

The parameters of cylindrical threads refer to the values measured radially or axially along the cross-section passing through the thread axis. As shown in Fig. 8-2 and Fig. 8-3, the main parameters are as follows.

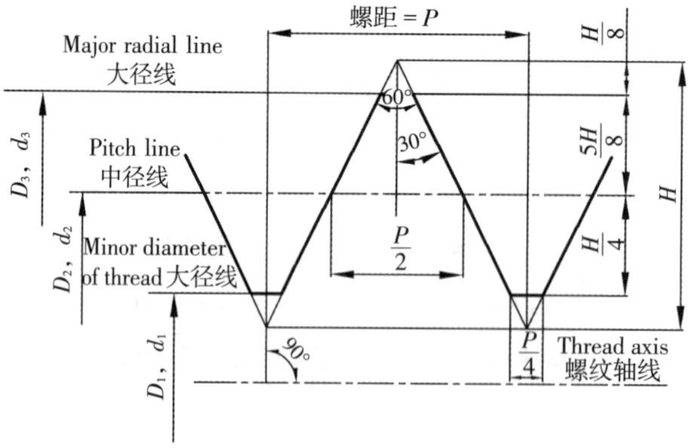

Fig. 8-2 Basic tooth profile of ordinary threads

图8-2 普通螺纹基本牙型

圆柱螺纹的几何参数是指在通过螺纹轴线的横截面上径向或轴向测量的值。参看图8-2和图8-3，主要参数如下。

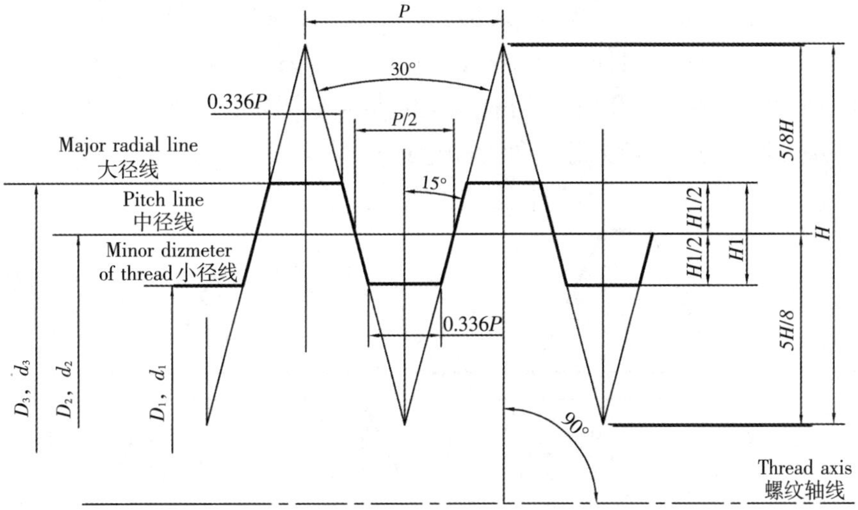

H—Initial angular height, 原始角形高度

Fig. 8-3 Basic tooth profile of trapezoidal thread

图8-3 梯形螺纹基本牙型

(1) Major diameter (*D* or *d*) refers to the diameter of an imaginary cylinder that is tangent to the top of an external thread or bottom of an internal thread. For external threads, the major diameter is the top diameter; For internal threads, the major diameter is the base diameter. The nominal diameter of a regular thread is the basic size of the major diameter of the thread

大径（*D*或*d*）：是指与外螺纹牙顶或内螺纹牙底相切的假想圆柱的直径。对于外螺纹，大径即为顶径；对于内螺纹，大径为底径。普通螺纹的公称直径即螺纹大径的基本尺寸。

(2) Minor diameter (*D* or *d*) refers to the diameter of an imaginary cylinder that is tangent to the base of an external thread or the top of an internal thread. For external threads, the minor diameter is the base diameter; For internal threads, the minor diameter is the top diameter.

小径（*D*或*d*）是指与外螺纹牙底或内螺纹牙顶相切的假想圆柱的直径。对外螺纹而言，小径为底径；对内螺纹而言，小径为顶径。

(3) Mid diameter (*D or d*) is the diameter of an imaginary cylinder, whose generatrix passes through a groove and the protrusion on the tooth profile where the width of the protrusion is equal, as shown in Fig. 8-4. The hypothetical cylinder is called a medium diameter cylinder.

中径（*D*或*d*）是一个假想圆柱的直径，该圆柱的母线通过牙型上沟槽和凸起宽度相等的地方，如图8-4所示。该假想圆柱称为中径圆柱。

(4) Pitch (*P*) refers to the axial distance between two adjacent teeth at the corresponding two points on the median diameter.

螺距（*P*）是指相邻两牙在中径线上对应两点间的轴向距离。

(5) The single pitch diameter is the diameter of an imaginary cylinder, whose generatrix passes through a groove on the tooth where the width of the groove is equal to half of the nominal pitch size, as shown in Fig.8-4. A single pitch diameter can be measured using the three needle method to represent the actual pitch diameter of a thread.

单一螺距直径是一个假想圆柱的直径，该圆柱的母线通过齿上的一个槽，该槽的宽度等于螺距公称尺寸的一半，如图8-4所示。单一螺距直径可以通过三针法来测量，以表示螺纹的实际。

（6）The tooth shape angle and tooth shape half angle refer to the angle between two adjacent tooth sides on the threaded tooth shape, and the tooth shape half angle is half of the tooth shape angle, as shown in Fig. 8-5. Symbols for tooth angles are α. The tooth profile angle of ordinary threads is 60°.

牙型角和牙型半角是指在螺纹牙型上两相邻齿侧间的夹角，牙型半角为牙型角的一半，如图8-5所示。牙型角用符号α表示。普通螺纹的牙型角为60°。

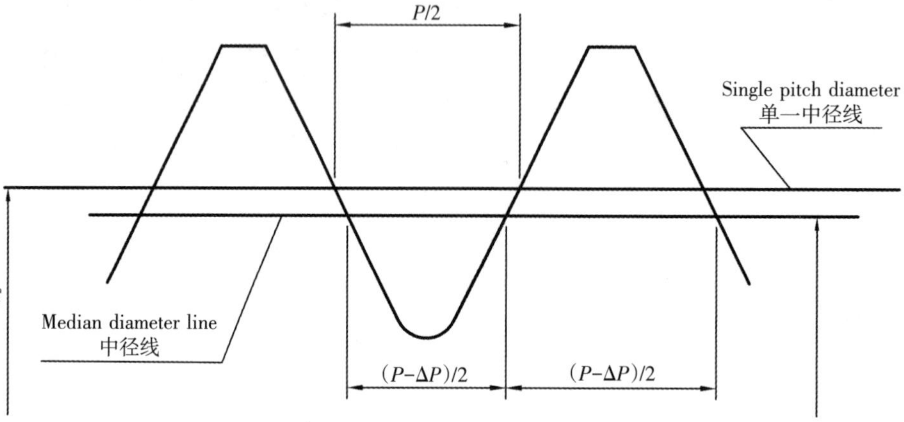

Fig. 8-4 Pitch diameter and single pitch diameter

图8-4 中径与单一中径

（7）The tooth flank angle refers to the angle between the tooth flank on the thread profile and the perpendicular line of the thread axis, as shown in Fig. 8-5. It is denoted by the symbol β_1 and β_2 symbols for the left and right tooth corners respectively . The basic value of tooth flank angle is equal to the half angle of the tooth shape, and the basic value of tooth flank angle for ordinary threads is 30 °.

牙侧角是指在螺纹牙型上牙侧与螺纹轴线的垂线间的夹角，如图8-5所示。左、右牙侧角分别用符号β_1和β_2表示。牙侧角基本值与牙型半角相等，普通螺纹牙侧角基本值为30°。

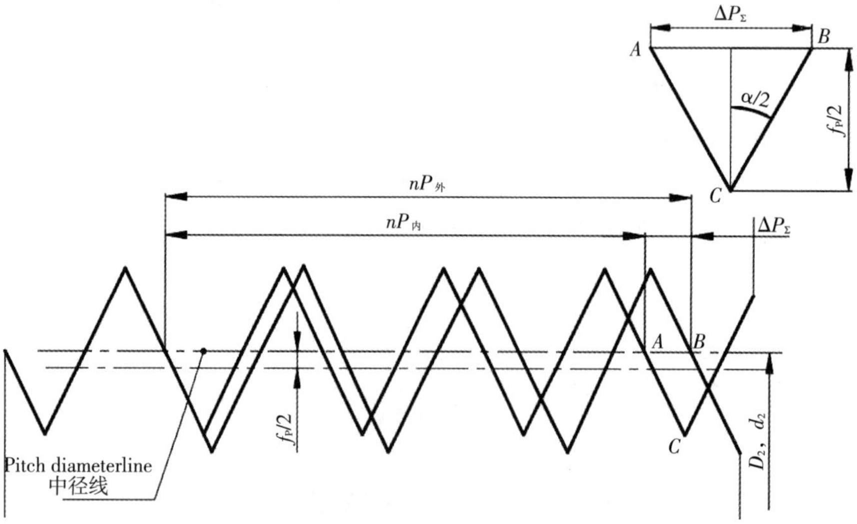

Fig. 8-5 Impact of accumulated pitch error on rotational fit

图8-5 螺距累积误差对旋合性的影响

（8）The contact height of threads refers to the distance between the overlapping tooth sides of two mating thread profiles in the direction perpendicular to the thread axis. The basic value of the contact height of ordinary threads is $5H/8$ ($3H/8+H/4=5H/8$), as shown in Fig. 8-2.

螺纹的接触高度是指两个相互配合的螺纹牙型在垂直于螺纹轴线方向上，其重叠牙侧之间的距离。普通螺纹的接触高度基本值为5H/8（3H/8+ H/4 = 5H/8），如图8-2所示。

（9）The length of thread engagement refers to the length of the portion of two threads that fit together along the axis of the thread.

螺纹旋合长度是指两个相互配合的螺纹沿螺纹轴线方向相互旋合部分的长度。

The basic size series of ordinary threads is shown in Tab. 8-1.

普通螺纹基本尺寸系列见表8-1。

Tab. 8-1 The basic size of ordinary threads（From GB/T 196-2003）

表8-1 普通螺纹的基本尺寸（摘自GB/T 196—2003）

单位：mm

Major diameter大径（D或d）			Prtch 螺距（P）	Medium diameter 中径（D_2或d_2）	Minor diameter 小径（D_1或d_1）
First series 第一系列	Second seies 第二系列	Third seies 第三系列			
6			**1**	5.350	4.917
			0.75	5.513	5.188
			（0.5）	5.675	5.459
	7		**1**	6.350	5.917
			0.75	6.513	6.188
			0.5	6.675	6.459
8			**1.25**	7.188	6.647
			1	7.350	6.917
			0.75	7.513	7.188
			（0.5）	7.675	7.459
		9	**（1.25）**	8.188	7.647
			1	8.350	7.917
			0.75	8.513	8.188
			（0.5）	8.675	8.459
10			**1.5**	9.026	8.376
			1.25	9.188	8.647
			1	9.350	8.917
			0.75	9.513	9.188
			（0.5）	9.675	9.459
		11	（1.5）	10.026	9.376
			1	10.350	9.917
			0.75	10.513	10.188
			（0.5）	10.675	10.459
12			**1.75**	10.853	10.106
			1.5	11.026	10.376
			1.25	11.188	10.647
			1	11.350	10.917
			（0.75）	11.513	11.188
			（0.5）	11.675	11.459
	14		**2**	12.701	11.835
			1.5	13.026	12.376
			（1.25）	13.188	12.647
			（1）	13.350	12.917
			（0.75）	13.513	13.188
			（0.5）	13.675	13.459
		15	1.5	14.026	13.376
			（1）	14.350	13.917

Tab. 8-1（Continued）表8-1（续）

Major diameter大径（D或d）			Prtch 螺距（P）	Medium diameter 中径（D_2或d_2）	Minor diameter 小径（D_1或d_1）
First series 第一系列	Second seies 第二系列	Third seies 第三系列			
16			**2**	14.701	13.835
			1.5	15.026	14.376
			1	15.350	14.917
			(0.75)	15.513	15.188
			(0.5)	15.575	15.459
		17	1.5	16.026	15.376
			1	16.350	15.917
	18		**2.5**	16.376	15.294
			2	16.701	15.835
			1.5	17.026	16.376
			1	17.350	16.917
			(0.75)	17.513	17.188
			(0.5)	17.675	17.459
20			**2.5**	18.376	17.294
			2	18.701	17.835
			1.5	19.026	18.376
			1	19.350	18.917
			(0.75)	19.513	19.188
			(0.5)	19.675	19.459
	22		**2.5**	20.376	19.294
			2	20.701	19.835
			1.5	21.026	20.376
			1	21.350	20.917
			(0.75)	21.513	21.188
			(0.5)	21.675	21.459
24			**3**	22.051	20.752
			2	22.701	21.835
			1.5	23.026	22.376
			1	23.350	22.917
			(0.75)	23.513	23.188
		25	2	23.701	22.835
			1.5	24.026	23.376
			(3)	24.350	23.917
		26	1.5	25.026	24.376
	27		**3**	25.051	23.752
			2	25.701	24.835
			1.5	26.026	25.376
			1	26.350	25.917
			(0.75)	26.513	26.188
		28	**2**	26.703	25.835
			1.5	27.026	26.376
			1	27.350	26.917
30			**3.5**	27.727	26.211
			(3)	28.051	26.752
			2	28.701	27.835
			1.5	29.026	28.376
			1	29.350	28.917
			(0.75)	29.513	29.188

Notes: ① The first preference for diameters is the first series, followed by the second series, and the third series should be avoided as much as possible;

② The pitch in parentheses should be avoided as much as possible;

③ The bold type in dicates coarse-threaded screws.

注：① 直径优先选用第1系列，其次是第2系列，第3系列尽可能不用；

② 括号内的螺距尽量不用；

③ 用黑体字表示的是粗牙螺纹。

8.3 The Influence of Geometric Parameter Errors on Thread Interchangeability 螺纹几何参数误差对螺纹互换性的影响

8.3.1 Impact of Thread Diameter Deviation 螺纹直径偏差的影响

The actual diameter of the thread directly affects the tightness of the thread connection. To ensure the rotational fit of the threaded joint, it is necessary to ensure that the actual diameter of the internal thread is greater than or equal to the actual diameter of the external thread. Since the basic diameter dimensions of the matching internal and external threads are the same, if the actual diameter of the internal thread is greater than or equal to its basic size (i.e. the actual deviation of the internal thread diameter is positive), and the actual diameter of the external thread is less than or equal to its basic size (i.e. the actual deviation of the external thread diameter is negative), the rotational fit of the internal and external thread combination can be guaranteed. However, the actual minor diameter of the internal thread cannot be too large, and the actual major diameter of the external thread cannot be too small, otherwise it will reduce the contact height of the thread and lead to insufficient thread connection strength. The actual pitch diameter of the internal thread cannot be too large, and the actual pitch diameter of the external thread cannot be too small, otherwise it will weaken the connection strength of the thread. So, it is necessary to limit the actual size of the thread diameter to be neither too large nor too small. Among the three diameter parameters of threads, the most important one is the actual size of the pitch diameter, which directly determines the fitting properties of the thread joint.

螺纹的实际直径大小直接影响螺纹连接的紧密程度。要保证螺纹连接的旋合性，就必须使内螺纹的实际直径大于或等于外螺纹的实际直径。由于相配合内、外螺纹的直径公称尺寸相同，如果使内螺纹的实际直径大于或等于其公称尺寸（即内螺纹直径实际偏差为正值），而外螺纹的实际直径小于或等于其公称尺寸（即外螺纹直径实际偏差为负值），就能保证内、外螺纹结合的旋合性。但是，内螺纹实际小径不能过大，外螺纹实际大径不能过小，否则会使螺纹接触高度减小，导致螺纹连接强度不足。内螺纹实际中径不能过大，外螺纹实际中径也不能过小，否则将削弱螺纹的联接强度。所以，必须限制螺纹直径的实际尺寸，使之不能过大，也不能过小。在螺纹的三个直径参数中，最主要的是中径的实际尺寸，它直接决定了螺纹结合的配合性质。

8.3.2 Influence of Pitch Deviation 螺距偏差的影响

Screw pitch deviation includes partial pitch deviation ΔP and pitchcumulative deviation ΔP_{Σ}. The partial pitch deviation ΔP is the maximum difference between any simple actual screw pitch and nominal screw pitch, it is maximum difference of screw pitch and it has no relationship with the length of thread engagement. The cumulative deviation ΔP_{Σ} is the maximum difference between several actual screw pitches and the nominal value, the cumulative deviation, and it affects thread interchangeability.

螺距偏差包括局部螺距偏差 ΔP 和螺距累积偏差 ΔP_{Σ}。局部螺距偏差是任意单个实际螺距值与公称螺距值之间的最大差值，是螺距的最大差值，与螺纹接合长度无关。螺距累积偏差 ΔP_{Σ} 则是多个实际螺距与公称值之间的最大差值，螺距累积偏差对螺纹互换性有更显著的影响。

Assuming the internal thread has an ideal profile, the pitch diameter and thread angle of the external thread are the same as those of the internal thread, yet there exists a discrepancy in the pitch, specifically, the pitch of the external thread is greater than that of the internal thread. Suppose that over the length of n thread pitches, the cumulative pitch error is ΔP_{Σ}. Obviously, this pair of threads will interfere and cannot be mated, as shown in Fig. 8-5.

假设内螺纹具有理想牙型，外螺纹中径及牙型角与内螺纹相同，但螺距有误差，并假设外螺纹的螺距比内螺纹的大。假定在 n 个螺牙长度上，螺距累积误差为 ΔP_{Σ}。显然，这对螺纹将发生干涉而无法旋合，如图8-5所示。

In practical manufacturing processes, to facilitate the mating of an external thread with pitch variations with a standard internal thread, the pitch diameter of the external thread may be decreased by a specific value f_p. f_p is professionally termed the pitch diameter compensation value for accommodating pitch errors.

在实际生产中，为了使有螺距误差的外螺纹能旋入标准的内螺纹，可把外螺纹中径减去一个数值 f_p，f_p 被称为螺距误差的中径补偿值。

Similarly, when the internal thread has only pitch deviation, the pitch diameter of the internal thread should be increased by a certain value f_p to ensure interchangeability .

同理，当内螺纹螺距有误差时，为了保证可旋合性，应把内螺纹的中径加大一个数值 f_p。

From Fig.8-5, the pitch diameter compensation value f_p can be obtained from $\triangle ABC$.

根据图8-5，螺距误差的中径补偿值 f_p 可以从 $\triangle ABC$ 中推导出来。

$$f_p = |\Delta P_{\Sigma}| \cot\frac{\alpha}{2} \tag{8-1}$$

For metric threads with $\alpha = 60°$ thread angle:

对于螺纹角 $\alpha = 60°$ 的公制螺纹：

$$f_p = 1.732|\Delta P_{\Sigma}| \tag{8-2}$$

From the above equation, it can be seen that if ΔP_{Σ} is too large, the pitch diameter of the innernal and external threads will increase or decrease significantly, respectively. Although it can ensure screwability, it will cause a decrease in the actual number of threads in contact, concentrating the load on the contact area of the thread contact surface, causing an increase in contact pressure on the thread contact surface, and subsequently reducing the connection strength of the thread.

由上式可知，如果 ΔP_{Σ} 过大，内、外螺纹中径要分别增大或减小许多，虽然可以保证旋合性，却会造成螺纹实际接触的螺牙数目减少，使载荷集中在螺牙接触面的接触部位，造成螺牙接触面接触压力增加，继而降低螺纹的连接强度。

8.3.3 The Impact of Flank Angle Deviation 牙侧角偏差的影响

The deviation of tooth flank angle refers to the difference between the actual value of tooth flank angle and its basic value, which includes the shape error of the thread flank and the position error of the flank relative to the thread axis.

螺纹牙侧角偏差是指螺纹牙侧角的实际值与其基本值之差，它包括螺纹牙侧的形状误差和牙侧相对于螺纹轴线的位置误差。

As shown in Fig. 8-6, assuming that the internal thread 1 has an ideal tooth shape（the size of the left and right tooth flank angles is both basic values of 30 °), the external thread 2 only has tooth flank angle deviation. In the figure, the deviation of the left lateral angle of the external thread is $\Delta\alpha_1$. If $\Delta\alpha_1 < 0$ and the deviation of the right tooth flank angle $\Delta\alpha_2 > 0$, interference will occur on the internal and external threads, making them unable to rotate. In order to eliminate interference and ensure rotational fit, it is necessary to move the external thread $\frac{f_\alpha}{2}$ towards the thread axis in the direction perpendicular to the thread axis, reaching the dashed line 3, that is, to reduce the pitch diameter of the external thread by one value f_α. f_α is called the pitch diameter equivalent of flank angle deviation. From Fig. 8-6, it can be concluded that the relationship between f_α and P is as follows:

如图8-6所示，假设内螺纹1具有理想牙型（左、右牙侧角的大小均为基本值30°），外螺纹2仅存在牙侧角偏差。图中，外螺纹左牙侧角偏差$\Delta\alpha_1 < 0$，右牙侧角偏差$\Delta\alpha_2 > 0$，则会在内、外螺纹牙侧产生干涉而不能旋合。为了消除干涉，保证旋合性，就必须将外螺纹、螺牙沿垂直于螺纹轴线的方向向螺纹轴线移动$\frac{f_\alpha}{2}$，到达虚线3处，即需将外螺纹中径减少一个数值f_α。f_α被称为牙侧角偏差的中径当量。由图8-6可得出f_α与P的关系如式（8-3）：

$$f_\alpha = 0.073P\left(K_1|\Delta\alpha_1| + K_2|\Delta\alpha_2|\right) \tag{8-3}$$

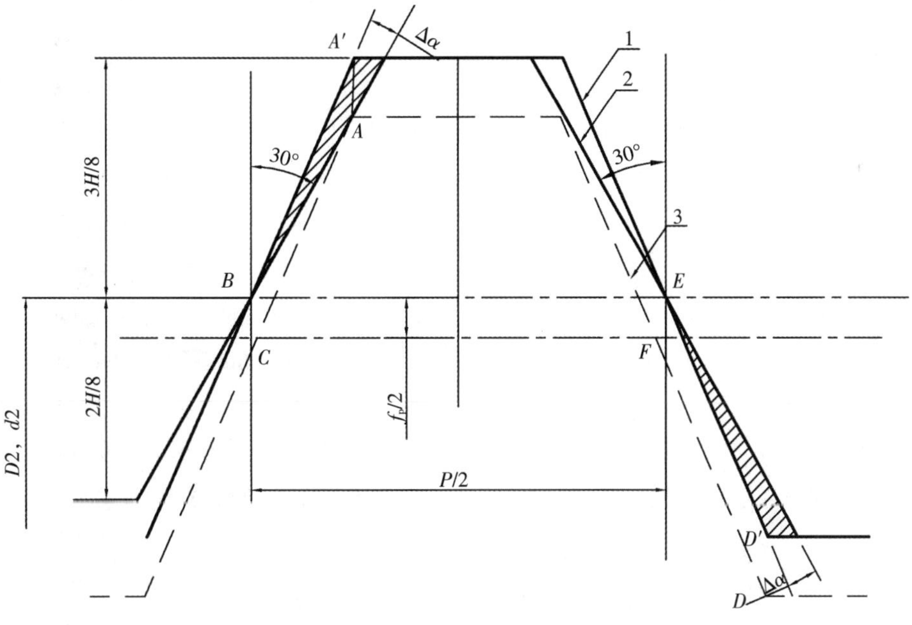

Fig. 8-6 The effect of thread flank deviation on screw-thread fit

图8-6 牙侧角偏差对旋合性的影响

For external threads, when $\Delta\alpha_1$(or $\Delta\alpha_2$) is positive, interference occurs on the tooth side between the median and minor diameters, with K_1(or K_2)taken as 2; When $\Delta\alpha_1$(or $\Delta\alpha_2$)is negative, interference occurs on the tooth side between the median and major diameters, and K_1(or K_2) is taken as 3; For internal threads, when $\Delta\alpha_1$(or $\Delta\alpha_2$)is positive, interference occurs on the tooth side between the median and major diameters, and K_1(or K_2)is taken as 3; When $\Delta\alpha_1$(or $\Delta\alpha_2$)is negative, interference occurs on the tooth side between the median and minor diameters, and K_1(or K_2)is taken as 2.

对于外螺纹，当$\Delta\alpha_1$（或$\Delta\alpha_2$）为正时，在中径与小径之间的牙侧产生干涉，K_1（或K_2）取2；当$\Delta\alpha_1$（或$\Delta\alpha_2$）为负时，在中径与大径之间的牙侧产生干涉，K_1（或K_2）取3；对于内螺纹，当$\Delta\alpha_1$（或$\Delta\alpha_2$）为正时，在中径与大径之间的牙侧产生干涉，K_1（或K_2）取3；当$\Delta\alpha_1$（或$\Delta\alpha_2$）为负时，在中径与小径之间的牙侧产生干涉，K_1（或K_2）取2。

When there is a deviation in the tooth flank angle of the thread, by reducing the pitch diameter of the external thread by a value f_α and increasing the pitch diameter of the internal thread by a value f_α, although the screwability can be ensured, the unequal tooth flank angles of the internal and external threads will reduce the contact area of the tooth flank and cause the load to be relatively concentrated at the contact area, resulting in an increase in contact pressure and a decrease in the strength of the thread connection.

当螺纹存在牙侧角偏差时，通过将外螺纹中径减小一个数值f_α，将内螺纹中径增大一个数值f_α，这样虽然可保证旋合性，但内、外螺纹的牙侧角不相等，会使牙侧接触面积减小，也会使载荷相对集中到接触部位，造成接触压力增加，降低螺纹连接强度。

8.4 Tolerance and Fit of Ordinary Threads 普通螺纹的公差和配合

8.4.1 Tolerance Zones for Ordinary Threads 普通螺纹的公差带

The tolerance zone of ordinary threads is the same as the tolerance zone of dimensions; its position is determined by the basic deviation, and its size is determined by the tolerance level. The national standard GB/T 197-2018 for ordinary threads specifies the tolerance zones for major, pitch, and minor diameters of threads.

普通螺纹的公差带与尺寸公差带一样，其位置由基本偏差决定，大小由公差等级决定。普通螺纹国家标准GB/T 197—2018规定了螺纹的大、中、小径的公差带。

8.4.1.1 Size and tolerance level of thread tolerance zone 螺纹公差带的大小和公差等级

The tolerance levels of threads are shown in Tab. 8-2. Among them, level 6 is the basic level; Grade 3 has the smallest tolerance value and the highest accuracy; Grade 9 has the lowest accuracy. The tolerance values for each level are shown in Tab. 8-3 and Tab. 8-4. Due to the difficulty in machining internal threads, the pitch diameter tolerance of internal threads within the same tolerance level is about 32% larger than that of external threads.

螺纹的公差等级见表8-2。其中6级是基本级；3级公差值最小，精度最高；9级精度最低。各级公差值见表8-3和表8-4。由于内螺纹加工比较困难，同一公差等级内螺纹中径公差比外螺纹中径公差大32%左右。

Tab. 8-2 Tolerance grades for threads

表8-2 螺纹的公差等级

Thread diameter 螺纹直径	Tolerance grade 公差等级	Thread diameter 螺纹直径	Tolerance grade 公差等级
Pitch diameter of external thread 外螺纹中径（d_1）	3，4，5，6，7，8	Pitch diameter of internal thread 内螺纹中径（D_1）	4，5，6，7，8
Major diameter of external thread 外螺纹大径（d）	4，6，8	Minor diameter of internal thread 内螺纹小径（D_2）	4，5，6，7，8

Tab. 8-3 Basic deviation and top diameter tolerance of ordinary thread

表8-3 普通螺纹的基本偏差和顶径公差

单位：mm

Screw pitch 螺距 (*P*)/mm	Basic deviation of internal thread 内螺纹基本偏差 (EI)/μm		Basic deviation of external thread 外螺纹的基本偏差(es)/μm				Minor diameter tolerance of internal thread 内螺纹小径公差 (T_{D_1})/μm					Major diameter tolerance of external thread 外螺纹大径公差(T_d)/μm		
	G	H	e	f	g	h	4	5	6	7	8	4	6	8
1	+26		−60	−40	−26		150	190	236	300	375	112	180	280
1.25	+28		−63	−42	28		170	212	265	335	425	132	212	335
1.5	+32		−67	−45	−32		190	236	300	375	485	150	236	375
1.75	+34		−71	−48	−34		212	265	335	425	530	170	365	425
2	+38	0	−71	−52	−38	0	236	300	375	475	600	180	380	450
2.5	+42		−80	−58	42		280	355	450	560	710	212	225	530
3	+48		−85	−63	−48		315	400	500	630	800	236	275	600
3.5	+53		90	−70	−53		355	450	560	710	900	265	425	670
4	+60		−95	75	−60		375	475	600	750	950	300	475	750

Tab. 8-4 Pitch diameter tolerances for ordinary threads

表8-4 普通螺纹的中径公差

单位：mm

Nominal diameter 公称直径 (*D*)/mm	Screw pitch 螺距 (*P*)/mm	Pitch diameter tolerance of internal thread 内螺纹中径公差(T_{D1})					Pitch diameter tolerance of external thread 外螺纹中径公差(T_{d1})						
		4	5	6	7	8	3	4	5	6	7	8	9
	0.5	71	90	112	140	—	42	53	67	85	106	—	—
	0.75	85	106	132	170	—	50	63	80	100	125	—	—
>5.6~11.2	1	95	118	150	190	236	56	71	90	112	140	180	224
	1.25	100	125	160	200	250	60	75	95	118	150	190	236
	1.5	112	140	180	224	280	67	85	106	132	170	212	295
	0.5	75	95	118	150	—	45	56	71	90	112	—	—
	0.75	90	112	140	180	—	53	67	85	106	132	—	—
	1	100	125	160	200	250	60	75	95	118	150	190	236
	1.25	112	140	180	224	280	67	85	106	132	170	212	265
>11.2~22.4	1.5	118	150	190	236	300	71	90	112	140	180	224	280
	1.75	125	160	200	250	315	75	95	118	150	190	236	300
	2	132	170	212	265	335	80	100	125	160	200	250	315
	2.5	140	180	224	280	355	85	106	132	170	212	265	335
	0.75	95	.118	150	190	—	56	71	90	112	140	—	—
	1	106	132	170	212	—	63	80	100	125	160	200	250
	1.5	125	160	200	250	315	75	95	118	150	190	236	300
	2	140	180	224	280	355	85	106	132	170	212	265	335
22.4 ~ 45	3	170	212	265	335	425	100	125	160	200	250	315	400
	3.5	180	224	280	355	450	106	132	170	212	265	335	425
	4	190	236	300	375	475	112	140	180	224	280	355	450
	4.5	200	250	315	400	500	118	150	190	236	300	375	475

Due to the fact that the minor diameter d_1 and pitch diameter d_2 of the external thread, as well as the major diameter D and pitch diameter D_1 of the internal thread, are simultaneously cut by the tool, their dimensions are naturally formed during the machining process and guaranteed by the tool. Therefore, the national standard does not specify specific tolerances for the major diameter of the internal thread and the minor diameter of the external thread. It only stipulates that any point on the actual contour of the inner and outer thread base cannot exceed the maximum solid tooth shape determined by the basic deviation.

由于外螺纹的小径d_1与中径d_2、内螺纹的大径D和中径D_1是同时由刀具切出的，其尺寸在加工过程中自然形成，由刀具保证。因此，国家标准中对内螺纹的大径和外螺纹的小径均不规定具体的公差值范围，只规定内、外螺纹牙底实际轮廓的任何点均不能超过基本偏差所确定的最大实体牙型。

8.4.1.2 Position and basic deviation of thread tolerance zone 螺纹公差带的位置和基本偏差

The tolerance zone of threads is arranged with the basic tooth profile as the zero line, and its position is shown in Fig.8-7. The basic tooth shape of a thread is the benchmark for calculating thread deviation.

螺纹的公差带是以基本牙型为零线布置的，其位置如图8-7所示。螺纹的基本牙型是计算螺纹偏差的基准。

The national standard only specifies two basic deviations G and H for internal threads. The basic deviation is the lower deviation EI. As shown in Fig. 8-7（a）and（b), four basic deviations e, f, g and h are specified for external threads, the basic deviation is the upper deviation es. As shown in Fig. 8-7（c）and（d). The basic deviation of H and h is zero, the basic deviation value of G is positive, and the basic deviation values of e, f and g are negative, as shown in Tab. 8-3.

国家标准中对内螺纹只规定了两种基本偏差G、H。基本偏差为下偏差EI。如图8-7（a）（b）所示。

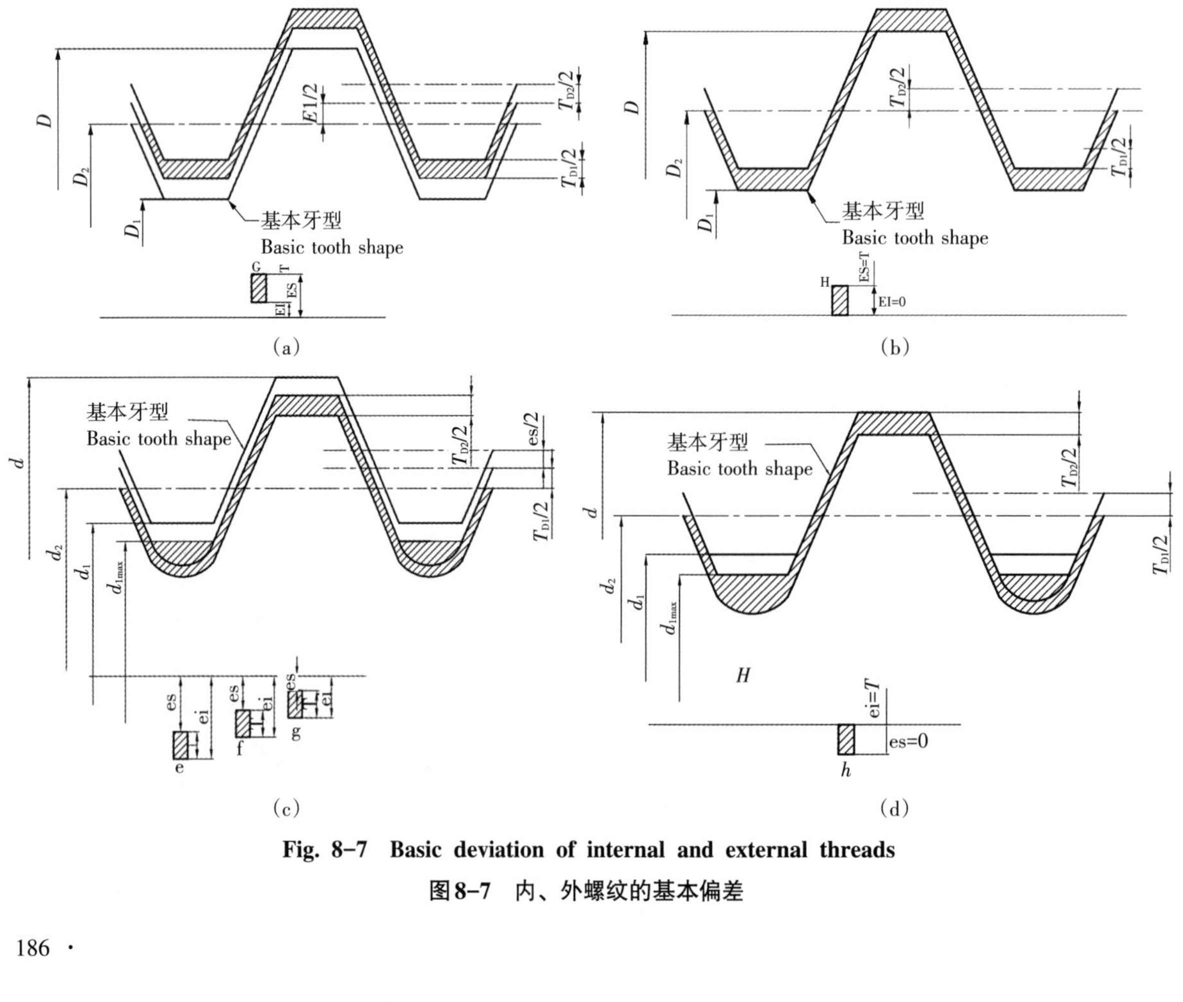

Fig. 8-7 Basic deviation of internal and external threads

图8-7 内、外螺纹的基本偏差

对外螺纹规定了四种基本偏差e，f，g，h，基本偏差为上偏差es，如图8-7（c）（d）所示。H和h的基本偏差为零，G的基本偏差值为正，e，f，g的基本偏差值为负，具体数值见表8-3。

According to the tolerance level and basic deviation of threads, many tolerance zones can be formed. The tolerance zone code for ordinary threads consists of numbers representing the tolerance level and basic deviation letters, such as 6h, 5G. Unlike general dimensional tolerance zone symbols, the tolerance level symbol comes first and the basic deviation code comes last.

按螺纹的公差等级和基本偏差，可以组成很多公差带。普通螺纹的公差带代号由表示公差等级的数字和基本偏差字母组成，如6h、5G。与一般的尺寸公差带符号不同，其公差等级符号在前，基本偏差代号在后。

8.4.2 Selection of Thread Tolerance Zones 螺纹公差带的选用

In order to reduce the specifications and types of cutting tools and measuring tools in production, the national standard specifies commonly used tolerance zones that can meet current needs in limited quantities, as shown in Tab. 8-5. The table specifies the priority, second-choice, and least used selection order. Except for special needs, tolerance zones other than those specified in the standard should generally not be selected.

为了在生产中减少切削刀具和测量工具的规格和种类，国家标准规定了既能满足当前需要而数量又有限的常用公差带，见表8-5。表中规定了优先，次选和尽可能不用的选用顺序。除了特殊需要之外，一般不应该选择标准规定以外的公差带。

Tab. 8-5 Selection of tolerance zones for ordinary threads

表8-5 普通螺纹选用公差带

Standard tolerance 公差精度	Internal thread tolerance zone 内螺纹公差带			External thread tolerance zone 外螺纹公差带		
	S	N	L	S	N	L
Precision 精密	4H	5H	6H	(3h4h)	**4h**	(5h4h) (5g4g)
Medium 中等	**5H** (5g)	**[6H]** **6G**	**7H** (7G)	(5g6g) (5h6h)	**6e** **6f** **[6g]** 6h	(7e6e) — (7g6g) (7h6h)
Coarse 粗糙	—	7H (7G)	8H (8G)	—	(8e) 8g	(9e8e) (9g8g)

Note: ① The order of selection is: bold font tolerance zone, general font tolerance zone, and tolerance zone within parentheses.

② Boxed bold font tolerance zones and mass-produced fastener threads.

③ The recommended tolerance zone also applies to thin coated threads, such as electroplated threads. The selected tolerance zone before coating should ensure that any point on the actual contour of the thread after coating does not exceed the maximum solid tooth shape determined by the tolerance zone position H or h.

注：① 选用顺序依次为：**公差带**、一般字体公差带、括号内的公差带。

② 带方框的粗字体公差带与大量生产的紧固件螺纹。

③ 推荐公差带也适用于薄涂镀层的螺纹，如电镀螺纹。所选择的涂镀前公差带应满足涂镀后螺纹实际轮廓上的任何点不超出公差带位置H或h确定的最大实体牙型。

8.4.2.1 Selection of fitting accuracy 配合精度的选用

According to GB/T 197-2018, the fitting accuracy of threads is divided into three levels: precision, medium, and roughness. Precision grade threads are mainly used for threads that require stable mating performance; Medium grade threads are for general purposes; Coarse grade is used for threads that are not important or difficult to manufacture, such as long blind hole tapping threads or threads on hot-rolled bars. Generally, a tolerance level of 6 with a moderate engagement length is used as the benchmark for moderate accuracy.

GB/T 197—2018规定螺纹的配合精度分为精密、中等和粗糙三个等级。精密级主要用于要求配合性能稳定的螺纹；中等级用于一般用途的螺纹；粗糙级用于不重要或难以制造的螺纹，如长盲孔攻螺纹或热轧棒上的螺纹。一般以中等旋合长度下的6级公差等级为中等精度的基准。

8.4.2.2 Determination of rotation length 旋合长度的确定

Due to the ease of processing and assembly of short parts and difficulty of processing and assembly of long parts, the length of engagement length affects the fitting accuracy and interchangeability of threaded connectors. The national standard specifies three types of thread connection lengths: short, medium, and long, represented by S, N, and L (see Tab. 8-6). Generally, medium thread connection lengths are preferred. From Tab. 8-5, it can be seen that for different screw lengths with the same accuracy, the tolerance level used for the pitchdiameter is also different. This is because different screw lengths have different effects on the cumulative error of the thread pitch.

由于短件易加工和装配，而长件难加工和装配，因此螺纹旋合长度影响螺纹连接件的配合精度和互换性。国家标准对螺纹连接规定了短、中等和长三种旋合长度，分别用S，N，L表示（见表8-6）。通常，优先选用中等旋合长度。从表8-5中可以看出，在同一精度中，对不同的旋合长度，其中径所采用的公差等级也不相同。这是因为不同旋合长度的螺纹对螺距累积误差有不同的影响。

Tab. 8-6 Lengths of thread engagement

表8-6 螺纹的旋合长度

单位：mm

Nominal diameter 公称直径(D或d)		Screw pitch 螺距(P)	Lengths of thread engagement 螺纹的旋合长度			
			S	N		L
>	≤		≤	>	≤	>
5.6	11.2	0.75	2.4	2.4	7.1	7.1
		1	3	3	9	9
		1.25	4	4	12	12
		1.5	5	5	15	15
11.2	22.4	1	3.8	3.8	11	11
		1.25	4.5	4.5	13	13
		1.5	5.6	5.6	16	16
		1.75	6	6	18	18
		2	8	8	24	24
		2.5	10	10	30	30

8.4.2.3 Determination of tolerance levels and basic deviations 公差等级和基本偏差的确定

Based on the fitting accuracy and engagement length, the tolerance level and basic deviation are selected from Tab. 8-5. The specific values are shown in Tab. 8-4 and Tab. 8-3.

根据配合精度和旋合长度，由表8-5中选定公差等级和基本偏差，具体数值见表8-4 和表8-3。

8.4.2.4 Selection of coordination 配合的选用

The tolerance zone of internal and external thread fit can be arbitrarily combined into multiple fits. In practical use, the thread fit is mainly selected according to the usage requirements. To ensure good coaxiality and sufficient connection strength after the nuts and bolts are tightened, a fit with a minimum clearance of zero（H/h）is selected; For the convenience of disassembly and improvement of thread fatigue strength, a small clearance fit（H/g and G/h）can be selected; The thread that needs to be coated with a protective layer has a gap size depending on the thickness of the coating, such as: 5μm selects 6H/6g; 10μm selects 6H/6e; For both internal and external coatings, 6G/6e should be selected.

内外螺纹配合的公差带可以任意组合成多种配合。在实际使用中，主要根据使用要求选用螺纹的配合。为保证螺母、螺栓旋合后同轴度较好且具有足够的连接强度，选用最小间隙为零的配合（H/h）；为了拆装方便和改善螺纹的疲劳强度，可选用小间隙配合（H/g和G/h）；对于需要涂镀保护层的螺纹，间隙大小取决于镀层厚度，如：镀层厚度为5μm则选用6H/6g；镀层厚度为10μm则选用6H/6e；内外均涂则选用6G/6e。

8.4.3 Marking of Ordinary Threads 普通螺纹的标记

The complete marking of threads consists of thread code M, thread tolerance zone code, and screw length code. The code for thread tolerance zone includes the code for pitch diameter tolerance zone and the code for major diameter（major diameter of external thread and minor diameter of internal thread）tolerance zone. The tolerance zone code consists of a tolerance level number representing its size and a basic deviation code representing its position. For a fine thread, the pitch also needs to be marked. An example of ordinary thread markings on part drawings:

螺纹的完整标记由螺纹代号M、螺纹公差带代号和旋合长度代号等组成。螺纹公差带代号包括中径公差带代号和顶径（外螺纹大径和内螺纹小径）公差带代号。公差带代号是由表示其大小的公差等级数字和表示其位置的基本偏差代号组成。对细牙螺纹，还需要标注出螺距。在零件图上的普通螺纹标记示例如下：

External thread 外螺纹：

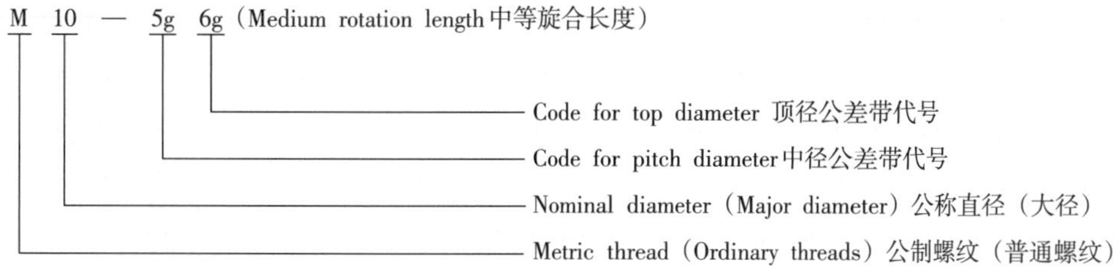

Internal thread 内螺纹：

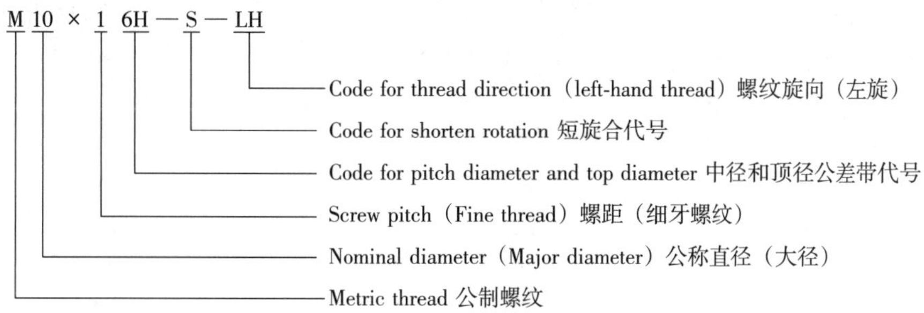

In the assembly drawing, the tolerance zone codes for internal and external threads are separated by diagonal lines, with the internal thread code on the left and the external thread code on the right. For example: M10×2—6H/5g6g. If necessary, add the screw thread length code S or L（medium screw thread length code N is not marked）after the thread tolerance zone code, such as M10-5g6g-S. When there is a special need, the numerical value of the screwing length can be marked, for example, the screwing length of the M10 — 5g6g — 25 thread is 25mm.

在装配图中，内、外螺纹公差带代号用斜线分开，左内右外，如 M10×2—6H/5g 6g。必要时，在螺纹公差带代号之后加注旋合长度代号S或 L（中等旋合长度代号N不标注），如M10—5g6g—S。特殊需要时，可以标注旋合长度的数值，如M10 — 5g6g — 25，代表螺纹的旋合长度为25 mm。

The markings that can be omitted for threads include:（1）the basic pitch value of coarse thread;（2）The code N for the length of medium engagement（3）Right rotation direction code RH;（4）Medium precision level threads refer to the codes for 6H and 6g tolerance zones with a nominal diameter of ≥ 1.6mm, or 5H and 6h tolerance zones with a nominal diameter of ≤ 1.4 mm.

螺纹的标记可以省略的标注有：（1）粗牙螺纹的螺距基本值；（2）中等旋合长度代号N；（3）右旋旋向代号RH；（4）中等精度级螺纹，即公称直径不小于1.6 mm的6H、6g公差带的代号，或公称直径不大于1.4 mm的5H、6h公差带的代号。

8.5 Description of Trapezoidal Threads 梯形螺纹简述

Various transmission threads often use trapezoidal threads in their thread profiles, such as machine tool screws, crane screws, etc. This is because trapezoidal threads have advantages such as high transmission efficiency, high accuracy, and convenient processing, and can meet the requirements of using transmission threads applications.

各种传动螺纹在其螺纹牙型上常采用梯形螺纹，如机床螺钉、起重螺钉等。这是因为梯形螺纹具有传动效率高、精度高、加工方便等优点，并能够满足传动螺纹的使用要求。

8.5.1 Basic Dimensions of Trapezoidal Threads 梯形螺纹基本尺寸

The trapezoidal thread specified by the national standard is formed by truncating the top and bottom of the original triangle, which is an isosceles triangle with a top angle equal to 30°. In order to ensure the flexibility of trapezoidal thread transmission, it is necessary to leave a guaranteed gap a_c between the major and minor diameters after the internal and external threads are matched. Therefore, a gap equal to a_c is given at the base of the internal and external threads by the basic tooth shape. As shown in Fig. 8-8。

国家标准规定的梯形螺纹是由原始三角形截去顶部和底部形成的，其原始三角形为顶角等于30°的等腰三角形。为了保证梯形螺纹传动的灵活性，必须使内、外螺纹配合后在大径和小径间保留一个保证间隙a_c。为此，分别在内外螺纹的牙底上，由基本牙形留出一个大小等于a_c的间隙。如图8-8所示。

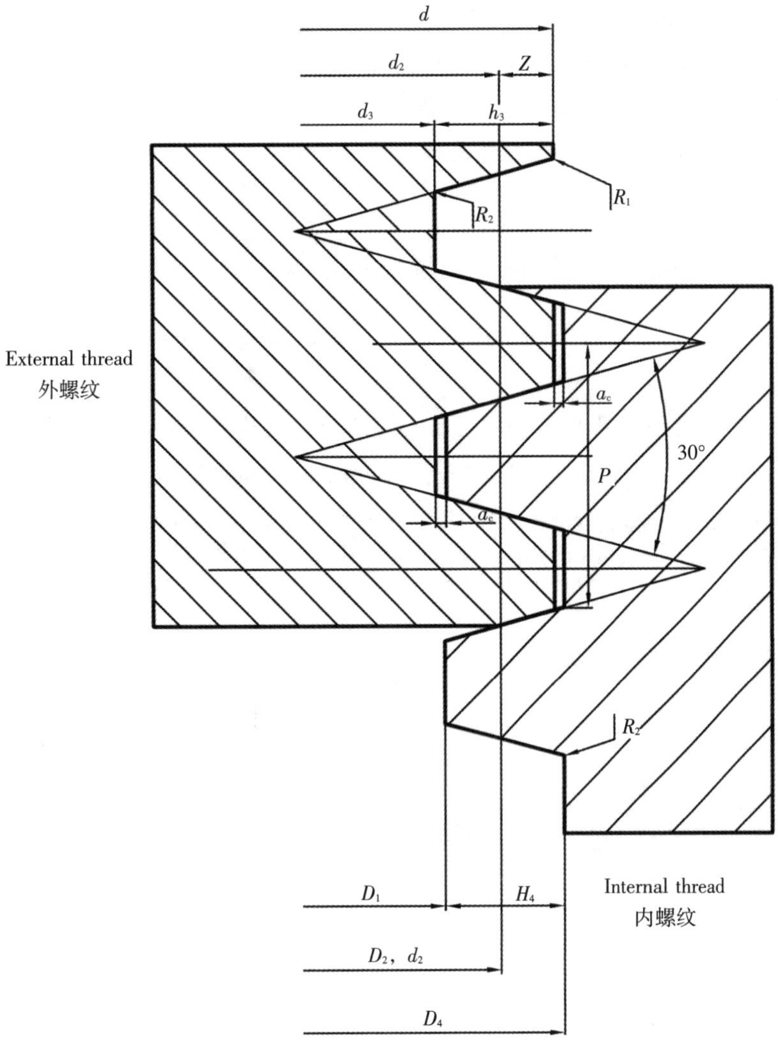

Fig. 8-8 Trapezoidal thread

图8-8 梯形螺纹

8.5.2 Tolerance of Trapezoidal Threads 梯形螺纹公差

In the trapezoidal thread standard, tolerance levels are specified for the major, pitch, and minor diameters of internal and external threads, as shown in Tab. 8-7 .

在梯形螺纹标准中，对内、外螺纹的大、中、小径分别规定了公差等级，见表8-7。

Tab. 8-7 Trapezoidal threads tolerance levels

表8-7 梯形螺纹公差等级

Diameter 直径	Tolerance grades 公差等级	Diameter 直径	Tolerance grades 公差等级
Minor diameter of external thread 外螺纹小径（d）	7, 8, 9	Minor diameter of internal thread 内螺纹小径（D）	4
Pitch diameter of external thread 外螺纹中径（d）	(6), 7, 8, 9	Pitch diameter of internal thread 内螺纹中径（D）	7, 8, 9
Major diameter of external thread 外螺纹大径（d）	4		

The standard only specifies one basic deviation H (with a value of zero) for the major diameter D_4, pitch diameter D_2, and minor diameter D_1 of internal threads. The pitch diameter d_2 of the external thread specifies two basic deviations, e and c. For the major diameter d and minor diameter d_3, a basic deviation h (with a value of zero) is specified and the basic deviation of e and c are negative.

标准对内螺纹大径D_4，中径D_2和小径D_1只规定了一种基本偏差H（其值为零）；对外螺纹的中径d_2规定了e，c两种基本偏差；对大径d和小径d_3规定了一种基本偏差h（其值为零），e和c的基本偏差为负值。

8.5.3 Marking of Trapezoidal Threads 梯形螺纹标记

The marking of trapezoidal threads consists of trapezoidal thread codes, tolerance zone codes and screw length codes.

梯形螺纹的标记由梯形螺纹代号、公差带代号及旋合长度代号组成。

When the engagement length is medium, do not label the twist length code. When the engagement length is the long engagement length, the group code L should be written after the tolerance code and separated by "–".

当旋合长度为中等旋合长度时，不标注旋合长度代号。当旋合长度为长旋合长度时，应将组别代号L写在公差代号的后面，并用"—"隔开。

In the assembly drawing, the tolerance zone of the trapezoidal thread should be marked with the tolerance zone codes of the internal and external threads respectively. The front is the tolerance code of the internal thread, and the back is the tolerance code of the external thread. The middle is separated by a diagonal line, for example: Tr 40 × 7 — 7H.

在装配图中，梯形螺纹的公差带要分别注出内外螺纹的公差代号。前面是内螺纹公差代号，后面是外螺纹公差代号，中间用斜线分开。例如：Tr 40 × 7 — 7H.

Internal thread 内螺纹：

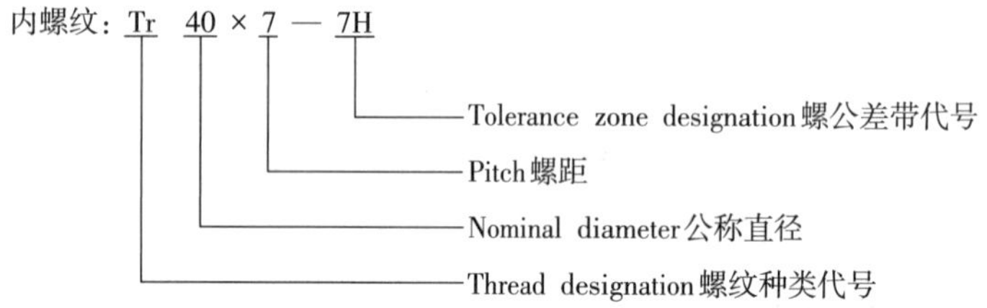

Multiple-thread多线螺纹：

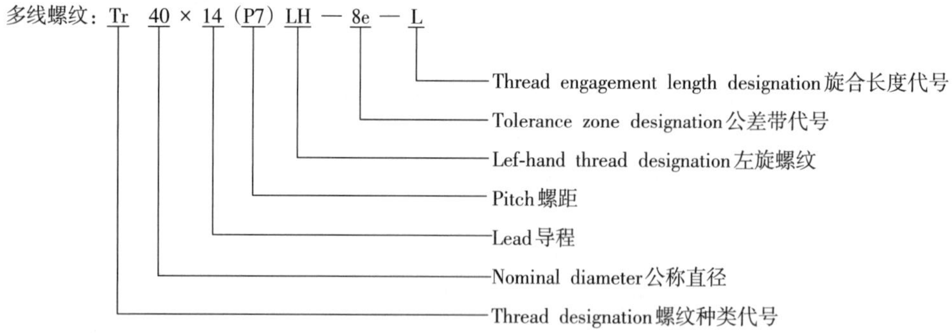

The top diameter tolerance and bottom diameter tolerance of multi thread are the same as those of single thread. The pitch diameter tolerance of multi-thread is obtained by multiplying the number of threads by a coefficient based on the pitch diameter tolerance of single thread.

多线螺纹的顶径公差和底径公差与单线螺纹相同。多线螺纹的中径公差是通过将单线螺纹的中径公差乘以一个系数获得的。

8.6 Testing of Ordinary Threads 普通螺纹的检测

According to usage requirements, thread inspection can be divided into two categories: comprehensive inspection and individual measurement.

根据使用要求，螺纹检测可分为综合检验和单项测量两类。

8.6.1 Comprehensive Inspection 综合检验

The comprehensive inspection of general threads refers to the use of gauges to test the comprehensive results of geometric parameter deviations that affect thread interchangeability. This includes: Using a regular thread gauge to check the effective pitch diameter and a single pitch diameter of the tested thread using a go gauge and a not-go gauge, respectively; Using a smooth limit gauge to verify the actual top diameter of the tested thread.

普通螺纹的综合检验是指使用量规来检验影响螺纹互换性的几何参数偏差的综合结果。其中包括：使用普通螺纹量规通规和止规分别对被测螺纹的作用中径（含底径）和单一中径进行检验；使用光滑极限量规对被测螺纹的实际顶径进行检验。

The thread gauge used to inspect internal threads is called a thread plug gauge, and the gauge used to inspect external threads is called a thread ring gauge. The design of thread gauges should comply with the Taylor Principle. As shown in Fig. 8-9, the thread gauge is designed to measure the maximum solid tooth shape of the thread to be tested. The effective pitch diameter of the tested thread is checked to see if it exceeds the pitch diameter of its maximum solid tooth shape, and the actual size of the base diameter is also checked to see if it exceeds its maximum solid size.

检验内螺纹用的螺纹量规称为螺纹塞规，检验外螺纹用的量规称为螺纹环规。螺纹量规的设计应符合泰勒原则。如图8-9所示。螺纹量规是为测量被测螺纹的最大实体牙型而设计的，用于检验被测螺纹的作用中径是否超出其最大实体牙型的中径，并同时检验底径实际尺寸是否超出其最大实体尺寸。

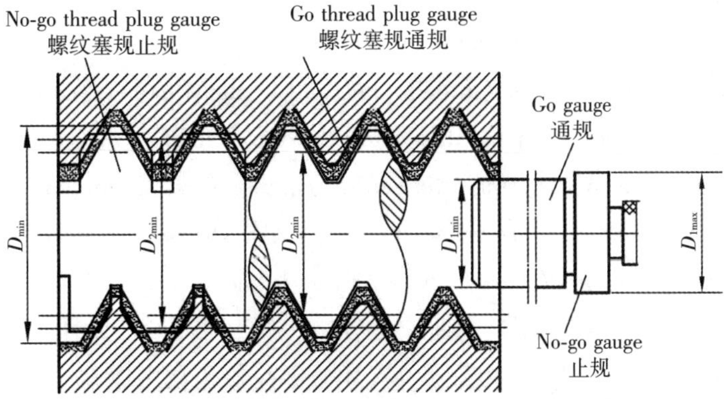

Fig. 8-9 Inspecting internal threads using a thread plug gauge and a plain limit plug gauge

图8-9 用螺纹塞规和光滑极限塞规检验内螺纹

Therefore, the gauge should have a complete tooth shape and the length of the thread should be equal to the length of the thread being tested. The not-go gauge is used to check whether the single pitch diameter of the tested thread exceeding the pitch diameter of its minimum solid tooth profile. As shown in Fig. 8-10, therefore, the not-go gauge should adopt a truncated tooth shape and only have a thread length of 2-3 pitches to reduce the influence of tooth flank angle deviation and pitch error on the inspection results.

因此，量规应具有完整的牙型，并且螺纹的长度等于被测螺纹的长度。止规用来检验被测螺纹的单一中径是否超出其最小实体牙型的中径。如图8-10所示。因此，止规应采用截短牙型，并且只有2~3个螺距的螺纹长度，以减少牙侧角偏差和螺距误差对检验结果的影响。

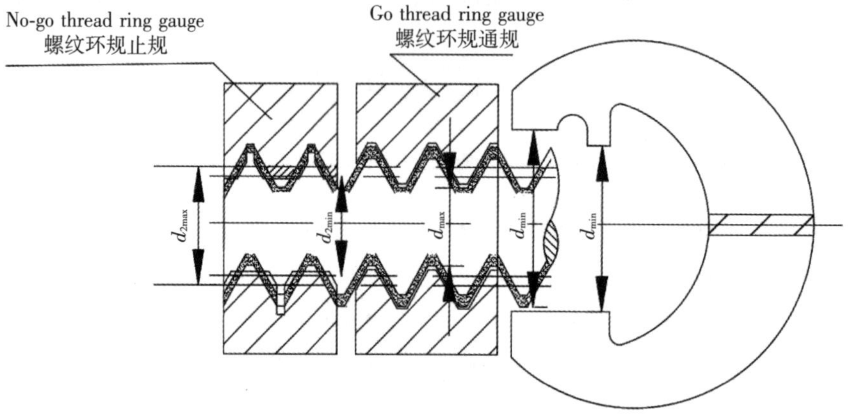

Fig. 8-10 Inspecting external threads using a thread ringgauge and a plain limit ring gauge

图8-10 用螺纹环规和光滑极限卡规检验外螺纹

If the tested thread can be screwed through by the thread gauge and not fully screwed through into the thread not-go gauge (the thread not-go gauge only allows screwing into both ends of the tested thread, and the screwing amount should not exceed two pitches), it indicates that the effective pitch diameter of the tested thread has not exceeded the pitch diameter of its maximum solid tooth shape, and a single pitch diameter has not exceeded the pitch diameter of its minimum solid tooth shape. This can ensure screwing and connection strength, and the measured thread pitch diameter is qualified. Otherwise, it is considered unqualified.

如果被测螺纹能够与螺纹通规旋合通过，而螺纹止规不能完全旋合通过（螺纹止规只允许与被测螺纹两端旋合，旋合量不得超过两个螺距），则表明被测螺纹的作用中径没有超出其最大实体牙型的中径，且单一中径没有超出其最小实体牙型的中径。那么就可以保证旋合性和连接强度，所测得的螺纹中径是合格的，否则视为不合格。

The basic dimensions and limit deviations of the pitch, major diameter, minor diameter, and tooth flank angle of the go and no go gauges of the thread plug gauge and thread ring gauge must be determined separately. The smooth limit gauges used for inspecting the top diameter of threads, including the go gauge and the no go gauge, should also determine the corresponding forming dimensions and limit deviations, similar to the smooth limit gauges used for inspecting holes and shafts. These are specified in GB/T 3934-2003 *Specification of Gauges for General Purpose Screw Threads* and its appendices.

螺纹塞规、螺纹环规的通规和止规的螺距、中径、大径、小径和牙侧角都要分别确定相应的基本尺寸及极限偏差。检验螺纹顶径用的光滑极限量规通规和止规也要分别确定相应的定形尺寸及极限偏差，与检验孔、轴用的光滑极限量规类似。这些在GB/T 3934—2003《普通螺纹量规技术条件》及其附录中都有具体规定。

8.6.2 Single Item Measurement 单项测量

The single item measurement of ordinary threads refers to measuring the various geometric parameters of the thread separately. Single item measurement is used for process analysis of threaded workpieces and measurement of thread gauges and thread cutting tools. There are several commonly used single thread measurement methods.

普通螺纹的单项测量是指分别对螺纹的各个几何参数进行测量。单项测量用于螺纹工件的工艺分析、螺纹量规和螺纹刀具的测量。常用的单项螺纹测量方法有以下几种。

8.6.2.1 Three needle method for measuring single pitch diameter of external threads 三针法测量外螺纹单一中径

As shown in Fig. 8-11, place three rigid cylindrical measuring needles of identical diameter, one in the tooth space of the thread being measured, and the other two in the adjacent tooth spaces on the opposite side. Then, utilizing a dial gauge or an equivalent indicating instrument, ascertain the pin-to-pin distance M. With this measurement in hand, coupled with the predefined nominal pitch size P of the thread, the theoretical half-angle $\alpha/2$ that characterizes the thread profile, and the diameter d of the gauge pins, proceed to compute the single pitch diameter d of the thread. As shown inthe following formula:

$$d_{2s}=M-d_{m}\left(1+1\bigg/\frac{\sin\alpha}{2}\right)+\frac{P}{2}\cot\frac{\alpha}{2} \tag{8-4}$$

如图8-11所示，将三根直径相同的刚性圆柱形量针放在被测螺纹对径位置的沟槽中，其中一根放在被测螺纹的牙槽中，另外两根放在对边相邻的两牙槽中，然后用指示量仪测出针距M值，并根据已知被测螺纹的螺距公称尺寸P，牙型角理论值$\alpha/2$和量针直径d的数值，计算出被测螺纹的单一中径d，即

For a standard thread, $\alpha/2=30°$, therefore the formula（8-4）simplifies to:

对于普通螺纹$\alpha/2=30°$，因此式（8-4）简化为：

$$d_{2s}=M-3d_{m}+0.886P \tag{8-5}$$

To avoid the influence of deviations in the flank angle on the measurement results and ensure that the contact points between the gauge pins and the thread flanks lie on the pitch diameter, the optimal diameter for the gauge pins should be determined as follows:

为了避免牙侧角偏差对测量结果的影响，使量针与牙侧的接触点落在中径上，此时最佳量针直径应为：

$$d_{m}=\frac{P}{2}\cos\frac{\alpha}{2} \tag{8-6}$$

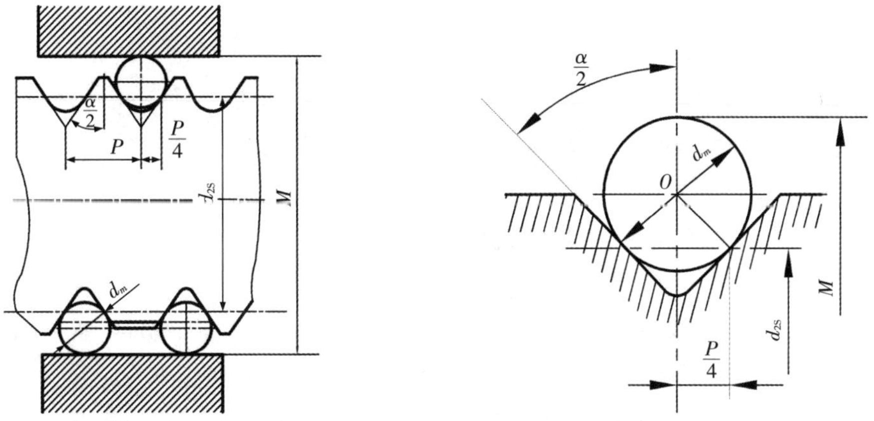

（a）Measure the pin-to-pin distance M
测出针距M

（b）Measure the pin-to-pin distance d_m
量针最佳直径d_m

Fig. 8-11 Three needle method for measuring single pitch diameter of external threads

图8-11 三针法测量外螺纹单一中径

8.6.2.2 Image method for measuring various geometric parameters of threads 影像法测量螺纹各几何参数

Image method for measuring threads refers to using a tool microscope to magnify and image the tooth profile of the tested thread, and measuring its pitch, tooth flank angle, and pitch diameter based on the image of the tested thread. It can also measure its major and minor diameters.

影像法测量螺纹是指用工具显微镜将被测螺纹的牙型轮廓放大成像，按被测螺纹的影像来测量其螺距、牙侧角和中径，也可测量其大径和小径。

8.6.2.3 Using a thread micrometer to measure the pitch diameter of the external threads 用螺纹千分尺测量外螺纹中径

A thread micrometer is a measuring tool for measuring high-precision threads. As shown in Fig.8-12, insert a pair of conical measuring heads 3 and V-groove measuring heads 2 that match the measured thread profile angle and pitch into the positions of the two micrometer anvils to measure the thread pitch.

螺纹千分尺是测量低精度螺纹的量具。如图8-12所示，将一对符合被测螺纹牙型角和螺距的锥形测头3和V形槽测头2分别插入千分尺两测砧的位置，以测量螺纹中径。

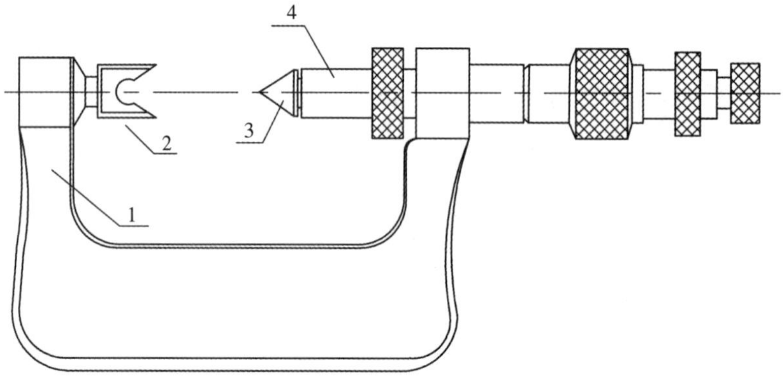

1—Micrometer body; 2—V-groove measuring head; 3—Cone measuring head; 4—Micrometer spindle

1—千分尺身；2—V形槽测头；3—锥形测头；4—测微螺杆

Fig. 8-12 Thread micrometer

图8-12 螺纹千分尺

In order to meet the needs of measuring threads with different pitches, the thread micrometer comes with a set of replaceable measuring heads of different specifications. By installing a conical measuring head and a V-groove measuring head on an inner diameter micrometer, it can also achieve the measurement of internal threads.

为了满足不同螺距的被测螺纹的需要，螺纹千分尺带有一套可更换的不同规格的测头。将锥形测头和V形槽测头安装在内径千分尺上，也能够实现对内螺纹的测量。

Exercise 8 习题8

8-1 What are the factors that affect the interchangeability of threads? How are these deviations controlled?

影响螺纹互换性的因素有哪些？对这些偏差是如何控制的？

8-2 What are the differences and relationships among the pitch diameter, single pitch diameter, and effective pitch diameter of a thread?

螺纹的中径、单一中径和作用中径三者有何区别和联系？

8-3 In a standard thread connection, how is the pitch diameter tolerance of internal and external threads constituted? How is the conformity of the pitch diameter assessed?

普通螺纹结合中，内、外螺纹中径公差是如何构成的?如何判断中径的合格性?

8-4 How many grades are there for the pitch diameter tolerance of standard threads? What are the differences between internal and external threads, and which grade is commonly used?

普通螺纹的中径公差分几级?内外螺纹有何不同，常用的是多少级?

8-5 Explain the meanings of the following thread designations:

试说明下列螺纹标记的含义：

（1）M20；（2）M24 × 2—5g6g—L；

（3）M6—7H/7g6g—L；（4）M6 × 0.75—5h6h—S—LH。

8-6 Using the three-wire method on a length measuring instrument, the measured wire spacing M for an external thread M10 is 10.3323 mm. If the diameter of the three wires d_m is 0.8660 mm, and the actual pitch P is 1.5004 mm, calculate the single pitch diameter d_{2a}.

在测长仪上用三针法测得外螺纹M10的针距M值为10.3323 mm，若三针直径d_m为0.8660 mm，实际螺距P为1.5004 mm，试计算其单一中径d_{2a}的大小。

Chapter 9 Tolerances and Fit of Keys and Splines 键和花键的公差与配合

Key and spline connections are widely used in connections between shaft and shaft parts (such as gear, pulley, coupling, hand wheel, etc.) to transfer torque and movement, and allow axial relative movement between fittings when required. Key and spline connections are detachable connections, often used for frequent disassembly and ease of assembly.

键连接和花键连接广泛应用于轴和轴上零件（如齿轮、带轮、联轴器、手轮等）之间的连接，用以传递扭矩和运动。需要时，配合件之间还可以有轴向相对运动。键和花键连接属于可拆卸连接，常用于需要经常拆卸和便于装配的场合。

9.1 Flat Key Tolerance and Coordination 平键公差与配合

Keys are usually called single keys, which are divided into flat keys, semi-round keys, tangential keys and wedge keys. Among them, the flat keys are the most widely used, which are divided into ordinary flat keys, guide flat keys and thin flat keys. The former is used for fixed connections, while the latter two are used for guiding connections.

键通常被称为单键，其分为平键、半圆键、切向键和楔键。其中，平键应用最广泛，又分为普通平键、导向平键和薄型平键。前者用于固定连接，而后两者用于导向连接。

A flat key transmits torque through the side of the key and the side of the keyway. The size between the two sides is an important size of the flat key connection, and its matching accuracy is high. Flat keys offer good alignment and are easy to install and remove. Guide flat keys are suitable for the shaft parts that can be moved along the axis; Thin flat keys are suitable for hollow shaft, thin-walled structure, low-torque, mainly transferring movement or other special occasions.

平键通过键的侧面和键槽侧面传递扭矩，两侧面间的尺寸是平键连接的重要尺寸，其配合精度较高。平键的对中性良好，拆装方便，导向平键适用于轴上零件可沿轴向移动的场合；薄型平键适用于空心轴、薄壁结构和传递扭矩小、主要传递运动的场合或其他特殊的场合。

The ordinary flat key connection is combined of three parts: Key, shaft and hub, which transfers the movement and torque through the side contact of the key with the shaft slot and the hub slot respectively. There is a certain gap between the upper surface of the key and the bottom surface of the hub slot. Therefore, the sides of the key and the shaft slot should have a large enough actual effective area to bear the load, and the key should be embedded in the shaft slot firmly and reliably to prevent loosening and falling off. Therefore, the key width b and key slot width B are the main parameters to determine the matching nature and matching accuracy. As the main matching size, the strict tolerance should be specified; while the key length L, key height h, shaft slot depth t and hub slot depth t_1 are non-matching size, so their accuracy requirements are low. The flat key connection mode and main parameters are shown in Fig. 9–1. The parameter values are shown in Tab. 9–1.

普通平键连接由键、轴和轮毂三部分组成，通过键的侧面分别与轴槽、轮毂槽的侧面接触来传递运动和转矩，键的上表面和轮毂槽底面留有一定的间隙。因此，键的侧面与轴槽应具有足够大的实际有效接触面积，以承受载荷，并且键应牢固可靠地嵌入轴槽中，以防止松动和脱落。所以，键宽b和键槽宽B是决定配合性质和配合精度的主要参数。作为主要配合尺寸，应规定较严的公差；而键长L、键高h、轴槽深t和轮毂槽深t_1为非配合尺寸，其精度要求较低。平键连接方式及主要参数如图9–1所示。平键连接的配合与应用见表9–1。

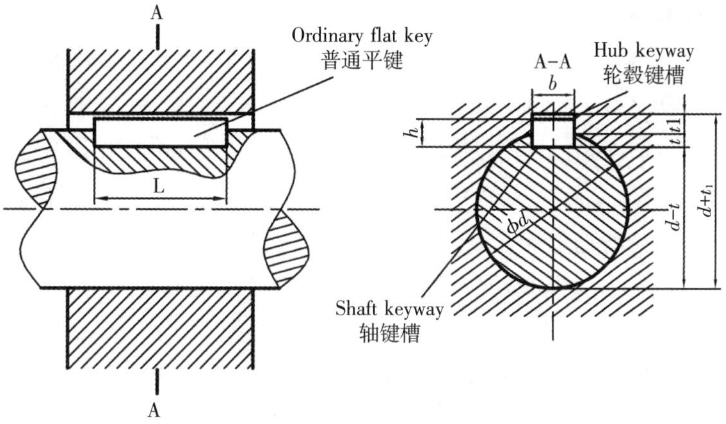

Fig. 9–1 Connecting structure of ordinary flat bonds

图9–1 普通平键的连接结构

Tab. 9–1 Coordination and application of flat key connection

表9–1 平键连接的配合与应用

Fit type 配合各类	Tolerance zone of size b 尺寸b的公差带			Appliction 应用
	Key 键	Shaft slot 轴槽	Hub slot 轮毂槽	
Loose key connection 松联接	h8	H9	D10	The key can slide both on the shaft and in the bub, primarily used for guiding flat keys, allowing the hub to move along the shaft. 键在轴上及轮毂中均能滑动，主要用于导向平键，轮毂可在轴上移动
Normal key connection 普通联接	h8	N9	JS9	The key is fixed in both the shaft slot and the hib slot, suitable for applications with1ow loads 键在轴槽中和轮毂槽中均固定，用于载荷不大的场合
Tight key connection 紧密联接		P9	P9	The key is securely fixed in both the shaft slot and the hub slot, providinga tighter fit then standard key joints, and is used in applications with high loads, impact, and bi directional torque transmission 键在轴槽中和轮毂槽中均牢固地固定，比一般键联接配合更紧，用于载荷较大、有冲击和双向传递扭矩的场合

9.1.1 Tolerance Zone and Coordination Type of Flat Key and Keyway Coordination Dimensions 平键和键槽配合尺寸的公差带与配合种类

In the coordination of the key and the key way, the key width is equivalent to the “shaft”in the broad sense, and the key way is equivalent to the “hole” in the broad serse. The key should cooperate with the shaft slot and the hub slot at the same time, and the fit properties are different. The flat key is a standard part, so the flat key fit adopts the shaft-basis system. The size of the key is selected according to the diameter of the shaft, as shown in Tab. 9–2.

在键与键槽的配合中，键宽相当于广义的“轴”，键槽相当于广义的“孔”。键同时与轴槽和轮毂槽配合，而且配合性质不同。平键是标准件，因此平键配合采用基轴制。键的尺寸根据轴的直径来选择，如表9–2所示。

In order to ensure the key is tightened on the shaft slot and easy to disassemble, at the same time the shaft slot and hub slot can use difference tolerance zone, to make the fit loose or tight, National Standard GB/T 1095–2003 *Square and Rectangular Keywords* stipulates three kinds of connection types, namely loose connection, normal connection and close connection; It also stipulates three tolerance zones for key width and the width of the hub slot, which constitute three different fit properties to meet the various needs different uses. The tolerance zone of fitting size (key and key slot width) are selected from GB/T 1801–2009 standard, as shown in Tab. 9–2.

为保证键在轴槽上紧固，同时又便于拆装，轴槽和轮毂槽可以采用不同的公差带，使其配合的松紧不同，国家标准GB/T 1095—2003《平键、键槽的剖面尺寸》对平键槽与键槽和轮毂槽的宽度规定了三种联接类型，即松联接、普通联接和紧密联接；该标准对轴和轮毂的键槽宽各规定了三种公差带，构成了三种不同性质的配合，以满足各种不同用途的需要。配合尺寸（键与键槽宽）的公差带均从GB/T 1800.1—2020标准中选取。见表9–2。

Tab. 9–2 Size and tolerance of ordinary flat key slots

表9–2 普通平键键槽尺寸与公差

单位：mm

Shaft 轴		Bond Groove 键槽												
Diameter 直径(d)		Key size 键尺寸 ($b \times h$)	Width 宽度(b)						Depth 深度				Radus 半径(r)	
			Nominal size 公称尺寸	Limit deviation 极限偏差					Shaft 轴		Hub 毂			
				Normal connection 正常联接		Tight connection 紧密联接	Loose connection 松联接		Nominal size 公称尺寸	Limit deviation 极限偏差	Nominal size 公称尺寸	Limit deviation 极限偏差		
大于	到			Shaft 轴N9	Hub 毂JS9	Shaft and Hub 轴和毂JS9	Shaft 轴N9	Hub 毂D10					min	max
=6	8	2×2	2	−0.004 −0.029	±0.0125	−0.006 −0.031	+0.025 0	+0.060 +0.020	12	+0.10 0	1.0	+0.10	0.08	0.06
8	10	3×3	3						18		1.4			
10	12	4×4	4	0 −0.030	±0.018	−0.012 −0.042	+0.030 0	+0.078 +0.30	2.5		1.8			
12	17	5×5	5						3.0		2.3		0.16	0.25
17	22	6×6	6						3.5		2.8			
22	30	8×7	8	0 −0.036	±0.018	−0.015 −0.051	+0.036 0	+0.098 +0.040	4.0	+0.2 0	3.3	+0.2 0		
30	38	10×8	10						5.0		3.3		0.25	0.40
38	44	12×8	12	0 −0.043	±0.0215	−0.018 −0.061	+0.043 0	+0.120 +0.050	5.0		3.3			
44	50	14×9	14						5.5		3.8			
50	58	16×10	16						6.0		4.3			
58	65	18×11	18						7.0		4.4			
65	75	20×12	20	0 −0.052	±0.026	−0.022 −0.074	+0.052 0	+0.149 +0.065	7.5		4.9		0.40	0.60
75	82	22×14	22						9.0		5.4			

Tab. 9–2 (Continued) 表9–2（续）

Shaft 轴		Bond Groove 键槽											Radus 半径(r)	
Diameter 直径(d)		Key size 键尺寸 ($b \times h$)	Widh 宽度(b)						Depth 深度					
			Nominal size 公称尺寸	Limit Deviation 极限偏差					Shaft 轴		Hub 毂			
				Normal connection 正常联接		Tight connection 紧密联接	Loose connection 松联接		Nominal size 公称尺寸	Limit deviation 极限偏差	Nominal size 公称尺寸	Limit deviation 极限偏差		
大于	到			Shaft 轴N9	Hub 毂JS9	Shaft and Hub 轴和毂JS9	Shaft 轴N9	Hub 毂D10					min	max
82	95	25 × 14	25	0 −0.052	±0.026	−0.022 −0.074	+0.052 0	+0.149 +0.065	9.0	+0.2 0	5.4	+0.2 0	0.40	0.60
95	110	28 × 16	28						10.0		6.4			
110	130	32 × 18	32	0 −0.062	±0.081	−0.026 −0.088	+0.062 0	+0.180 +0.080	11.0	+0.2 0	7.4	+0.2 0	0.40	0.60
130	150	36 × 20	36						12.0	+0.3 0	8.4	+0.3 0	0.70	1.00
150	170	40 × 22	40						13.0		9.4			
170	200	45 × 25	45						15.0		10.4			
200	230	50 × 25	50						17.0		11.4			
230	260	60 × 32	56	0 −0.074	±0.037	−0.032 −0.106	+0.074 0	+0.220 +0.100	20.0		12.4		1.20	1.60
260	290	63 × 32	63						20.0		12.4			
290	330	70 × 36	70						20.0		14.4			
330	380	80 × 40	80						25.0		15.4		2.00	2.50
380	440	90 × 45	90	0 −0.087	±0.0435	−0.037 −0.124	+0.087 0	+0.260 +0.10	28.0		17.4			
440	500	100 × 50	100						31.0		19.5			

9.1.2 The Tolerance for Non-fitting Sizes 非配合尺寸的公差

The non-mating dimensions of a plain key and a half-round key connection are shown in Fig. 9–2.
普通平键和半圆键连接的非配合尺寸如图9–2所示。

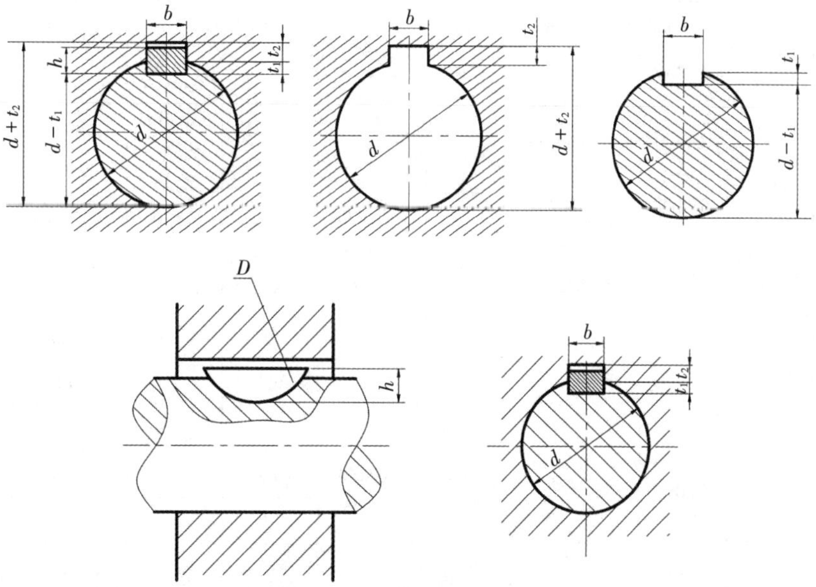

Fig. 9–2 Non-mating dimensions of a plain key and a half-round key connection
图9–2 普通平键和半圆键连接的非配合尺寸

The tolerances for non-fitting sizes are specified as follows:

非配合尺寸公差规定如下：

t_1（shaft slot depth）and t_2（hub slot depth）can be found in Tab. 9−2. The size tolerance zone for L（the shaft slot length）is designated as H14; The size tolerance zone for L（the key length）is designated as h14; The size tolerance zone for h（key height）is specified as h11; The size tolerance zone for d_1（a half-round key）is specified as h12.

t_1（轴槽深）、t_2（轮毂槽深）见表9−2；L（轴槽长）的公差带为H14；L（键长）的公差带为h14，h（键高）的公差带为h11；d_l（半圆键直径）的公差带为h12。

The tolerances for various elements of the key are shown in Tab. 9−3 to Tab. 9−5.

键的各要素公差见表9−3～表9−5。

Tab. 9−3 Tolerance of plain flat key（From GB/T 1096−2003）

表9−3 普通平键的公差（摘自 GB/T 1096—2003） 单位：mm

Hub width 键宽(b)	Nominal size 公称尺寸		2	3	4	5	6	8	10	12	14	16	18	20	22	25	28
	Limit devation 极限偏差（b8）		0 −0.014		0 −0.018			0 −0.022		0 −0.027				0 −0.033			
Hub hight 键高(h)	Nominal size 公称尺寸		2	3	4	5	6	7	8	8	9	10	11	12	14	16	18
	Limit devation 极限偏差	Rectangle（h11）矩形（h11）						0 −0.090					0 −0.110				
		Square（h8）方形（h11）	0 −0.014		0 −0.018												

Tab. 9−4 Dimensions and tolerances of half-round keyways（From GB/T 1098−2003）

表9−4 半圆键键槽的尺寸与公差（摘自GB/T 1098—2003） 单位：mm

Key size 键尺寸 ($b \times h \times D$)	Keyways 键槽										Radials 半径(R)	
	Width 宽度(b)						Depth 深度					
	Nominal size 公称尺寸	Limits of deviation 极限偏差					Shaft 轴(t_1)		Hubs 毂(t_2)			
		Normal connection 正常联接		Tight connection 紧密联接	Loose connection 松联接		Nominal size 公称尺寸	Limits of deviation 极限偏差	Nominal size 公称尺寸	Limits of deviation 极限偏差		
		Shaft N9 轴N9	Hubs JS9 毂JS9	Shaft and hubs P9 轴和毂P9	Shaft H9 轴 H9	Hubs D10 毂D10					max	min
3 × 5 × 13	3	−0.004 −0.029	±0.0125	−0.006 −0.031	+0.025 0	+0,060 +0.020	3.8	+0.2 0	1.4	+0.1 0	0.16	0.08
3 × 6.5 × 16	3						5.3		1.4			
4 × 6.5 × 16	4	0 −0.030	±0.015	−0.012 −0.042	+0.030 0	+0.078 +0.030	5.0		1.8		0.25	0.16
4 × 7.5 × 19	4						6.0		1.8			
5 × 6.5 × 16	5						4.5		2.3			
5 × 7.5 × 19	5						5.5		2.3			
5 × 9 × 22	5						7.0	+0.3 0	2.3			
6 × 9 × 22	6						6.5		2.8			
6 × 10 × 25	6						7.5		2.8	+0.2 0		

Tab. 9–5 Dimensions and tolerances of half-round key (From GB/T 1099.1–2003)

表9–5 半圆键的尺寸与公差（摘自GB/T 1099.1—2003） 单位：mm

键尺寸 ($b \times h \times D$)	Width 宽度b		Height 高度h		Diameter 直径D	
	Nominal size 公称尺寸	Limits of deviation 极限偏差	Nominal size 公称尺寸	Limits of deviation 极限偏差（$h12$）	Nominal size 公称尺寸	Limits of deviation 极限偏差（$h12$）
3 × 5 × 13	3		5		13	
3 × 6.5 × 16	3		6.5		16	0 −0.180
4 × 6.5 × 16	4		6.5		16	
4 × 7.5 × 19	4		7.5		19	0 −0.210
5 × 6.5 × 16	5	0 −0.025	6.5	0 −0.15	16	0 −0.180
5 × 7.5 × 19	5		7.5		19	
5 × 9 × 22	5		9		22	0 −0.210
6 × 9 × 22	6		9		22	
6 × 10 × 25	6		10		25	

9.1.3 Geometric Tolerance of Key and Keyway 键与键槽的几何公差

To ensure sufficient contact area between the key and the keyway and to avoid assembly difficulties, national standards also specify the symmetry tolerance of the axis of the shaft key way relative to the axis of shaft, and the axis of the hub key way relative to the hole's depth, as well as the parallelism tolerance of the two fitting sides of the key. See Tab. 9–2. Symmetry tolerance grades 7 ~ 9 for the shaft key slot and hub key slot are chosen according to GB/T 1184–1996, titled *Shape and Positional Tolerance, Without Individual Tolerance Indications*. When the ratio of the key's length L to its width b is greater than or equal to 8, the parallelism of the key's two sides must adhere to GB. Specifically, for keys with a width b of 6 mm or less, the parallelism tolerance is selected according to Grade 7; for keys with a width between 8 mm and 36 mm, Grade 6 is applied; and for keys with a width of 40 mm or greater, Grade 5 is utilized.

为了保证键和键槽之间具有足够的接触面积，并避免装配困难，国家标准还规定了轴键槽对轴线和轮毂键槽对孔槽深的轴线的对称度公差，同时规定了键的两个配合侧面的平行度公差，见表9–2。轴键槽和轮毂键槽的对称度公差等级按GB/T 1184—1996《形状和位置公差、未注公差值》选择7 ~ 9级。当键长L与键宽b之比大于或等于8时，键的两侧面的平行度必须遵守国家标准的规定。当键宽$b \leq$ 6 mm时平行度公差按7级选取，当键宽b为8 ~ 36 mm时按6级选取，当键宽$b \geq$ 40 mm时按5级选取。

9.1.4 Surface Roughness of Flat and Keyway 平键和键槽的表面粗糙度

At the same time, it is also stipulated that the upper limit of the surface roughness Ra value for the two sides of the width b of shaft keyway and hub keyway is generally taken as 1.6 ~ 3.2 μm, and the upper limit of the surface roughness Ra value for the bottom surface of the shaft keyway and the hub keyway is taken as 6.3 μm.

同时规定轴键槽、轮毂键槽宽b的两侧面的表面粗糙度Ra值的上限值一般取为1.6 ~ 3.2 μm，轴键槽底面、轮毂键槽底面的表面粗糙度Ra值的上限值取为6.3 μm。

When the control of form errors can be guaranteed by the process, tolerances need not be specified on the drawing.

当形状误差的控制可由工艺保证时，图样上可不给出公差。

The profile size of the shaft key slot and key slot of the hub and their tolerance zone, the key slot geometric tolerance and surface roughness requirements are marked on the drawing as shown in Fig. 9-3, and the symmetry tolerance in the figure adopts the independent principle.

轴键槽和轮毂键槽剖面尺寸及其公差带、键槽的几何公差和表面粗糙度要求。在图样上的标注，如图9-3所示，图中的对称度公差采用独立原则。

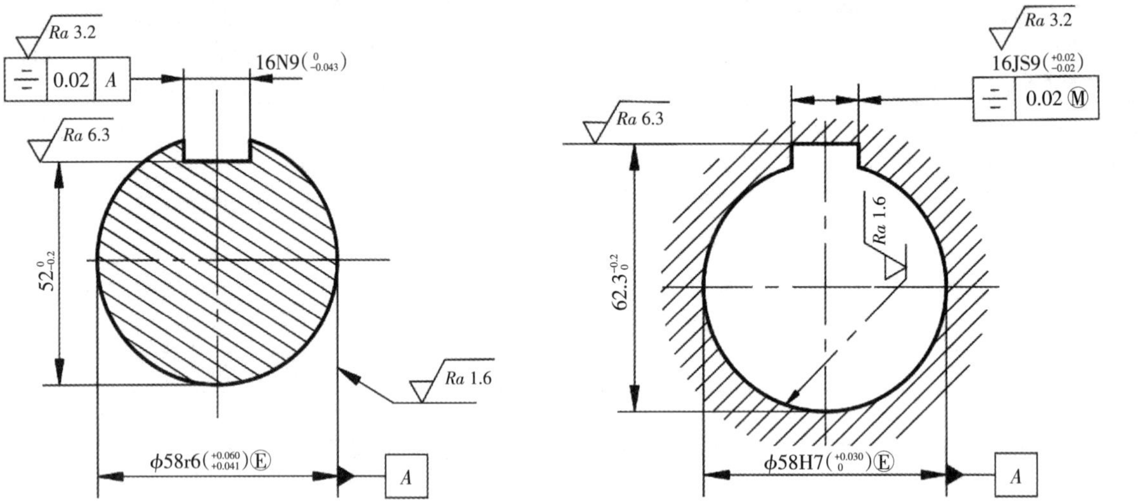

(a) The symmetry tolerance adopts the independence principle (b) The symmetry tolerance adopts the maximum entity requirement

(a) 对称度公差采用独立原则 (b) 对称度公差采用最大实体要求

Fig. 9-3 Note of key slot dimension and tolerance

图9-3 键槽尺寸与公差标注

9.2 Rectangular Splines 矩形花键

When the heavy torque is transferred and the centering precision is high, the single key connection cannot meet the requirements, so the spline connection is needed. Spline connection is the combination of two parts, spline shaft and spline hole. Splines can be used as a fixed connection or as a sliding connection.

当传递较大的转矩，定心精度又要求较高时，单键联接满足不了要求，需采用花键联接。花键联接是花键轴、花键孔两个零件的结合。花键可用作固定联接，也可用作滑动联接。

Compared with the flat key connections, the spline connections offer obvious advantages: the axis alignment precision (centering precision) and good guidance, the load distribution on the shaft and hub is relatively uniform, so it can transfer a large torque, and the strength is high, the connection is more reliable.

花键联接与平键联接相比具有明显的优势：孔、轴的轴线对中精度（定心精度）高，导向性好，轴和轮毂上承受的负荷分布比较均匀，因而可以传递较大的转矩，而且强度高，连接更可靠。

Splines are divided into rectangular spline and involute spline. Compared with the involute spline connection, the rectangular spline connection has higher strength and stronger bearing capacity, and has the advantages of high precision, good tooth surface contact, automatic centering and convenient processing.

花键分为矩形花键和渐开线花键两种。与渐开线花键联接相比，矩形花键联接的强度更高，承载能力更强，并且具有精度高、齿面接触良好、能自动定心、加工方便等优点。

9.2.1 Basic Dimensions of the Rectangular Splines 矩形花键的基本尺寸

GB/T 1144-2001 specifies three main dimensions for rectangular splines: mayor diameter (D), minor path (d), key width (B), and keyway width B, as shown in Fig. 9-4. To facilitate processing and measurement, the number of keys for rectangular splines is specified as even, with six, eight and ten. Based on load carrying capacity, the nominal size of the rectangular spline is divided into two series; Medium and Light series. Medium series has strong carrying capacity, and it is mostly used in automobile and tractor manufacturing; light series has relatively low carrying capacity and is mostly used in machine tool manufacturing.

GB/T 1144—2001规定了矩形花键的三个主要尺寸参数为大径D、小径d、键宽和键槽宽B，如图9-4所示。为了便于加工和测量，矩形花键的键数规定为偶数，有6键、8键和10键三种。按照承载能力不同，矩形花键分为中系列、轻系列两个系列。中系列的承载能力强，多用于汽车、拖拉机等制造业；轻系列的承载能力相对低，多用于机床制造业。

The light series and medium series of the same trail have the same number of keys, and the key width and keyway width are the same, only the large diameter varies. Medium series has large key height size and strong carrying capacity; light series has small key height size and low carrying capacity. The basic size series of rectangular splines is shown in Tab. 9-6.

同一规格（或称为同一轨迹）的轻系列和中系列的键数相同，键宽和键槽宽也相同，仅大径不同。中系列花键的键高尺寸较大，承载能力强；轻系列花键的键高尺寸较小，承载能力较低。矩形花键的基本尺寸系列见表9-6。

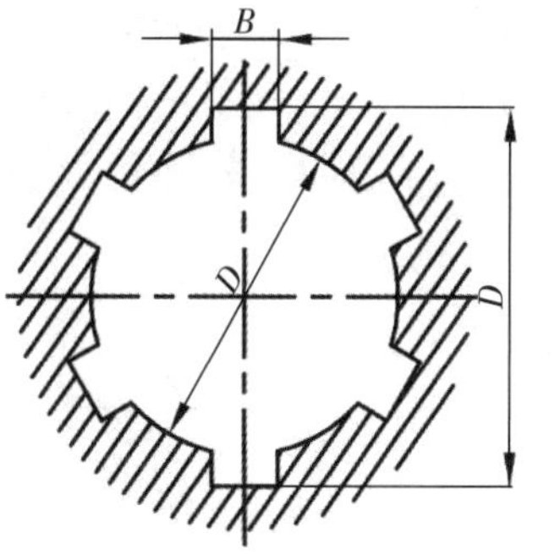

(a) Internal spline
内花键

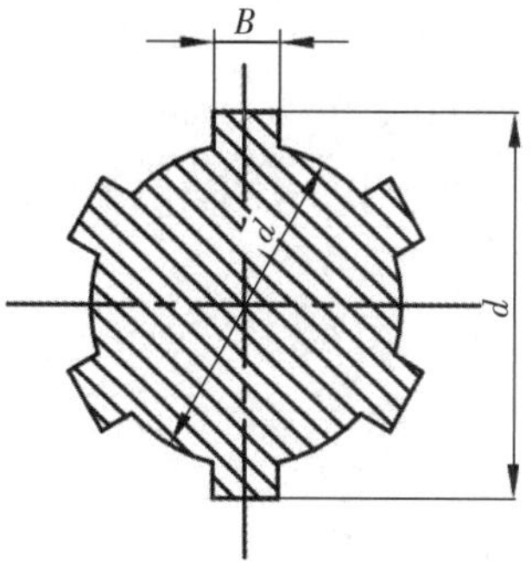

(b) External spline
外花键

Fig. 9-4 Main dimensions of the rectangular splines

图9-4 矩形花键的主要尺寸

Tab. 9-6 Basic size series of rectangular sines (GB/T 1144-2001)

表9-6 矩形花键的基本尺寸系列（GB/T 1144—2001）

Minor diameter 小径 d/mm	Light series 轻系列				Medium series 中系列			
	Specification 规格 $N\times d\times D\times B$	Number of key 键数 N	Major diameter 大径 D/mm	Key width 键宽 B/mm	Specification 规格 $N\times d\times D\times B$	Number of key 键数 N	Major diameter 大径 D/mm	Key width 键宽 B/mm
23	6×23×26	6	26	6	6×23×28	6	28	6
26	6×26×30	6	30	6	6×26×31	6	32	6
28	6×28×32	6	32	7	6×28×34	6	34	7
32	8×32×36	8	36	6	8×32×38	8	38	6
36	8×36×40	8	40	7	8×36×42	8	42	7
42	8×42×46	8	46	8	8×42×48	8	48	8
46	8×46×50	8	50	9	8×46×54	8	54	9
52	6×52×58	8	58	10	8×52×60	8	60	10
56	8×56×62	8	62	10	8×56×65	8	65	10
62	8×62×67	8	68	12	8×62×72	8	72	12
72	10×72×78	10	78	12	10×72×82	10	82	12

Notes: N—the numbers of keys; d—small diameter; D—big diameter; B—key (key slot) width.

注：N—键数；d—小径；D—大径；B—键（键槽）宽。

9.2.2 Spline Centering Way 花键定心方式

The main requirement of spline connection is to ensure a high coaxial degree after the internal and external spline bonding, and to transfer the torque. Splines have three main size parameters: large diameter *D*, small path *d* and key（slot）width *B*. If these three sizes are required to fit and concentration, it is difficult to ensure the coaxiality of internal and external splines, and it is not necessary. Therefore, in order to improve its processing technology, only the dimensions *B* and *D* or *d* are made more accurately, so that it plays a clearance role in the fit, while the other size *d* or *D* is processed with a lower accuracy, and give a larger clearance.

花键联接的主要要求是保证内、外花键联接后具有较高的同轴度，并能传递扭矩。花键有大径*D*、小径*d*和键（槽）宽*B*三个主要尺寸参数，若要求这三个尺寸同时配合紧密，很难保证内、外花键同轴度，而且也没有必要。因此，为了改善其加工工艺性，只需将尺寸*B*和*D*或*d*加工得较准确，使其起配合定心作用，而另一尺寸*d*或*D*则按较低精度加工，并给予较大的间隙。

Since torque transfer is achieved through key and keyway sides, key and slot width require high dimensional precision, whether they serve as centering sizes or not.

由于扭矩的传递是通过键和键槽两侧面来实现的，因此，键宽和槽宽不论是否作为定心尺寸，它们都需要较高的尺寸精度。

According to the different centering elements, there are three centering modes:（1）centering by large diameter *D*;（2）centering by small path *d*;（3）centering by key width *B*. As shown in Fig. 9-5.

根据定心要素的不同，分为三种定心方式：（1）按大径*D*定心；（2）按小径*d* 定心；（3）按键宽*B*定心。如图9-5所示。

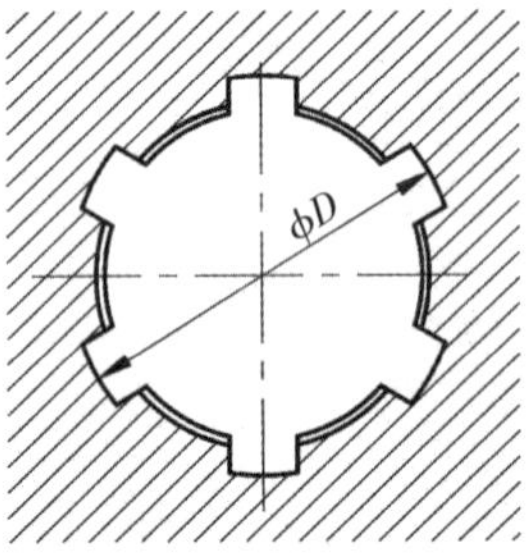

（a）Major diameter centering
大径定心

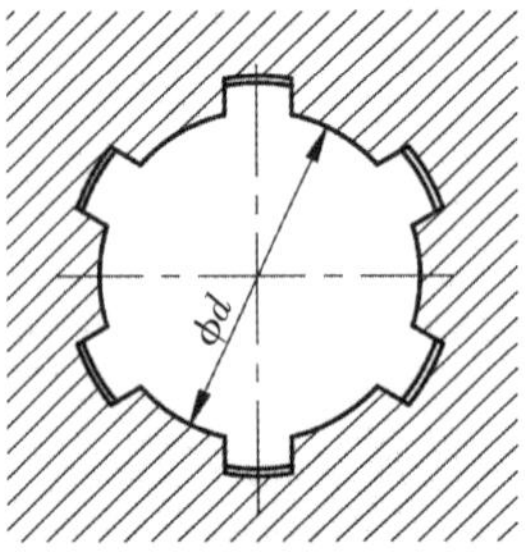

（b）Miner path centering
小径定心

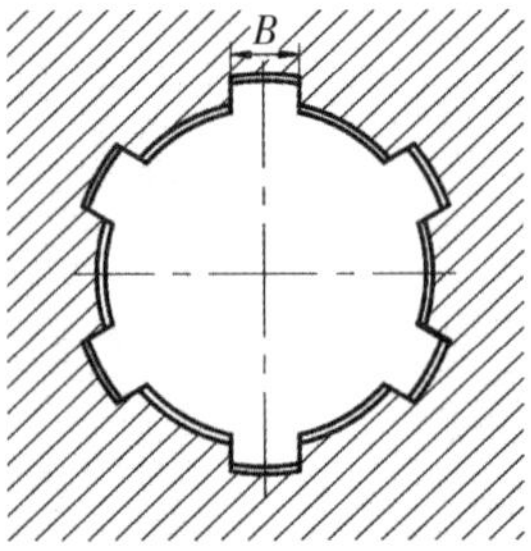

（c）Key width centering
键宽定心

Fig. 9-5 The centering mode of the rectangular spline connection

图9-5 矩形花键连接的定心方式

The national standard of the rectangular spline（GB/T 1144-2001）stipulates that the rectangular spline is fixed by a minor diameter, because it has a series of advantages. When centering with large diameter, the accuracy of the internal spline centering surface is guaranteed by the pull knife. When the surface hardness of internal spline centering is high（above HRC40), the deformation after heat treatment is difficult to correct by push knife; When the surface roughness of internal spline centering is high（$Ra < 0.63$ μm), it is difficult to guarantee the drawing process; In single piece and small batch production and large-size spline, it is difficult to use because this processing method is not economical. When using small path centering, the deformation after heat treatment can be repaired by the internal mill, and the internal mill can meet higher dimensional precision and higher surface roughness requirements. Therefore, the small path centering accuracy is higher, with better centering stability and longer service life, which is conducive to the improvement of product quality. The accuracy of the external spline path can be guaranteed by form grinding.

关于矩形花键的国家标准（GB/T 1144—2001）规定，矩形花键以小径定心，因为小径定心有一系列优点。当用大径定心时，内花键定心表面的精度由拉刀保证。而当内花键定心表面硬度要求高(HRC40以上）时，热处理后的变形难以用推刀修正；当内花键定心表面粗糙度要求高（$Ra<0.63$ μm）时，用拉削工艺也难以保证；在单件、小批生产及大规格花键中，内花键也难以用拉削工艺，因为该种加工方式不经济。采用小径定心时，热处理后的变形可用内圆磨修复，而且内圆磨可达到更高的尺寸精度和更高的表面粗糙度要求。因此，小径定心的定心精度更高，定心稳定性较好，使用寿命长，有利于产品质量的提高。外花键小径精度可由成形磨削保证。

9.2.3 Tolerance and Fit of the Rectangular Spline Connection 矩形花键连接的公差与配合

9.2.3.1 Size tolerance of the rectangular splines 矩形花键的尺寸公差

The size tolerance zones of the internal and external spline centering diameters(d), non-centering major diameter(D) and key width or keyway width(B) are classifid into General Use and Precision Transmission categories. The size tolerance zones of the internal and external spline are shown in Tab. 9-7. In order to reduce the number of special cutting tools and measuring tools, spline connection adopts a hole-basis system.

Tab. 9-7 Size tolerance band of rectangular spline（GB/T 1144-2001）

表9-7 矩形花键的尺寸公差带（GB/T 1144—2001）

Internal spline 内花键				External spline 外花键			Assembly type 装配形式
Minor diameter 小径（*d*）	Major diameter 大径（*D*）	Keyway width 键槽宽（*B*）		Minor diameter 小径（*d*）	Major diameter 大径（*D*）	Keyway width 键宽（*B*）	
		No heat treatment after broaching 拉削后不热处理	Heat treatment after broaching 拉削后热处理				
General used 一般用							
H7	H10	H9	H11	f7	a11	d10	滑动
				g7		f9	紧滑动
				h7		h10	固定
Precision transmission used 精密传动用							
H5	H10	H7，H9		f5	a11	d8	滑动
				g5		f7	紧滑动
				h5		h8	固定
H6				f6		d8	滑动
				g6		f7	紧滑动
				h6		d8	固定

Notes：① When the internal spline for precision transmission is needed to control the key side coordination gap, H7 can be used for the slot width, and H9 can be used in general.

② When the internal spline tolerance zone is H6 and H7, coordination with the higher external splines is allowed.

注：① 当精密传动用的内花键需要控制键侧配合间隙时，槽宽可选用H7，一般情况可选用H9。

② 当内花键公差带为 H6 和 H7 时，允许与高一级的外花键配合。

内、外花键定心小径、非定心大径和键宽或键槽宽（*B*）的尺寸公差带分一般用和精密传动两类。其内、外花键的尺寸公差带见表9-7。为减少专用刀具和量具的数量，花键连接采用基孔制配合。

As can be seen from Tab. 9-7: two kinds of tol-erance zones are specified for the general slot width processing, with the tolerance zone is H9; For heat-treatment after processing, the tolerance zone is H11; For precision transmission, when the key side fit clearance gap is small, H7 is the slot width tolerance zone, and H9.

从表9-7中可以看出，对于一般槽宽的加工，规定了两种公差带，其中公差带为H9；对于加工后的热处理，公差带为H11；对于精密传动，当键侧配合间隙较小时，槽宽公差带为H7，也有使用H9的情况。

9.2.3.2 Geometric tolerance of rectangular spline connections 矩形花键连接的几何公差

Internal and external splines are connecting components with complex surfaces, and the ratio of key length to key width is relatively large. Therefore, geometric tolerances must be specified separately for internal and external splines to ensure the precision and strength requirements of spline connections. To guarantee fitting properties, GB/T 1144-2001 stipulates that the relationship between the form tolerance and size tolerance of the minor diameter centering surfaces for internal and external splines shall comply with the envelope requirement. To ensure uniform distribution of spline teeth (or splines slots) around the circumference, during batch production, the position tolerance of the key width (see Tab. 9-8) and the size tolerance of the minor diameter centering surface shall adhere to the maximum material requirement (MMR), as is shown in Fig.9-6.

内、外花键是具有复杂表面的联接件，且键长与键宽的比值较大，因此对内、外花键必须分别规定几何公差，以保证花键键联接精度和强度的要求。为了保证配合性质，GB/T1144—2001规定内、外花键小径定心表面的形状公差与尺寸公差的关系采用包容要求。为了保证花键(或花键槽)在圆周上分布的均性，批量生产时，键宽的位置度公差(见表9-8)与小径定心表面的尺寸公差关系应符合最大实体要求，图样标注如图9-6所示。

Tab. 9-8 Position tolerance of rectangular spline（GB/T 1144-2001）

表9-8 矩形花键的位置度公差（GB/T 1144—2001） 单位：mm

Keyways width or key width 键槽宽或键宽（B）		3	3.5 ~ 6	7 ~ 10	12 ~ 18
Keyways width 键槽宽		0.010	0.015	0.020	0.025
Key width	Slide & fixed 滑动、固定	0.010	0.015	0.020	0.025
	Close slide 紧滑动	0.006	0.010	0.013	0.016

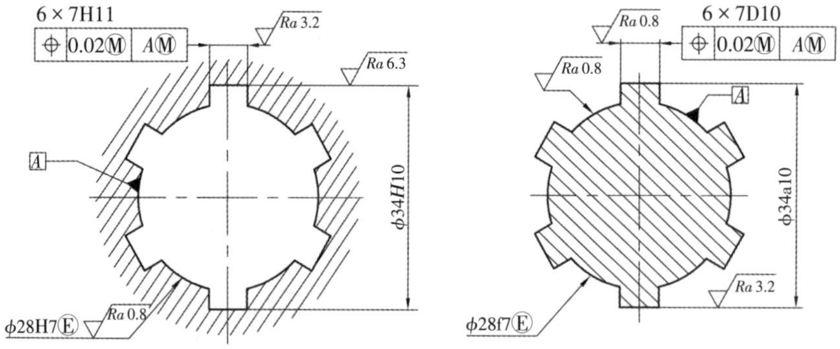

(a) Internal spline
内花键

(b) External spline
外花键

Fig. 9-6 Annotation of spline position tolerance

图9-6 花键位置度公差的标注

For single-piece or small-batch production,the relationship between the symmetry tolerance of the key width (see Tab. 9-9) and the size tolerance of the minor diameter centering surface should comply with the independent principle,and the drawing annotation is shown in Fig.9-7. Additionally,for longer splines,parallelism tolerance of the key sides relative to the axis may be specified according to product performance requirements.

单件小批生产时，键宽的对称度公差（见表9–9）与小径定心表面的尺寸公差关系则应遵守独立原则，图样标注如图9–7所示。另外，对于较长花键，可根据产品性能自行规定键侧对轴线的平行度公差。

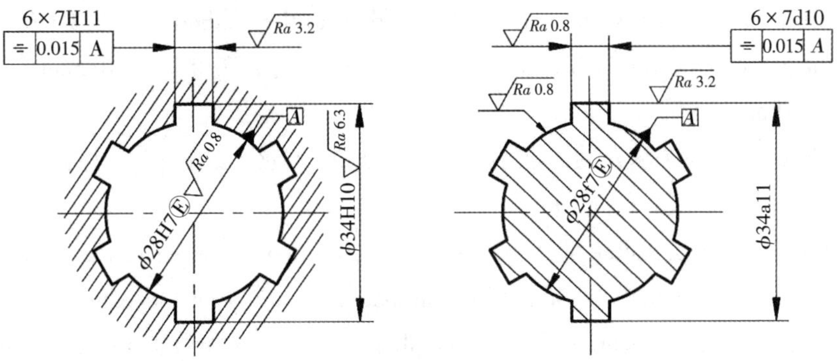

（a）Internal spline

（a）内花键

（b）External spline

（b）外花键

Fig. 9–7 Annotation of the spline symmetry tolerance

图9–7 花键对称度公差的标注

Tab. 9–9 Symmetry tolerance of rectangular spline（GB/T 1144–2001）

表9–9 矩形花键的对称度公差（GB/T 1144—2001）

Keyways width or key width 键槽宽或键宽（*B*）	3	3.5～6	7～10	12～18
General purpose 一般用	0.010	0.012	0.015	0.018
Precision transmission 精密传动用	0.006	0.008	0.009	0.011

9.2.3.3 Surface roughness of the rectangular splines 矩形花键的表面粗糙度

The upper limits of the surface roughness parameter *Ra* of the rectangular spline are recommended as follows:

矩形花键的表面粗糙度参数*Ra*的上限值推荐如下：

（1）Internal spline: the surface of small diameter is not greater than 0.8 μm, the side of the slot is not greater than 3.2 μm, the big diameter surface is not greater than 6.3 μm.

内花键：小径表面不大于0.8 μm，键槽侧面不大于3.2 μm，大径表面不大于6.3 μm。

（2）External spline: the small diameter surface is not greater than 0.8 μm, the side of the key slot is not greater than 0.8 μm, the big diameter surface is not greater than 3.2 μm。The sample annotation is shown in Fig. 9–7. Recommended values for spline surface roughness are provided in Tab. 9–10.

外花键：小径表面粗糙度不大于0.8 μm，键槽侧面不大于0.8 μm，大径表面不大于3.2 μm。样例标注如图9–7所示。花键表面粗糙度推荐值见表9–10。

Tab. 9–10 Surface roughness of spline

表9–10 花键表面粗糙度推荐值

Surface 加工表面	Internal splines 内花键	External splines 外花键
	Ra/μm	
Small diameter 小径	1.6	0.8
Big diameter 大径	6.3	3.2
Key sides 键侧	6.3	1.6

9.2.4 Mark of the Rectangular Splines on the Pattern 矩形花键连接在图样上的标注

Rectangular spline specifications are expressed according to the method of $N \times d \times D \times B$, (e.g. 8 × 52 × 58 × 10), where the number of key is 8 and small diameter is 52 mm, the big diameter is 58 mm, the key (key way) width is 10 mm.

矩形花键规格按$N \times d \times D \times B$的方法表示，如8 × 52 × 58 × 10，依次表示键数为8，小径为52 mm，大径为58 mm，键（键槽）宽为10 mm。

The mark of the rectangular spline shall be written in the order specified in the spline specification, and the fit or tolerance zone symbol shall be added, which shall be marked on the pattern as shown in Fig. 9-8. Fig. 9-8 (a) is a spline pair, indicating that the number of splines is 6, the small diameter fit is 23H7/f7, the large diameter fit is 28H10/a11, and the key width fit is 6H11/d10. On the part diagram, the spline tolerance zone can still be nated in the sequence of the spline specification , as shown in Fig. 9-8 (b) and (c).

矩形花键的标记按花键规格所规定的顺序书写，另需加上配合或公差带代号，其在图样上的标注如图9-8所示。图9-8（a）为一花键副，表示花键数为6，小径配合为23H7/f7，大径配合为28H10/a11，键宽配合为6H11/d10；在零件图上，花键公差带可仍按花键规格按顺序注出，如图9-8（b）、（c）所示。

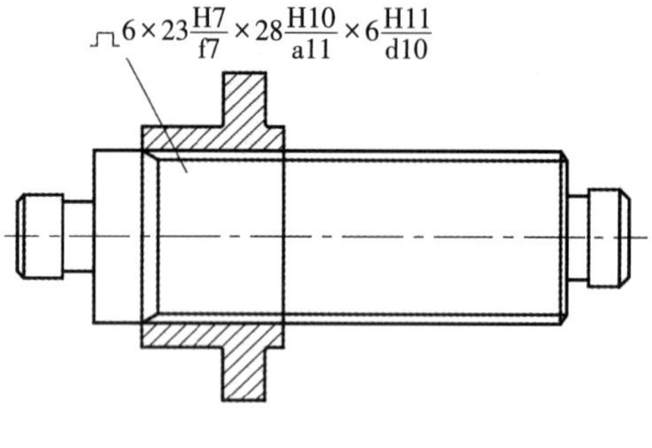

(a) Marking of on the assembly drawing

(a) 在装配图样上的标注

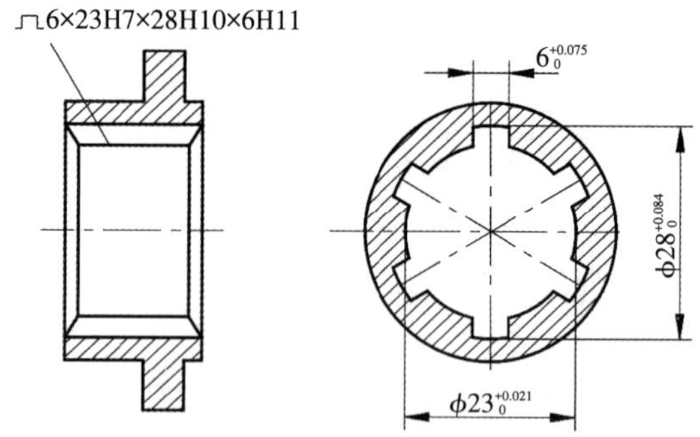

(b) Marking of the internal spline

(b) 内花键的标注

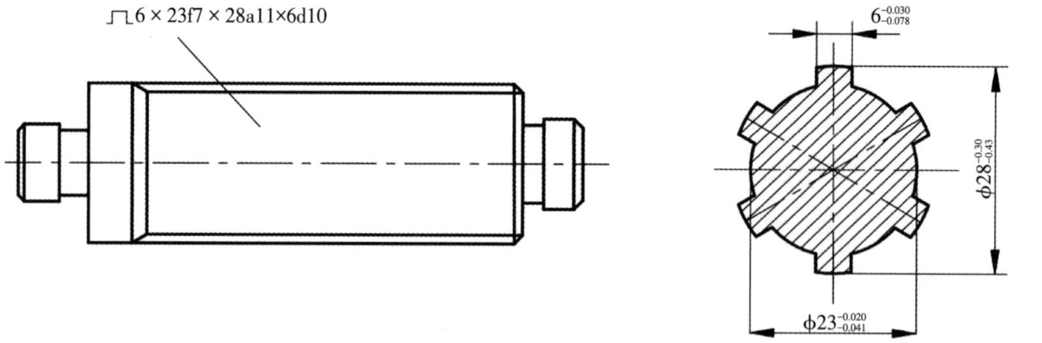

(c) Marking of the external spline

(c) 外花键的标注

Fig. 9-8 Marking of fit and tolerance of rectangular spline

图9-8 矩形花键配合及公差标注

Exercise 9 习题9

9-1 What are the characteristics of flat key connections? What are the main geometric parameters?

平键连接的特点是什么？主要几何参数有哪些？

9-2 Why is a stricter tolerance only specified for the key（slot）width in flat key connections?

平键连接为什么只对键（槽）宽规定较严的公差？

9-3 What basis fit system is used for flat key connection fits? What basis fit system is used for spline connections?

平键连接的配合采用何种基准制？花键连接采用何种基准制？

9-4 What is spline centering? Why do rectangular splines generally use minor diameter centering? What are the advantages and disadvantages of minor diameter centering?

什么是花键定心？矩形花键为什么一般采用小径定心?小径定心有什么优缺点？

9-5 A reducer transmits general torque, and a gear on one shaft is connected to the shaft via a flat key. Given the key width b=10mm, determine the fit designation for the key width b, find its limit deviations, and draw the tolerance zone diagram.

某减速器传递一般扭矩，其中一轴上齿轮与轴的连接采用平键。已知键宽$b=10$ mm，试确定键宽b的配合代号，查出其极限偏差值，并画出公差带图。

Chapter 10 Accuracy and Detection of the Cylindrical Gear 圆柱齿轮的精度与检测

10.1 Overview 概述

Gear is an important transmission part, which is widely used in various fields, such as automobiles, agricultural machinery, machine tools, mining machinery, military engineering, instruments and home appliances and other machinery and equipment. The purpose of gear accuracy design is to ensure the accuracy and interchangeability of gear transmission. When the manufacturing accuracy grade of a gear varies by one grade, the carrying capacity difference is by 20% ~ 30%, the noise difference is by 2.5 ~ 3.0 decibels, and the manufacturing cost difference is by 60% ~ 80%. The design, process, manufacturing, inspection, sales and procurement of gears are all based on the gear accuracy standards.

齿轮是一种重要的传动零件，被广泛地应用于各种领域，如汽车、农业机械、机床、矿山机械、军事工程、仪器和家电等的机器设备中。齿轮精度设计的目的是保证齿轮传动的准确性和互换性。齿轮的制造精度等级每提高一级，其承载能力差异为20% ~ 30%，噪声差异为2.5 ~ 3.0 dB，而制造成本差异则高达60% ~ 80%。齿轮的设计、工艺、制造、检验以及销售和采购，都以齿轮精度标准为重要的依据。

Gear transmission application requirements can be divided into transmission accuracy and tooth side clearance. According to the functional characteristics of gear transmission, the transmission accuracy can be divided into the accuracy of transmission motion, the stability of transmission motion and the uniformity of load distribution. Therefore, in general, the application requirements of gear transmission can be divided into the following four aspects.

齿轮传动应用要求可分为传动精度和齿侧间隙两个方面。根据齿轮传动的功能特性，传动精度可分为传递运动的准确性、传递运动的平稳性和载荷分布的均匀性三个方面。因此，一般情况下，齿轮传动的应用要求可分为以下4个方面。

10.1.1 Accuracy of the Transmission Motion 传递运动的准确性

The accuracy of transmission motion means that over one complete revolution of the gear, the maximum angular error does not exceed a certain limit.

传递运动的准确性是指齿轮在一转范围内，最大转角误差不超过一定的限度。

10.1.2 Stability of the Transfer Motion 传递运动的平稳性

The stability of the transfer movement means that the gear is within the range of one too the pitch, the instantaneous transmission ratio does not exceed a certain range.

传递运动的平稳性是指要求齿轮在转一齿范围内，瞬时传动比变化不超过一定的范围。

10.1.3 Uniformity of the Load Distribution 载荷分布的均匀性

The uniformity of load distribution means that the working tooth surface should ensure good contact, avoid stress concentration, reduce tooth surface wear, and improve the strength and life of the tooth surface.

载荷分布的均匀性是指要求一对齿轮在啮合时，工作齿面要保证接触良好，避免应力集中，减少齿面磨损，提高齿面强度和寿命。

10.1.4 Rationality of the Tooth Side Clearance 齿侧间隙合理性

Tooth side clearance, or gear side clearance, refers to the requirement that the non-working flanks of a meshing gear pair have proper clearance. Backlash is naturally formed when gears, shafts, and other parts are assembled into a gearbox. reducer, or other transmission device. Appropriate gear side clearance is used to store lubricants, compensate for thermal and elastic deformation of the gears during operation, and prevent tooth surface ablation or jamming, ensuring the gear pair operates normally.

齿轮齿侧间隙是指齿轮副啮合时非工作齿面之间需保持适当间隙的要求。齿侧间隙是在齿轮、轴以及其他部件组装成箱式减速器、齿轮箱或其他传动装置时自然形成的。适当的齿轮齿侧间隙可用于储存润滑油，补偿齿轮在工作中的热变形和弹性变形，防止齿面发生胶合或卡死现象，从而确保齿轮副能够正常工作。

High-speed heavy-duty gears, such as those used in steam turbine reducers, are characterized by extremely high rotational speeds and substantial power transmission. These gears require superior transmission smoothness to minimize vibration. impact, and noise. Meanwhile, strict requirements are also imposed on the uniformity of tooth surface contact to ensure reliable load distrbution. In contrast, low-speed heavy-duty gears employed in steel rolling mills, mining machinery, cranes, and similar equipment are designed for high-power transmission at low speeds. Here, the primary focus are on uniform tooth surface contact and high load-carrying capacity, with relatively lower demands for transimission accuracy and smoothness. Regarding gear pair backlash, regardless of the application, a specific amount of backlash is indispenable to guarantee flexible meshing and prevent jamming during transmission.

对于不同用途的齿轮及其传动，这四个方面要求的程度并不完全一致。例如，分度齿轮和读数齿轮，突出的要求是分度正确和示值正确，因此要求较高的传递运动准确性，而其他方面精度可相应低些。对于高速重载齿轮，如汽轮机减速器中的齿轮，其特点是转速很高、传动功率也较大，因此要求较高的传动平稳性，使振动、冲击和噪声尽量小些。同时，对齿面接触均匀性也应有较高的要求。对于低速重载齿轮，如轧钢机、矿山机械以及起重机用的齿轮，其特点是传动功率相当大、转速比较低，这时主要的要求是齿面接触均匀、承载能力高，而对传递运动的准确性、传动平稳性要求均可低些。至于齿轮副侧隙，无论任何齿轮，为保证其传动灵活，都必须留有一定的侧隙。

10.2 Evaluation Parameters of Gear Accuracy 齿轮精度的评定参数

10.2.1 Check the Required Deviation Items for Assessing the Gear Accuracy 评定齿轮精度的必检偏差项目

In order to assess the three accuracy requirements of gears, according to the GB/T 10095.1-2022 standard, mandatory deviations (individual pitch deviation, cumulative pitch deviation, and total cumulative pitch deviation), tooth profile deviation, and helix deviation must be measured. To assess the backlash, it is customary to measure either tooth thickness deviation or common normal deviation length.

为了评定齿轮的三项精度要求，根据GB/T 10095.1—2022标准，需检测强制性偏差（单个齿距偏差、齿距累积偏差以及齿距累积总偏差）、齿廓偏差和螺旋线偏差。为了评定齿轮的侧隙大小，通常会检测齿厚偏差或公法线长度偏差。

10.2.1.1 Required inspection parameters for transmission accuracy 传动准确性的必检参数

(1) Total cumulative pitch deviation (ΔF_P) 齿距累积总偏差（ΔF_P）。

Total cumulative pitch deviation (ΔF_P), refers to the maximum cumulative deviation in any arc segment ($k=1$ to $k=z$) on the same side of the gear. It appears as the total amplitude of the cumulative deviation curve, as shown in Fig. 10-1.

齿距累积总偏差（ΔF_P）指的是齿轮同侧齿面任意弧段（$k=1$至$k=z$）内的最大齿距累积偏差。它表现为齿距累积偏差曲线的总幅值，如图10-1所示。

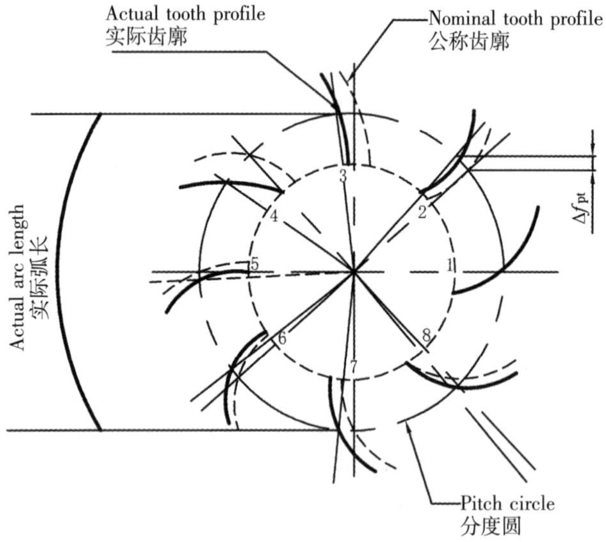

(a) Cross section error diagram
截面误差图

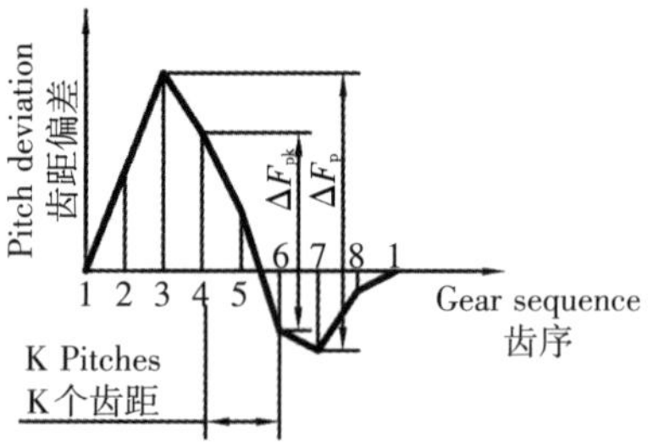

(b) Gear cumulative deviation curve
齿轮累积偏差曲线图

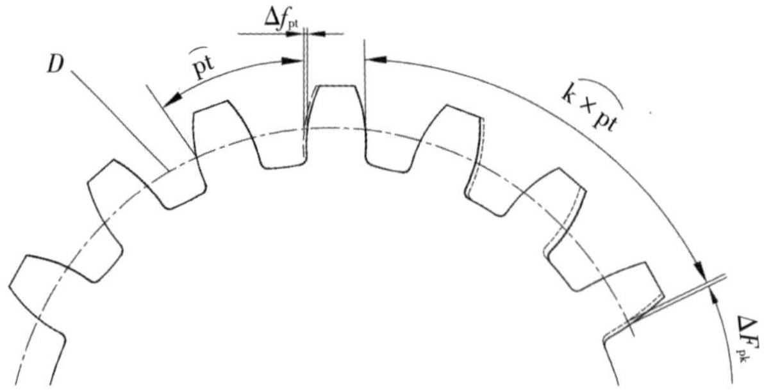

(c) Single pitch deviation (Δf_{pt}) and cumulative pitch deviation (ΔF_{pk})
单个齿距偏差（Δf_{pt}）与齿距累积偏差（ΔF_{pk}）

Fig. 10-1 Pitch deviation and cumulative pitch deviation

图10-1 齿距偏差与齿距累积偏差

In the process of gear processing, there will inevitably be geometric eccentricity and motion eccentricity, resulting in the occurrence of uneven tooth distance and gear cumulative deviation. The cumulative total deviation of the tooth distance can reflect the corner error caused by the eccentric error during a single revolution of the gear, and F_P can be used instead as an index for assessing the accuracy of gear movement.

齿轮在加工过程中，不可避免地会出现几何偏心和运动偏心，从而使齿距不均匀，产生齿轮累积偏差。齿距累积总偏差能反映齿轮一转中偏心误差引起的转角误差，因此，F_P可作为评定齿轮传动精度的指标。

Because ΔF_P can be measured by commonly available instruments such as gear tooth pitch gauges and universal gear measuring instruments, ΔF_P is a kind of gear motion accuracy evaluation index commonly used in factoies at present.

由于ΔF_P的测量可用常用的齿距仪、万能测齿仪等仪器进行测量，因此它是目前工厂中常用的一种齿轮传动精度的评定指标。

（2）Cumulative pitch deviation（ΔF_{pk}）齿距累积偏差（ΔF_{pk}）。

In the process of measurement,it is necessary to control cumulative pitch deviationΔF_{pk} across k tooth pitches, and ΔF_{pk} is equal to the algebraic sum of deviations of k teeth, usually $k = z/8$.

在测量的过程中，必要时还要控制k个齿距的累积偏差ΔF_{pk}，通常ΔF_{pk}等于k个齿的齿距偏差的代数和，通常 $k = z/8$ 。

The cumulative pitch error of measuring tooth distance is usually measured by relative method, which can be measured by universal gear meter or gear meter. Fig. 10-2 shows the gear pitch measuring diagram of the universal gear tester.

测齿距的累积误差通常采用相对法，可用万能齿轮测量仪或齿轮测量仪进行测量。图10-2为万能测齿仪测齿距测量图。

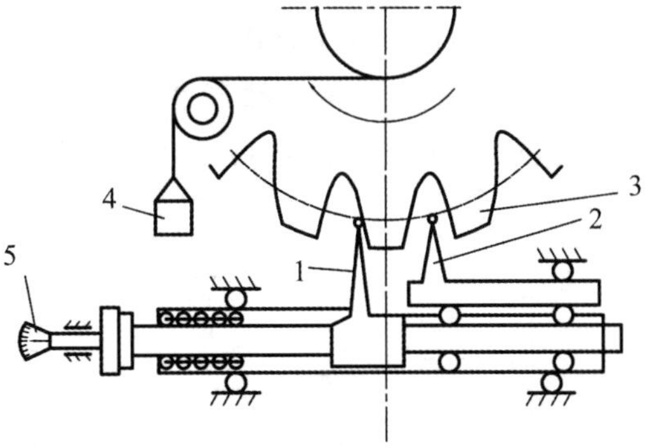

1—movable head; 2—fixed head; 3—wheel gear; 4—heavy hammer; 5—indicator table

1—活动测头；2—固定测头；3—轮齿；4—重锤；5—指示表

Fig. 10-2 Brief map of universal gear tester

图10-2 万能测齿仪测齿距简图

Firstly, the actual pitch of the measured gear is taken as the reference; the instrument indicator table is adjusted to zero, and then the difference between the other actual pitch and the reference pitch around the entire gear circle（called the relative pitch deviation）is measured, and ΔF_{pk} is obtained after data processing. At the same time, a single pitch deviation can also be obtained.

首先，以被测齿轮上任一实际齿距作为基准，将仪器指示表调至零位，然后沿整个齿圈依次测出其他实际齿距与作为基准的齿距的差值（称为相对齿距偏差），经过数据处理求出ΔF_{pk}，同时可求得单个齿距偏差。

10.2.1.2 Required inspection parameters of transmission stability 传动平稳性的必检参数

（1）Single pitch deviation $\left[\Delta f_{pt}(\pm f_{pt})\right]$ 单个齿距偏差 $\left[\Delta f_{pt}(\pm f_{pt})\right]$。

$\Delta f_{pt}(\pm f_{pt})$ refers to the algebraic difference between the actual pitch and the theoretical pitch on a circle concentric with the gear axis, and is located near the middle of the tooth height on the gear end plane. In Fig. 10-1, the value Δf with the largest absolute value is taken $\Delta f_{pt\max}$ to assess the value.

Δf_{pt} 是指在齿轮端平面上，在接近齿高中部的一个与齿轮轴线同心的圆上，实际齿距与理论齿距的代数差。在图 10-1 中，取其中绝对值最大的数值 $\Delta f_{pt\max}$ 为评定值。

When there is a pitch deviation in the gear, whether it is positive negative values when one pair of teeth finishes engagement, the driving teeth and an other pair enters engagment, the driving teeth and the driven teeth collide, which affects the stability accuracy of the gear transmission.

当齿轮存在齿距偏差时，无论是正值还是负值，都会在一对齿啮合完毕而另一对齿进入啮合时，主动齿与被动齿发生冲撞，影响齿轮传动的平稳性精度。

（2）Total profile deviation［$\Delta F_\alpha(F_\alpha)$］ 齿廓总偏差［$\Delta F_\alpha(F_\alpha)$］。

ΔF_α refers to the normal distance between the two design tooth profiles containing the actual profile working part with the smallest possible distance. As shown in Fig. 10-3（a), it is measured in the gear end plane along the direction perpendicular to the involute tooth profile.

ΔF_α是指包容实际齿廓工作部分的两条设计齿廓之间的法向距离，且距离为最小。如图 10-3（a）所示，该距离在齿轮端平面内，沿垂直于渐开线齿廓的方向上测量。

Design tooth profile refers to the tooth profile conforming to the design requirements, usually the involute tooth profile. Considering the manufacturing error and the elastic deformation of the teeth after load, in order to reduce the noise and reduce the influence of dynamic load, the design tooth profile also includes the modified involute（crowded teeth), such as convex teeth and trim teeth.

设计齿廓是指符合设计规定的齿廓，通常为渐开线齿廓。考虑制造误差和轮齿受载后的弹性变形，为了降低噪声和减小动载荷的影响，设计齿廓也包括修形的渐开线齿廓（修形齿廓），如凸齿廓、修缘齿廓等。

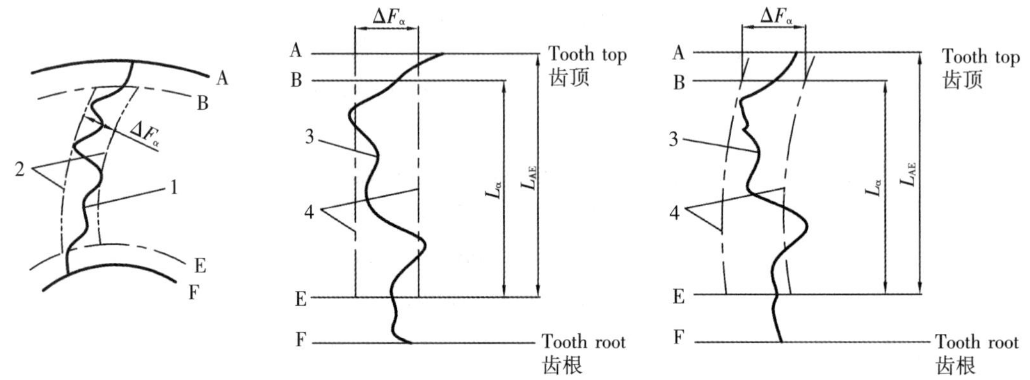

（a）Total deviation of tooth profile 齿廓总偏差
（b）Deviation measurement record of unmodified shape 未修形的偏差测量记录图
（c）Deviation measurement record of revised shape 修形的偏差测量记录图

1—actual tooth profile; 2—design tooth profile; 3—actual tooth profile trace; 4—design tooth profile trace; A—top circle; B—starting circle of tooth top value range; E—starting circle of effective tooth root; F—root circle; B ~ E—working part; L—tooth profile effective length; L_α—tooth profile value range

1—实际齿廓；2—设计齿廓；3—实际齿廓迹线；4—设计齿廓迹线；
A—齿顶圆；B—齿顶计值范围的起始圆；E—齿根有效齿廓的起始圆；F—齿根圆；
B ~ E—工作部分；L—齿廓有效长度；L_α—齿廓计值范围

Fig. 10-3 Record diagram of total deviation and measured tooth deviation

图 10-3 齿廓总偏差与测量齿廓偏差记录图

The profile deviation curve on the record map of measuring the profile deviation is called the tooth profile trace, as shown in Fig. 10-3（b）and（c）. Total profile deviation ΔF_α is indicated in the value range L, the distance between the two design tooth traces that enclose the actual tooth trace with the smallest possible distance.

在测量齿廓偏差时的记录图上，齿廓偏差曲线称为齿廓迹线，如图10-3（b）和（c）所示。齿廓总偏差ΔF_α是指在计值范围L内，包容实际齿廓迹线的两条设计齿廓迹线间的距离。

If the design tooth profile is a theoretical involute, the design tooth profile trace is a straight line, as shown in line 4 in Fig. 10-3（b）; if the design tooth profile is trimmed（convex）, the design tooth profile trace is a curve, as shown in curve 4（convex）in Fig. 10-3（c）.

若设计齿廓是理论渐开线，则设计齿廓迹线为一条直线，如图10-3（b）中直线4所示；如果设计齿廓采用修形（凸齿廓）齿廓，则设计齿廓迹线为一曲线，如图10-3（c）中曲线4（凸形）所示。

Total profile deviation ΔF_α mainly affects the gear transmission stability, because there is ΔF_α, the gear cannot guarantee a constant instantaneous transmission ratio,making it prone to vibration and noise.

齿廓总偏差ΔF_α主要影响齿轮传动平稳性，因为有齿廓总偏差ΔF_α的齿轮无法保证瞬时传动比恒定，易产生振动与噪声。

When judging Δf_{pt} andΔF_α, the qualification conditions are that all the single tooth pitch deviation（Δf_{pt}）of the tested gear must be confined within the allow able range of single tooth pitch distance deviation a$\pm f_{pt}$（$-f_{pt} \leq \Delta f_{pt} \leq +f_{pt}$）; The total deviation of tooth profile ΔF_α of the left and right tooth surfaces of three or more teeth, measured evenly over the circumference, must satisfy the maximum value $\Delta F_{\alpha max}$ not greater）than the total allowable deviation of tooth profile F_α（$\Delta F_{\alpha max} \leq F_\alpha$）。

评定Δf_{pt}和ΔF_α时，它们的合格条件是：被测齿轮所有的Δf_{pt}都在单个齿距偏差允许值$\pm f_{pt}$范围内（$-f_{pt} \leq \Delta f_{pt} \leq +f_{pt}$）；三个或更多齿的左右齿面的齿廓总偏差$\Delta F_\alpha$应在被测齿轮的圆周上均匀测量，其中的最大值$\Delta F_{\alpha max}$不大于齿廓总偏差允许值$F_\alpha$（$\Delta F_{\alpha max} \leq F_\alpha$）。

10.2.1.3 The required parameters of load distribution uniformity——The total helix deviation［ΔF_β（F_β）］载荷分布均匀性的必检参数——螺旋线总偏差［ΔF_β（F_β）］

The parameter required for load distribution uniformity in the tooth width direction is helix deviation ΔF_β, and the parameter in gear-height direction is related to transmission stability.

载荷分布均匀性的必检参数在齿宽方向是螺旋线总偏差ΔF_β（F_β）；在齿高方向上，它是传动平稳性的必检参数。

ΔF_β is indicated within the value range（L_β）, the distance between the two design helix traces that enclose the actual helical trace（Fig. 10-4）. This deviation mainly affects the contact accuracy of the tooth surface.

ΔF_β是指在计值范围（L_β）内，包容实际螺旋线迹线的两条设计螺旋线迹线间的距离（图10-4）。该项偏差主要影响齿面接触精度。

The helix deviation of the non-modified helix is measured using the helix detector. The principle is to compare the actual helix of the tested gear with the standard theoretical helix point by point, and draw the deviation curve on the recording paper, as shown in Fig. 10-4. A helix without helix deviation should be a straight line（design helix trace）, such as 1（1-design helix trace）in Fig. 10-4. If there is no ΔF_β, the instrument pen should step out of a line with 1; when ΔF_β exists in its trace, a curve 2（actual helix trace）. The axial distance of the gear from the base surface Ⅰ to the non-base surface Ⅱ is the tooth width b. The gauge value range of L_β is determined by subtracting , at both ends of the gear, the smaller of two vallces: the 5% of tooth width value or a length value equal to one modulus.

在螺旋线检查仪上测量非修形螺旋线的斜齿轮螺旋线偏差，原理是将被测齿轮的实际螺旋线与标准的理论螺旋线逐点进行比较并将所得的差值在记录纸上画出偏差曲线图，如图10-4所示。没有螺旋线偏差的螺旋线展开后应该是一条直线（设计螺旋线迹线），即图中的1。如果无ΔF_β偏差，仪器的记录笔应该走出一条与1重合的直线，而当存在ΔF_β偏差时，则走出一条曲线2（实际螺旋线迹线）。齿轮从基准面Ⅰ到非基准面Ⅱ的轴向距离为齿宽b。螺旋线计值范围L_β为在轮齿两端处各减去下面两个数值中较小的一个后的迹线长度，即一个是5%的齿宽数值，另一个是等于一个模数的长度数值。

When judging ΔF_β condition, it should be noted is that the total helix deviation ΔF_β of the left and right tooth flanks of the three or more teeth on the circumference of the tested gearshould be measured, and the maximum value $\Delta F_{\beta\max}$ shall not exceed the allowable total helix deviation ΔF_β. ($\Delta F_{\beta\max} \leq F_\beta$)。

During evaluation, the qualification criteria are: the total helix deviation of the left and right tooth flanks of three or more teeth the circumference of the gear under test shall be measured, and the maximum value $\Delta F_{\beta\max}$ shall not exceed the allowable value.

评定ΔF_β时，它的合格条件是：应在被测齿轮圆周上均布测量三个轮齿或更多的轮齿左、右齿面的螺旋线总偏差ΔF_β，其中最大值$\Delta F_{\beta\max}$不大于螺旋线总偏差允许值ΔF_β（$\Delta F_{\beta\max} \leq F_\beta$）。

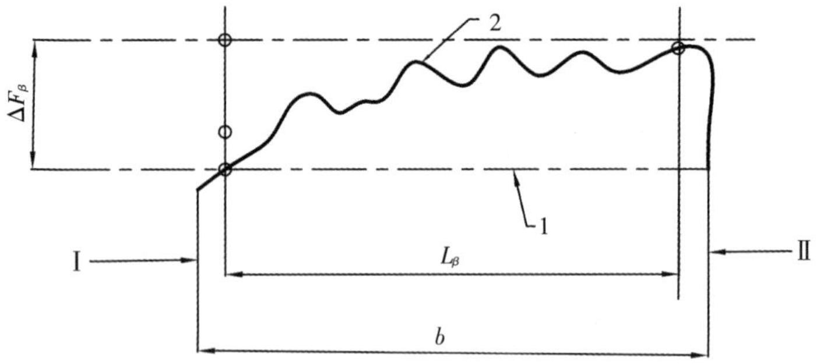

Ⅰ, Ⅱ—both sides of the wheel; 1—design helix trace; 2—actual helix trace; b—tooth width; L_β—helix gauge value range

Ⅰ，Ⅱ—轮齿两端面；1—设计螺旋线迹线；2—实际螺旋线迹线；b—齿宽；L_β—螺旋线计值范围

Fig. 10-4 Measure the helix deviation record graph

图10-4 测量螺旋线偏差记录图

10.2.1.4 Required inspection parameters of the gear side gap 齿轮侧隙的必检参数

(1) Tooth thickness deviation ΔE_{sn} (E_{sns}, ΔE_{sni}) 齿厚偏差ΔE_{sn}（E_{sns}, ΔE_{sni}）。

For spur gears, the ΔE_{sn} denotes the discrepancy between the actual tooth thickness and the nominal (or theoretical) tooth thickness measured on the pitch cylinder. In the case of helical gears, the ΔE_{sn} signifies the deviation between the normal actual tooth thickness and the nominal tooth thickness.

对于直齿轮，ΔE_{sn}是指在分度圆柱面上，实际齿厚与公称齿厚（齿厚理论值）之差。对于斜齿轮，ΔE_{sn}指法向实际齿厚与公称齿厚之差。

(2) Common normal length deviation ΔE_{bn} (E_{bns}, E_{bni}) 公法线长度偏差ΔE_{bn}（E_{bns}, E_{bni}）。

The common normal length refers to the arc of the base circle enclosed between the opposite-facing tooth profiles at the ends of several teeth on a gear, that is, the length of the tangent line segment of the base circle between these opposite-facing tooth profiles.

公法线长度是指齿轮上几个轮齿的两端异向齿廓间所包含的一段基圆圆弧，即该两端异向齿廓间基圆切线线段的长度。

ΔE_w refers to the difference between the actual common normal length（w_k）and its nominal value（w）.

ΔE_w是指实际公法线长度（w_k）与公称公法线长度（w）之差。

As mentioned above, the gear accuracy standard defines the essential deviation parameters for assessing the precision, including pitch deviation, profile deviation, helix deviation, and tooth thickness deviation. The latter is specifically used to assess backlash.

由上述可见，齿轮精度标准规定评定齿轮精度的必检偏差项目为齿距偏差、齿廓偏差和螺旋线偏差以及评定侧隙的齿厚偏差四项。

10.2.2 Optional Deviation Items for Evaluating the Gear Accuracy 评定齿轮精度的可选用偏差项目

When using a certain tooth cutting method to producing the first batch of gears, it is crucial to inspect the mandatory deviation items to ensure that the machined gear accuracy fulfills the design requirements. Upon successful completion of these inspections, and assuming the process conditions remain consistent, for the purpose of continued production of identical gears or for applications in error analysis and research, GB/T 10095.1—2022 and GB/T10095.2—2023 permit the use of non-mandatory parameters to further assess transmission motion precision and stability.

用某种切齿方法生产第一批齿轮时，为了掌握该齿轮加工后的精度是否达到设计要求，需要按上述强制性的必检偏差项目进行检测。检测合格后，在工艺条件不变的情况下继续生产同样的齿轮或用作误差分析研究时，GB/T 10095.1—2022和GB/T 10095.2—2023规定也可采用下列非强制性的参数来评定齿轮传递运动准确性和传动平稳性的精度。

10.2.2.1 Select the parameters for transmitting the movement accuracy 传递运动准确性的可选用参数

（1）Tangential comprehensive total deviation ［$\Delta F_i'(F_i')$］ 切向综合总偏差 ［$\Delta F_i'(F_i')$］。

$\Delta F_i'$ refers to the maximum deviation between the effective and theoretical circumferential displacements at the reference circle of the gear under inspection, when the tested grearis meshed with a master gear and turned through one complete revolution.

$\Delta F_i'$是指被测齿轮与测量齿轮（基准）单面啮合检验时，在被测齿轮一转内，被测齿轮分度圆上实际圆周位移与理论圆周位移的最大差值。

Fig. 10-5 shows the tangential comprehensive deviation curve drawn on the single-sided engagement meter（One turn of the inspected gear（2π）), with the abscissa indicating the rotation angle and the ordinate indicating the deviation. If the gear has no deviation, the curve shall be a straight line overlapping with the abscissa. In the range of gear rotation, if the highest curve and the lowest point are two straight lines parallel to the abscissa, then the distance betweem the highest and lowest parallel lines within the rotation range is $\Delta F_i'$ value.

图10-5为在单面啮合测量仪上画出的切向综合偏差曲线图（被检验齿轮的一转（2π)），横坐标表示被测齿轮转角，纵坐标表示偏差。如果齿轮没有偏差，偏差曲线应是与横坐标重合的直线。在齿轮一转范围内，过曲线最高、最低点作与横坐标平行的两条直线图10-5，则此平行线间的距离即为$\Delta F_i'$值。

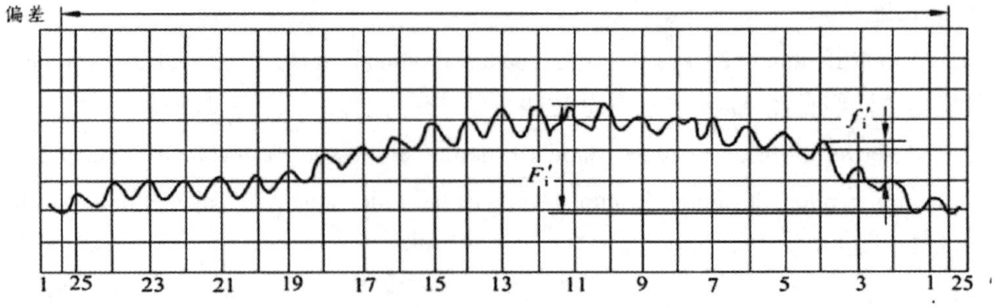

Fig. 10-5 Tangential comprehensive deviation

图10-5 切向综合偏差

(2) Radial run-out [$\Delta F_r(F_r)$] 径向跳动 [$\Delta F_r(F_r)$]。

Radial run-out ΔF_r denotes the variation between the maximum and minimum radial distances from a probe (spherical, cylindrical, or anvil-shaped) to the gear axis, as it is sequentially positioned within each tooth space.

径向跳动ΔF_r是指测头（球形、圆柱形、砧形）相继置于被测齿轮的每个齿槽内时，从它到齿轮轴线的最大和最小径向距离之差。

Radial run-out can be measured by a gear rim radial run-out measuring instrument. The probe, which can be fashioned in either a spherical or conical shape for insertion into the tooth space, or alternatively, a V-shaped probe designed to clamp securely onto the gear teeth is employed, as shown in Fig. 10-6. During one full revolution of the tested gear, with the probe maintained in double-sided contact at the mid-height of the tooth, the total fluctuation in radial distance relative to the axis is recorded, which precisely corresponds to the gear's radial run-out.

径向跳动可用齿圈径向跳动测量仪进行测量。测头做成球形或圆锥形插入齿槽中，也可做成V形卡在轮齿上，如图10-6所示；与齿高中部双面接触，被测齿轮一转所测得的相对于轴线径向距离的总变动幅度值，即齿轮的径向跳动。

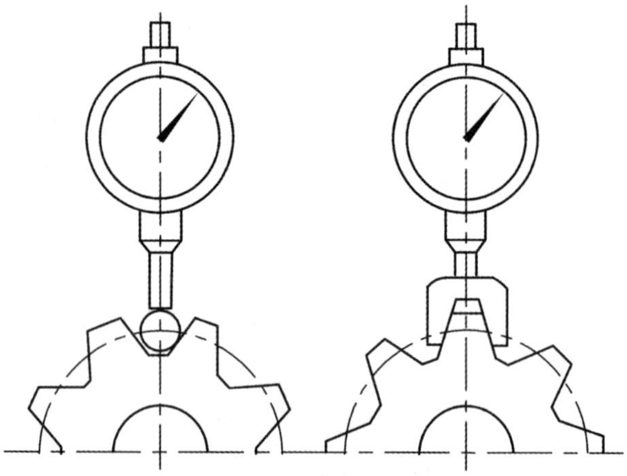

Fig. 10-6 Radial beating

图10-6 径向跳动

In radial run-out measurement, using the gear hole axis as the reference, radial run-out measurement only detects radial errors. The maximum error in one revolution appears only once, which is a long period error which serves as a period error definitive indicator of radial transmission motion. Hence, it is essential to couple this indicator with one that represents tangential errors to fully assess the accuracy of transmission motion.

径向跳动测量中，以齿轮孔的轴线为基准，只反映径向误差。齿轮一转中最大误差只出现一次，是长周期误差。它仅作为影响传递运动准确性的径向性质的单项指标。因此，采用这一指标必须与能表示切向误差的单项性指标组合，才能评定传递运动的准确性。

(3) Total radial composite deviation (F_i'') 径向综合总偏差 (F_i'')。

Total radial composite deviation F_i'' is the difference between the maximum and minimum values of centre distance which occur during a radial (double-flank) composite test, when the product gear with its right and left flanks simultaneously in contact with those of the master gear, is turned through one complete revolution, Fig. 10-7 shows an example of a relevant diagram.

径向综合总偏差 F''_i 是指在径向（双面）综合测试中，当产品齿轮的右齿面和左齿面同时与主齿轮的相应齿面接触并旋转一整圈时，所出现的中心距最大值与最小值之间的差异。图 10-7 展示了相关图示的一个例子。

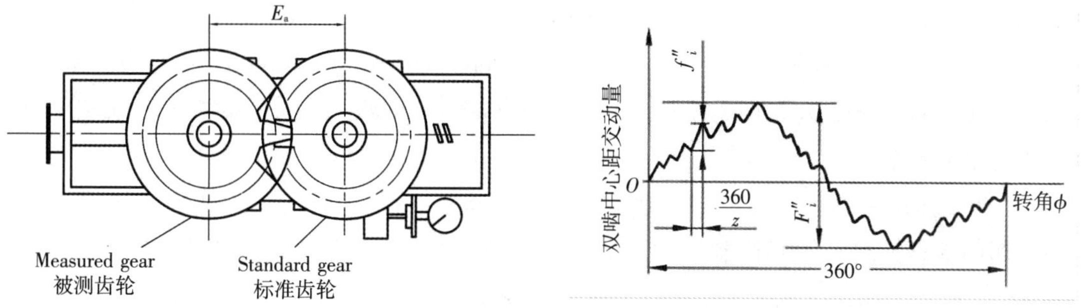

(a) (b)

Fig. 10-7 Combined radial error with double rodent

图 10-7 用双啮仪测径向综合误差

The radial composite deviation is measured using a double-flank gear meshing tester (also known as a double-flank tester). The radial composite deviation includes the total radial composite deviation F''_i and the single-tooth radial composite deviation F''_i. During the radial composite deviation test, the measurement gear and the gear under test are mounted on the double-flank tester, with one gear fixed on a stationary shaft and the other gear mounted on a shaft with a slide that incorporates a spring mechanism, allowing the two gears to mesh tightly in the radial direction. The total radial composite deviation F''_i of the tested gear is equal to the maximum variation in center distance during one complete revolution of the gear. The single-tooth radial composite deviation F''_i is equal to the variation in center distance when the gear rotates through one tooth pitch angle, with the maximum variation taken as the single-tooth radial composite deviation F''_i. As shown in Fig. 10-7 (a) is the double-flank gear meshing tester, and (b) is the data curve obtained after measurement.

径向综合偏差用齿轮双面啮合检查仪（双啮仪）测量。径向综合偏差包括径向综合总偏差 F''_i 和一齿径向综合偏差 F''_i。在径向综合偏差检测中，测量齿轮和被测齿轮安放在双啮仪上，其中一个齿轮装在固定的轴上，另一个齿轮则装在带有滑道的轴上，该滑道带一弹簧装置，从而使两个齿轮在径向能紧密地啮合。被检测齿轮径向综合总偏差 F''_i 等于齿轮旋转一整周中最大的中心距变动量，一齿径向综合偏差 F''_i 等于齿轮转过一个齿距角时中心距的变动量，取其中的最大变动量为一齿径向综合偏差 F''_i。如图 10-7 所示，(a) 为双面啮合综合检查仪，(b) 为测量后的数据曲线。

10.2.2.2 Optional parameters of transmission stability 传动平稳性的可选参数

(1) The tangential comprehensive deviation of one tooth $\left[\Delta f'_i(f'_i)\right]$ 一齿切向综合偏差 $\left[\Delta f'_i(f'_i)\right]$。

The tangential comprehensive deviation of one tooth $\Delta f'_i(f'_i)$ refers to the tangential comprehensive total deviation of the gear in a tooth pitch angle, that is, the maximum amplitude of the small ripple on the record curve of the tangential comprehensive total deviation. As shown in Fig. 10-5, every gear turns a tooth pitch angle, the angle error, that is, many small peaks and valleys appear. In these short period errors, the maximum amplitude of peak valley is the tangential comprehensive deviation $\Delta f'_i$. The f'_i reflects both the tangential error of short period and the radial error of short period, which is a comprehensive index to assess the stability of gear transmission.

一齿切向综合偏差 $\Delta f'_i(f'_i)$ 是指齿轮在一个齿距角内的切向综合总偏差，即在切向综合总偏差记录曲线上小波纹的最大幅度值，如图10-5所示齿轮每转过一个齿距角，都会引起转角误差，即出现许多小的峰谷。在这些短周期误差中，峰谷的最大幅度值即为一齿切向综合偏差 f'_i。f'_i 既反映了短周期的切向误差，又反映了短周期的径向误差，是评定齿轮传动平稳性较全面的综合性指标。

Obviously, the greater the tangential comprehensive deviation of one tooth, the higher the frequency, the less stable the transmission is. Therefore, the tangential comprehensive tolerance f'_i must be used according to the use requirements of gear drive. The f'_i, like the tangent synthesis total deviation, can be measured with a single-sided instrument.

显然，一齿切向综合偏差越大，频率越高，则传动越不平稳。因此，必须根据齿轮传动的使用要求，用一齿切向综合公差 f'_i 加以限制。f'_i 与切向综合总偏差一样，可以用单啮仪进行测量。

(2) The radial comprehensive deviation of one tooth $\left[\Delta f''_i(f''_i)\right]$ 一齿径向综合偏差 $\left[\Delta f''_i(f''_i)\right]$。

The $\Delta f''_i(f''_i)$ is the combined radial overall deviation within a pitch angle (360°/z) when the measured gear is engaged with the measuring gear.

$\Delta f''_i(f''_i)$ 是指当被测齿轮与测量齿轮啮合一整圈时，对应一个齿距角（360°/z）内的径向综合偏差值。

In order to ensure transmission stability and mitigate tangential error, the standard of the radial comprehensive tolerance should be multiplied by 0.8 as a correction factor. The $\Delta f''_i(f''_i)$ can be measured on a double-sided mesh comprehensive detector.

为了满足传动平稳性的要求，防止测不出切向误差部分的影响，应将标准规定的一齿径向综合公差乘以0. 8加以缩小。$\Delta f''_i(f''_i)$ 可采用双面啮合综合检查仪测量。

10.2.3 Precision Grade of Gear (Drawing Marking) 齿轮的精度等级（图样标注）

GB/T 10095.1-2022 and GB/T 10095.2-2023 establish 13 distinct precision grades for a series of gear parameters, including the total cumulative pitch deviation ΔF_p, the cumulative pitch deviation ΔF_{pk}, the single pitch deviation $\Delta f_{pt}(\pm f_{pt})$, the total profile deviation $\Delta F_\alpha(F_\alpha)$, the total helix deviation $\Delta F_\beta(F_\beta)$, the tangential comprehensive total deviation $\Delta F'_i(F'_i)$, the radial run-out $\Delta F_r(F_r)$, the radial comprehensive deviation of one tooth $\Delta F''_i(F''_i)$ and the total radial composite deviation F''_i. These grades, denoted by Arabic numerals from zero to twelve, represent a range from the highest to the lowest precision levels. Additionally, for the total radial composite deviation F''_i and the single-tooth radial composite deviation f''_i, the standard specifies nine precision grades spanning from grade four (highest) to grade twelve (lowest).

GB/T10095.1—2022和GB/T10095.2—2023对齿距累积总偏差ΔF_P、齿距累积偏差ΔF_{pk}、单个齿距偏差 $\Delta f_{pt}(\pm f_{pt})$、齿廓总偏差 $\Delta f_\alpha(\pm f_\alpha)$、螺旋线总偏差 $\Delta f_\beta(\pm f_\beta)$、切向综合总偏差 $\Delta F'_i(F'_i)$、径向跳动 $\Delta f_r(\pm f_r)$、一齿径向综合偏差 $\Delta F'_i(F'_i)$、径向综合总偏差 F''_i 分别规定了13个精度等级，从高到低分别用阿拉伯数字0，1，2，…，12表示；对径向综合总偏差 F''_i 和一齿径向综合偏差 f''_i 分别规定了9个精度等级（4，5，6，…，12），其中4级最高，12级最低。

Grade 5 precision is the basic level, from which allowable deviation values are calculated for all other grades. For grades 0 ~ 2, the requirements are very high, and there are almost no manufacture and measurement metheds available, so they are to be developed; grades 3 ~ 5 are high accuracy; grades 6 ~ 8 are medium accuracy (most used); and grade 9 is low accuracy and grades 10 ~ 12 are very low precision. A comprehensive and detailed list of numerical values for various gear deviations is provided in Tab. 10-1 ~ Tab. 10-3.

5级精度为基本等级，它是计算其他等级偏差允许值的基础。0 ~ 2级齿轮要求非常高，目前几乎没有能够制造和测量的手段，因此属于有待发展的展望级；3 ~ 5级为高精度等级；6 ~ 8级为中等精度等级（用得最多）;9级为较低精度等级；10 ~ 12级为低精度等级。齿轮各项偏差数值见表10-1 ~ 表10-3。

Tab. 10-1 Deviation allowable value of $\pm f_{pt}$, F_p, $\pm F_{pk}$, F_α, f'_i, F'_i, F_r, F_w (GB/T 10095.1-2020)

表10-1 齿轮各项偏差数值（$\pm f_{pt}$，F_p，$\pm F_{pk}$，F_α，f'_i，F'_i，F_r，F_w）（GB/T 10095.1—2020） 单位：μm

Pitch diameter 分度圆直径 d/mm	Deviation item 偏差项目	Individual tooth pitch deviation 单个齿距偏差 $\pm f_{pt}$				Total cumulative tooth pitch deviation 齿距累积总偏差 F_p				Total tooth profile deviation 齿廓总偏差 F_a				Radial run-out tolerance 径向跳动公差 F				f/K Value f/K值				Tolerance for variation in length of common normal 公法线长度变动公差 F_w			
	Accuracy grade 精度等级	5	6	7	8	5	6	7	8	5	6	7	8	5	6	7	8	5	6	7	8	5	6	7	8
$5 \leqslant d \leqslant 20$	$0.5 \leqslant m \leqslant 2$	4.7	6.5	9.5	13.0	11.0	16.0	23.0	32.0	4.6	6.5	9.0	13.0	9.0	13	18	25	14.0	19.0	27.0	38.0	10	14	20	29
	$2 < m \leqslant 3.5$	5.0	7.5	10.0	15.0	12.0	17.0	23.0	33.0	6.5	9.5	13.0	19.0	9.5	13	19	27	16.0	23.0	32.0	45.0				
$20 < d \leqslant 50$	$0.5 \leqslant m \leqslant 2$	5.0	7.0	10.0	14.0	14.0	20.0	29.0	41.0	5.0	7.5	10.0	15.0	11	16	23	32	14.0	20.0	29.0	41.0	12	16	23	32
	$2 < m \leqslant 3.5$	5.5	7.5	11.0	15.0	15.0	21.0	30.0	42.0	7.0	10.0	14.0	20.0	12	17	24	34	17.0	24.0	34.0	48.0				
	$3.5 < m \leqslant 6$	6.0	8.5	12.0	17.0	15.0	22.0	31.0	44.0	9.0	12.0	18.0	25.0	12	17	25	35	19.0	27.0	38.0	54.0				
$50 < d \leqslant 125$	$0.5 \leqslant m \leqslant 2$	5.5	7.5	11.0	15.0	18.0	26.0	37.0	52.0	6.0	8.5	12.0	17.0	15	21	29	42	16.0	22.0	31.0	44.0	14	19	28	37
	$2 < m \leqslant 3.5$	6.0	8.5	12.0	17.0	19.0	27.0	38.0	53.0	8.0	11.0	16.0	22.0	15	21	30	43	18.0	25.0	36.0	51.0				
	$3.5 < m \leqslant 6$	6.5	9.0	13.0	18.0	19.0	28.0	39.0	55.0	9.5	13.0	19.0	27.0	16	22	31	44	20.0	29.0	40.0	57.0				
$125 < d \leqslant 280$	$0.5 \leqslant m \leqslant 2$	6.0	8.5	12.0	17.0	24.0	35.0	49.0	69.0	7.0	10.0	14.0	20.0	20	28	39	55	17.0	24.0	34.0	49.0	16	22	31	44
	$2 < m \leqslant 3.5$	6.5	9.0	13.0	18.0	25.0	35.0	50.0	70.0	9.0	13.0	18,0	25.0	20	28	40	56	20.0	28.0	39.0	56.0				
	$3.5 < m \leqslant 6$	7.0	10.0	14.0	20.0	25.0	36.0	51.0	72.0	11.0	15.0	21.0	30.0	20	29	41	58	22.0	31.0	44.0	62.0				
$280 < d \leqslant 560$	$0.5 \leqslant m \leqslant 2$	6.5	9.5	13.0	19.0	32.0	46.0	64.0	91.0	8.5	12.0	17.0	23.0	26	36	51	73	19.0	27.0	39.0	54.0	19	26	37	53
	$2 < m \leqslant 3.5$	7.0	10.0	14.0	20.0	33.0	46,0	65.0	92.0	10.0	15.0	21.0	29.0	26	37	52	74	22.0	31.0	44.0	62.0				
	$3.5 < m \leqslant 6$	8.0	11.0	16.0	22.0	33.0	47.0	66.0	94.0	12.0	17.0	24.0	34.0	27	38	53	75	24.0	34.0	48.0	68.0				

Tab. 10-2 Total helix deviation F_β (From GB/T 10095.1-2020)

表10-2 螺旋线总偏差 F_β（摘自GB/T 10095.1—2020） 单位：μm

Pitch circle diameter 分度圆直径 d/mm	Tooth width 齿宽 b/mm	Precision grade 精度等级			
		5	6	7	8
		Total helix deviation 螺旋线总偏差 F_β			
$5 \leqslant d \leqslant 20$	$4 \leqslant b \leqslant 10$	6.0	8.5	12.0	17.0
	$10 \leqslant b \leqslant 20$	7.0	9.5	14.0	19.0
$20 \leqslant d \leqslant 50$	$4 \leqslant b \leqslant 10$	6.5	9.0	13.0	18.0
	$10 \leqslant b \leqslant 20$	7.0	10.0	14.0	20.0
	$20 \leqslant b \leqslant 40$	8.0	11.0	16.0	23.0
$50 \leqslant d \leqslant 125$	$4 \leqslant b \leqslant 10$	6.5	9.5	13.0	19.0
	$10 \leqslant b \leqslant 20$	7.5	11.0	15.0	21.0
	$20 \leqslant b \leqslant 40$	8.5	12	17	24
	$40 \leqslant b \leqslant 80$	10.0	14.0	20.0	28.0

Tab. 10-2 (Continued) 表 10-2（续）

Pitch circle diameter 分度圆直径 d/mm	Tooth width 齿宽 b/mm	Precision grade 精度等级			
		5	6	7	8
		Total helix deviation 螺旋线总偏差 F_β			
$125 \leqslant d \leqslant 280$	$4 \leqslant b \leqslant 10$	7.0	10.0	14.0	20.0
	$10 \leqslant b \leqslant 20$	8.0	11.0	16.0	22.0
	$20 \leqslant b \leqslant 40$	9.0	13.0	18.0	25.0
	$40 \leqslant b \leqslant 80$	10.0	15.0	21.0	29.0
	$80 \leqslant b \leqslant 160$	12.0	17.0	25.0	35.0
$280 \leqslant d \leqslant 560$	$10 \leqslant b \leqslant 20$	8.5	12.0	17.0	24.0
	$20 \leqslant b \leqslant 40$	9.5	13.0	19.0	27.0
	$40 \leqslant b \leqslant 80$	11.0	15.0	22.0	31.0
	$80 \leqslant b \leqslant 160$	13.0	18.0	26.0	36.0
	$160 \leqslant b \leqslant 250$	15.0	21.0	30.0	43.0

Tab. 10-3 Total radial composite deviation and single-tooth radial composite deviation (From GB/T 10095.2-2020)

表 10-3 径向综合总偏差和一齿径向综合偏差（摘自 GB/T 10095.2—2020） 单位：μm

Pitch circle diameter 分度圆直径 d/mm	Module 模数 m_n/mm	Precision grade 精度等级							
		5	6	7	8	5	6	7	8
		Total radial composite deviation 径向综合总偏差(F''_i)				Single-tooth radial composite deviation 一齿径向综合偏差(f''_i)			
$5 \leqslant d \leqslant 20$	$0.2 \leqslant m_n \leqslant 0.5$	11	15	21	30	2.0	2.5	3.5	5.0
	$0.5 < m_n \leqslant 0.8$	12	16	23	33	2.5	4.0	5.5	7.5
	$0.8 < m_n \leqslant 1.0$	12	18	25	35	3.5	5.0	7.0	10
	$1.0 < m_n \leqslant 1.5$	14	1+	27	38	4.5	6.5	9.0	13
$20 < d \leqslant 50$	$0.2 \leqslant m_n \leqslant 0.5$	13	19	26	37	2.0	5	3.5	5.0
	$0.5 < m_n \leqslant 0.8$	14	20	28	40	2.5	4.0	5.5	7.2
	$0.8 < m_n \leqslant 1.0$	15	21	30	42	3.5	5.0	7.0	10
	$1.0 < m_n \leqslant 1.5$	16	23	32	45	4.5	6.5	9.0	13
	$1.5 < m_n \leqslant 2.5$	18	26	37	52	6.5	9.5	13	19
$50 < d \leqslant 125$	$1.0 < m_n \leqslant 1.5$	19	27	39	55	4.5	6.5	9.0	13
	$1.5 < m_n \leqslant 2.5$	22	31	43	61	6.5	9.5	13	19
	$2.5 < m_n \leqslant 4.0$	25	36	51	72	10	14	20	29
	$4.0 < m_n \leqslant 6.0$	31	44	62	88	15	22	31	44
	$6.0 < m_n \leqslant 10$	40	57	80	114	24	34	48	67
$125 < d \leqslant 280$	$1.0 < m_n \leqslant 1.5$	24	34	48	68	4.5	6.5	9.0	13
	$1.5 < m_n \leqslant 2.5$	26	37	53	75	6.5	9.5	16	19
	$2.5 < m_n \leqslant 4.0$	30	43	61	86	10	15	21	29
	$4.0 < m_n \leqslant 6.0$	36	51	72	102	15	22	31	44
	$6.0 < m_n \leqslant 10$	45	64	90	127	24	34	48	67
$280 < d \leqslant 560$	$1.0 < m_n \leqslant 1.5$	30	43	61	86	4.5	6.5	9.0	13
	$1.5 < m_n \leqslant 2.5$	33	46	65	92	6.5	9.5	13	19
	$2.5 < m_n \leqslant 4.0$	37	52	73	104	10	15	21	29
	$4.0 < m_n \leqslant 6.0$	42	60	84	119	15	22	31	44

10.3 Accuracy Design of Cylindrical Gear 圆柱齿轮精度设计

In order to ensure the use requirements of gear transmission, the precision design of gear mainly includes the following contents:① Determing the precision grade of the gears, the deviation items that must be inspected, and their allowable values;② Specifying the deviation items related to the gear backlash and their upper and lower deviations; ③ Defining the precision of gear pairs and gear blanks.

为了保证齿轮传动的使用要求，齿轮的精度设计主要有下列内容：① 确定齿轮的精度等级和必须检测的偏差项目及其允许值；② 确定齿轮侧隙的偏差项目及其上、下偏差；③ 确定齿轮副和齿坯精度。

10.3.1 Determine the Gear Accuracy Level 确定齿轮精度等级和必检的偏差项目

10.3.1.1 Determine the accuracy level of the gear 确定齿轮的精度等级

The selection of accuracy level is mainly based on the use of the gear, use requirements and working conditions. The main two methods to choose are calculation method and empirical method（analogy method）.

选择精度等级的主要依据是齿轮的用途、要求和工作条件。选择的方法主要有计算法和经验法（类比法）两种。

The calculation method is primarily employed in the design of precision transmission chains, which can also be utilized according to the precision requirements of the transmission chain. For instance, it is necessary to first calculate the permissible angular error to ensure the accuracy of motion transmission, in order to select an appropriate precision grade.

计算法主要用于精密传动链设计，可按传动链精度要求。例如，传递运动准确性要求先计算出允许的转角误差大小，以便选择适宜的精度等级。

The empirical method determines the precision grade of a gear by referencing the gear precision grade of similar products and taking into account the specific requirements of the gear being designed.

经验法是参考同类产品的齿轮精度等级，结合所设计齿轮的具体要求来确定精度等级。

Tab. 10-4 summarizes the approximate accuracy grade range of gears for various gear applications based on production practice, serving as a reference for designers.

表10-4 为从生产实践中搜集到的各种用途齿轮的大致精度等级范围，可供设计者参考。

Tab. 10-4 Application of accuracy level（for reference）

表10-4 精度等级的应用（供参考）

Gear use 齿轮用途	Accuracy grade 精度等级	Gear use 齿轮用途	Accuracy grade 精度等级	Gear use 齿轮用途	Accuracy grade 精度等级
Prototype gear 测量齿轮	3 ~ 5	Light car 轻型汽车	5 ~ 8	Tractor, steel rolling mill 拖拉机、轧钢机	6 ~ 10
Steam turbine reducer 汽轮机减速器	3 ~ 6	Lorry 载重汽车	6 ~ 9	Crane 起重机	7 ~ 10
Metal cutting machine 金属切削机床	3 ~ 8	General reducer 一般减速器	6	Mine winch 矿山绞车	8 ~ 10
Aeroplane engine 航空发动机	3 ~ 7	Locomotive 机车	6 ~ 7	Enterprise machinery 企业机械	8 ~ 11

In mechanical transmissions, the gears most frequently applied are capable of transmitting both motion and power. Their precision grades are tightly correlated with the circumferential speed. Therefore, it is possible to determine the maximum circumferential speed of these gears through calculation. Please refer to Tab. 10-5 for further details.

在机械传动中，应用得最多的齿轮是既传递运动又传递动力的齿轮，其精度等级与圆周速度密切相关。因此，可计算出齿轮的最高圆周速度。参考表10-5确定齿轮精度等级。

Tab. 10-5 Selection of gear stability accuracy grade (for reference)

表10-5 齿轮平稳性精度等级的选用（供参考）

Accuracy grade 精度等级	Circumferential velocity 圆周速度（m/s）		Applications 应用范围
	Spur gear 直齿	Helix gear 斜齿	
3	≤40	≤75	Gear of very very high precision or work most stable and no voice 极高精度或运行超稳定、无噪声的齿轮
4	≤35	≤70	Gear of very high precision indexing structure or work most stable and no voice 具备超高精度分度结构或运行超稳定、无噪声的齿轮
5	≤20	≤40	Gear of high precision indexing structure or work at very stable and no noise 具有高精密分度机构或运行超稳定、无噪声的齿轮
6	≤16	≤30	Gear works maximum efficiency, stable and no voice 运行于高效率、稳定、无噪声的齿轮
7	≤10	≤15	Gear for speed reducer, feed mechanism of machine tools, aircraft and car 用于减速器、机床进给机构、航空器和汽车的齿轮
8	≤6	≤10	Gear for general manufacture mechanism, hosting mechanism, agriculture machinery 用于通用制造机械、主机传动系统和农业机械的齿轮
9	≤2	≤4	Gear used for low precision 用于精度要求低的齿轮

10.3.1.2 Determine the required deviation items of the gear and its allowable value 确定齿轮的必检偏差项目及其允许值

As specified by GB/T 10095.1-2022, the mandatory gear accuracy deviations are the cumulative total pitch deviation F_p, the single tooth deviation $\pm f_{pt}$, total. Profile F_α and the total helix deviation F_β. Their allowable values are listed in Tab. 10-1 and Tab. 10-2.

由GB/T 10095.1—2022 规定，评定齿轮精度的必检偏差为齿距累积总偏差F_p、单个齿距偏差$\pm f_{pt}$、齿廓总偏差F_α和螺旋线总偏差F_β，其允许值见表10-1和表10-2。

10.3.2 Determination of the Minimum Lateral Gap and Tooth Thickness Deviation 最小侧隙和齿厚偏差的确定

During gear meshing and motion transmission, it is crucial to maintain a clearance between the meshing tooth profiles. This clearance, known as backlash, serves to facilitate the formation of a lubricating oil film and to prevent seizure caused by the frictional heating and subsequent expansion of the teeth. The control of this backlash is meticulously governed by two key factors: the center distance between the gear pair and the tooth thickness, or alternatively, the length of the common normal line of each individual gear.

齿轮啮合传动时，为了在啮合齿廓之间形成润滑油膜，避免因轮齿摩擦发热膨胀而卡死，齿廓之间必须留有间隙，此间隙称为齿侧间隙，简称侧隙。侧隙是由齿轮副中心距以及单个齿轮的齿厚或公法线长度来控制的。

For spur gears, tooth thickness deviation E_{sn} is the difference between the actual tooth thickness S_{na} and the nominal tooth thickness S_n on the pitch cylinder, as shown in Fig. 10-8.

对于直齿轮，齿厚偏差E_{sn}为在分度圆柱上，实际齿厚S_{na}与公称齿厚S_n之差，如图10-8所示。

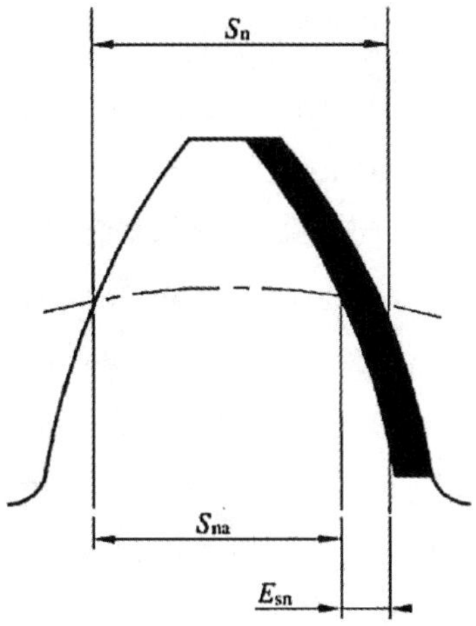

Fig. 10-8 Tooth thickness deviation

图10-8 齿厚偏差

The chordal tooth thickness is usually measured using either a vernier caliper or an optical caliper, with measurements taken in reference to the chordal tooth height, as shown in Fig. 10-9. Given that the chordal tooth thickness is the focus of measurement and the top land cylinder of the gear functions as the datum for these measurements, both the actual pitch diameter deviation and the radial run-out of the top land cylinder in relation to the gear ' s datum axis exert a considerable influence on the precision of the tooth thickness measurements.

弦齿厚通常用游标测齿卡尺或光学测齿卡尺来测量，以弦齿高为依据来测量，如图10-9所示。由于测量弦齿厚以齿轮齿顶圆柱面作为测量基准，因此，齿顶圆直径的实际偏差和齿顶圆柱面对齿轮基准轴线的径向圆跳动都对齿厚测量精度产生较大的影响。

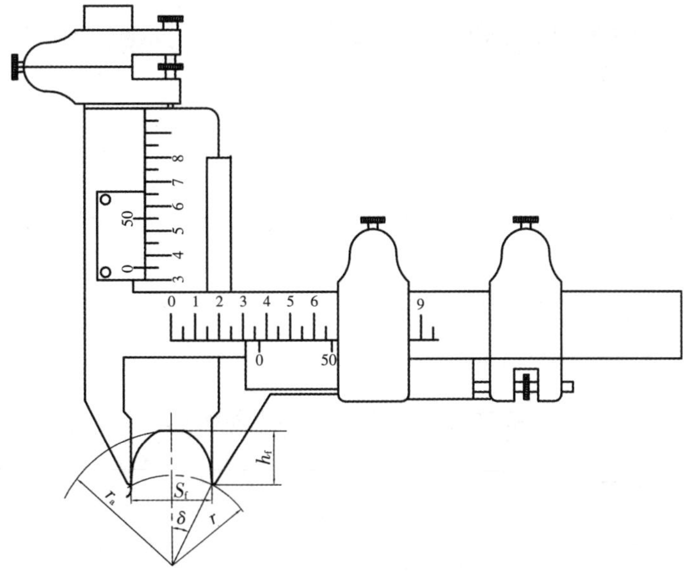

Fig. 10-9 Measurement of chordal tooth thickness on the pitch circle

图10-9 分度圆弦齿厚的测量

To constrain the actual deviation of tooth thickness, it is essential to specify the upper deviation (E_{sns}) and lower deviation (E_{sni}) of tooth thickness during the design stage.

为了限制齿厚的实际偏差，设计时应规定齿厚的上偏差E_{sns}和下偏差E_{sni}。

10.3.2.1 Determination of the minimum normal backlash ($j_{bn\,min}$) 最小法向侧隙（$j_{bn\,min}$）的确定

There are typically two methods used to represent tooth side clearance, namely circumferential backlash j_{wt} and normal backlash j_{bn}. As shown in Fig. 10-10, the circumferential backlash refers to the maximum arc length of the pitch circle that the other gear can rotate through when one of the gears in an installed gear pair is fixed.

齿侧间隙通常有两种表示方法，即圆周侧隙 j_{wt} 和法向侧隙 j_{bn} ，如图10-10所示。圆周侧隙 j_{wt} 是指安装好的齿轮副，当其中一个齿轮固定时，另一齿轮所能转过的节圆弧长的最大值。

The normal backlash j_{bn} is defined as the minimal separation between the non-functional tooth surfaces when the functional tooth surfaces of a mounted gear pair are in engagement.

法向侧隙 j_{bn} 是指安装好的齿轮副的工作齿面接触时，非工作齿面之间的最短距离。

This measurement is taken along the tangent to the base circle, which aligns with the direction of the line of action. Typically, this can be accurately assessed using the lead wire compression technique, involving the insertion of a lead wire between the teeth during the meshing process. Following meshing, the compressed lead wire is extracted and its thickness, indicative of the normal backlash, is measured. Alternatively, a feeler gauge offers a direct method for measuring normal backlash j_{bn}, ensuring precision and efficiency in the assessment of gear alignment and clearance. Theoretically, there is a relationship Eq. 10-1 between j_{wt} and j_{bn}.

测量时需在基圆切线方向，也就是在啮合线方向上测量。一般可以通过压铅丝的方法实现精准测量，具体操作是齿轮在啮合过程中将铅丝置入齿间，啮合完成后，取出被压扁了的铅丝并测量其厚度。该厚度值可反映法向侧隙。此外，还有一种塞尺法替代方案，就是直接测量法向侧隙j_{bn}，可确保齿轮对中状态及啮合间隙评估的精确性与高效性。理论上 j_{wt} 与 j_{bn} 的关系见式（10-1）：

$$j_{bn}=j_{wt}\cos\alpha_{wt}\cdot\cos\beta_b \tag{10-1}$$

Where, α_{wt} —end face working pressure angle;

β_b —base circle helix angle.

式中，α_{wt} ——端面工作压力角；

β_b ——基圆螺旋角。

Minimum normal backlash is determined by allow temperature (the standard temperature 20°), lubrication method and circumferential velocity during transfer of the gear pairs, usually the minimum normal backlash is calculated by Eq. (10-2).

最小法向侧隙$j_{bn\,min}$可以根据传动时允许的工作温度（标准温度20°）、润滑方法及齿轮的圆周速度等工作条件确定，计算公式见式（10-2）：

$$j_{bn\,min}=\frac{2}{3}(0.06+0.0005a+0.03m_n) \tag{10-2}$$

Where, α—the minimum center distance;

M_n—normal modulus.

式中，α——最小中心距；

M_n——法向模数。

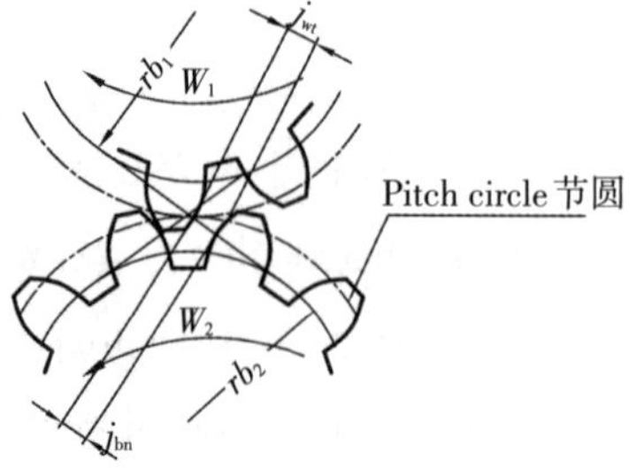

Fig. 10-10 Gear side gap

图10-10 齿轮侧隙

By the above calculation, the recommended data shown in Tab. 10-6.

按上式计算可以得出如表10-6所示的推荐数据。

Tab. 10-6 Minimum normal side gap of medium and large modular gears $j_{bn\,min}$ recommended data

表10-6 中、大模数齿轮最小法向侧隙$j_{bn\,min}$推荐数据

Module 模数(m_n)	Center distance 中心距(α)/mm					
	50	100	200	400	800	1600
1.5	0.09	0.11	—	—	—	—
2	0.1	0.12	0.15	—	—	—
3	0.12	0.14	0.17	0.24	—	—
5	—	0.18	0.21	0.28	—	—
8	—	0.24	0.27	0.34	0.47	—
12	—	—	0.35	0.42	0.55	—
18	—	—	—	0.54	0.67	0.94

(1) Upper deviation of tooth thickness E_{sns} 齿厚上偏差E_{sns}。

The tooth thickness shall ensure the minimum thinning, which is formed by the deviation on the rounded tooth thickness, as shown in Fig. 10-11. The reduction in normal backlash must also be considered. The calculation formula is as formula (10-3).

齿厚应保证有最小减薄量，它是由分度圆齿厚上偏差形成的，如图10-11所示。还要考虑齿轮法向侧隙减小量 j_{bn} 。计算公式如（10-3）所示：

$$j_{bn}=\sqrt{f_{pt1}^2+f_{pt1}^2+2(F_\beta\cos\alpha_n)^2+(f_{\Sigma\delta}\sin\alpha_n)^2+(f_{\Sigma\beta}\cos\alpha_n)^2}$$

$$J_{bn}=\sqrt{0.88(f_{pt1}^2+f_{pt2}^2)+[1.77+0.34(L/b)^2]F_\beta^2} \tag{10-3}$$

The calculation formuld for upper deviation of tooth thickness E_{sns} is as formula (10-4).

齿上偏差E_{sns}计算公式为（10-4）：

$$E_{sns}=-\left(\frac{j_{bn\,min}+J_{bn}}{2\cos\alpha_n}+|f_\alpha|\tan\alpha_n\right) \tag{10-4}$$

Where, f_{pt1}, f_{pt2} —individual tooth pitch deviation of the large and small gears;

L —bearing span of the gear pair;

b —tooth width;

α_n —normal pressure angle of the gear.

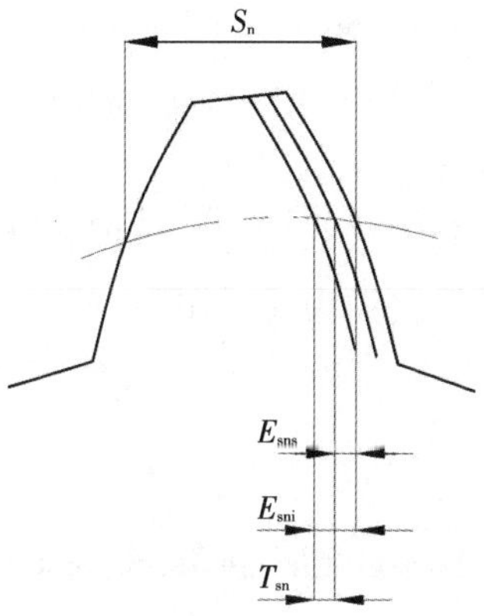

Fig. 10-11 Deviation tooth thickness

图10-11 分度圆齿厚上偏差

式中，f_{pt1}，f_{pt2}——大，小齿轮的单个齿距偏差；

L——齿轮副轴承跨距；

b——齿宽；

α_n——齿轮的法向压力角。

(2) The inferior tooth thickness deviation齿厚下偏差。

The lower deviation of tooth thickness (E_{sni}) is obtained from the upper deviation of tooth thickness

(E_{sns}) and the normal tooth thickness tolerance (T_{sn}).

齿厚下偏差E_{sni}由齿厚上偏差E_{sns}和法向齿厚公差T_{sn}求得。即

$$E_{sni} = E_{sns} - T_{sn} \tag{10-5}$$

The normal tooth thickness tolerance (T_{sn}) is largely unrelated to gear accuracy. Nonetheless, when specific requirements for maximum backlash limits are imposed, calculations become necessary. The choice of tooth thickness tolerance must be judicious; An overly restrictive tolerance can significantly elevate gear manufacturing costs. Conversely, an excessively large tolerance will lead to an increase in backlash, resulting in excessive lost motion during both forward and reverse gear rotation. The tooth thickness tolerance (T_{sn}) can be determined using the formula (10-6).

法向齿厚公差T_{sn}大体上与齿轮精度无关，如对最大侧隙有要求时，就必须进行计算。齿厚公差的选择要适当，公差过小，势必增加齿轮制造成本；公差过大，会使侧隙加大，使齿轮正、反转时空行程过大。齿厚公差T_{sn}可按式（10-6）求得。

$$T_{sn} = \sqrt{F_r^2 + b_r^2} \cdot 2\tan\alpha_n \tag{10-6}$$

Where, b_r—the radial cutter tolerance can be selected according to Tab.10-7.

式中，b_r——齿轮径向进刀公差，可按表10-7选取。

Tab. 10-7 Feed tolerance during cutting tooth b_r

表10-7 切齿径向进刀公差b_r值

The accuracy grade of gears 齿轮精度等级	4	5	6	7	8	9
Value b_r b_r值	1.26IT7	IT8	1.26IT8	IT9	1.26IT9	IT10

10.3.2.2 Common normal length deviation公法线长度偏差

The upper and lower deviations of the base pitch length (E_{ws}, E_{wi}) are calculated from the upper and lower deviations of the tooth thickness (E_{sns}, E_{sni}), respectively. For external gears, the conversion relationship is as follows.

公法线长度上、下偏差（E_{bns}，E_{bni}）分别由齿厚上、下偏差（E_{sns}，E_{sni}）换算得到。对外齿轮它们的换算关系为式（10-7）和（10-8）。

$$E_{ws} = E_{sns}\cos\alpha_n - 0.72F_r\sin\alpha_n \tag{10-7}$$

$$E_{wi} = E_{sni}\cos\alpha_n + 0.72F_r\sin\alpha_n \tag{10-8}$$

For internal gears, the conversion relationship is

对内齿轮，他们的换算关系式为式（10-9）和（10-10）：

$$E_{ws} = -E_{sns}\cos\alpha_n - 0.72F_r\sin\alpha_n \tag{10-9}$$

$$E_{wi} = -E_{sni}\cos\alpha_n + 0.72F_r\sin\alpha_n \tag{10-10}$$

The calculation of the nominal common normal length of a spur gear W_k and the number of teeth spanned k during measurement:

直齿轮的公称公法线长度W_k和测量时跨齿数k的计算:

$$W_k = m\cos\alpha[\pi(k-0.5) + z\cdot \mathrm{inv}\alpha] + 2xm\sin\alpha \tag{10-11}$$

Where, m, z, α, x—Module, Number of teeth, Standard pressure angle, Addendum modification coefficient;

invα —Involute Function, inv20° = 0.014904;

k—Number of teeth in span, for standard gears: $k = z\alpha/180^{\circ} + 0.5$.

式中，m，z，α，x——齿轮的模数、齿数、标准压力角、变位系数；

invα ——渐开线函数，inv20° = 0.014904；

k——跨齿数，对于标准齿轮：$k = z\alpha/180^{\circ} + 0.5$。

(3) Common normal length deviation ΔE_w (E_{ws}, E_{wi}) 公法线长度偏差ΔE_w (E_{ws}, E_{wi})

The common normal length refers to the arc of the base circle enclosed between the opposite-facing tooth profiles at the ends of several teeth on a gear, that is, the length of the tangent line segment of the base circle between these opposite-facing tooth profiles.

公法线长度是指齿轮上几个轮齿的两端异向齿廓间所包含的一段基圆圆弧，即该两端异向齿廓间基圆切线线段的长度。

For medium and small modular gears, the common normal length deviation is usually replaced to facilitate convenient measurement.

对于中、小模数齿轮，为测量方便，通常用公法线长度偏差代替齿厚偏差。

Common normal length deviation (ΔE_w) refers to the difference between the actual common normal length (W_k) and its nominal value (W).

公法线长度偏差（ΔE_w）是指实际公法线长度与公称公法线长度之差。$\Delta E_w = W_k - W$。

When assessing the acceptability of ΔE_{sn} and ΔE_w, standards are as follows: ΔE_{sn} must be within the limits defined by the upper and lower deviations of the tooth thickness ($E_{sni} \leqslant \Delta E_{sn} \leqslant E_{sns}$); similarly, ΔE_w must also fall within the specified range defined by the upper and lower deviations of the common normal length ($E_{wi} \leqslant \Delta E_w \leqslant E_{ws}$).

评定ΔE_{sn}和ΔE_w时，它们的合格条件是：ΔE_{sn}在齿厚上、下偏差的范围内（$E_{sni} \leqslant \Delta E_{sn} \leqslant E_{sns}$）；$\Delta E_{bn}$在公法线长度上、下偏差的范围内（$E_{wi} \leqslant \Delta E_{bn} \leqslant E_{ws}$）。

10.3.3 Determination of the Gear Pair and Gear Accuracy 齿轮副和齿轮精度的确定

A pair of gears that engage with each other to facilitate motion transfer forms a gear pair, the main evaluation indicators of the gear pair are center distance deviation, axis parallelism deviation, contact spots and so on.

一对齿轮相啮合进行传动就构成了齿轮副。齿轮副的主要评定指标有中心距极限偏差、轴线平行度偏差、接触斑点等。

10.3.3.1 Limit deviation of gear pair center distance $\left[\Delta f_{\alpha}(\pm f_{\alpha})\right]$ 齿轮副中心距极限偏差 $\left[\Delta f_{\alpha}(\pm f_{\alpha})\right]$

Center distance deviation (Δf_{α}) refers to the difference between the actual center distance and the nominal center distance. In the center distance deviation of the gear pair, the side gap of the gear pair will be affected. Select the center distance limit deviation $\pm f_{\alpha}$. See Tab. 10–8.

中心距偏差Δf_{α}是指实际中心距与公称中心距的差值。齿轮副存在中心距偏差时，会影响齿轮副的侧隙。选择中心距极限偏差 $\pm f_{\alpha}$ 时可参考表10–8。

Tab. 10-8 Center distance from the limit deviation $\pm f_\alpha$ (GB/T 10095.1-2022)

表10-8 中心距极限偏差$\pm f_\alpha$摘自 GB/T10095.1—2022）

Center distance of the gear pair 中心距 a/mm	Gear precision grade 齿轮精度等级	
	5,6	7,8
≥6 ~ 10	7.5	11
> 10 ~ 18	9	13.5
> 18 ~ 30	10.5	16.5
> 30 ~ 50	12.5	19.5
> 50 ~ 80	15	23
> 80 ~ 120	17.5	27
> 120 ~ 180	20	31.5
> 180 ~ 250	23	36
> 250 ~ 315	26	40.5
> 315 ~ 400	28.5	44.5
> 400 ~ 500	31.5	48.5

10.3.3.2 Axis parallelism deviations $\Delta f_{\Sigma\delta}$ and $\Delta f_{\Sigma\beta}$ ($f_{\Sigma\delta}$ and $f_{\Sigma\beta}$) 轴线平行度偏差 $\Delta f_{\Sigma\delta}$ 和 $\Delta f_{\Sigma\beta}$ ($f_{\Sigma\delta}$ 和 $f_{\Sigma\beta}$)

Since the influence of axis parallelism deviation is related to the direction of its vector, standard specifies two types of axis parallelism deviations: parallelism deviation ($\Delta f_{\Sigma\delta}$) in the axial plane and parallelism deviation ($\Delta f_{\Sigma\beta}$) in the perpendicular plane, see Fig. 10-12.

由于轴线平行度偏差的影响与其向量的方向有关，因此，标准规定了轴线两种平行度偏差：轴线平面上的平行度偏差 $\Delta f_{\Sigma\delta}$ 和垂直平面上的平行度偏差 $\Delta f_{\Sigma\beta}$，如图10-12 所示。

The axial axis plane is a plane that contains the datum axis and is defined by the intersection of the measured axis with the intermediate plane of a bearing.

轴线平面是指包含基准轴线并通过被测轴线与其一个轴承中间平面的交点所确定的平面。

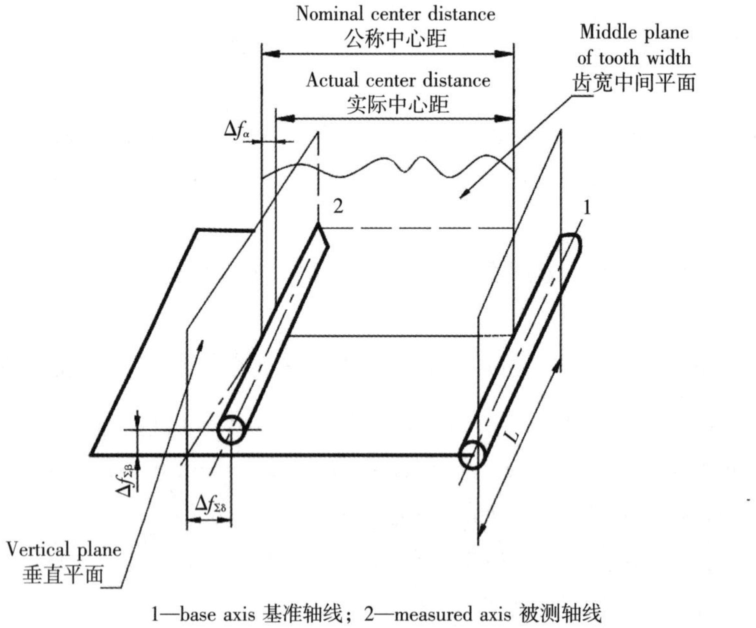

1—base axis 基准轴线；2—measured axis 被测轴线

Fig. 10-12 Axis parallelism deviation

图10-12 轴线平行度偏差

Perpendicular plane is a plane determined by the above intersection as perpendicular to the axis plane and parallel to the reference axis.

垂直平面是指通过上述交点确定的垂直于轴线平面，并且平行于基准轴线的平面。

The datum axis generally selects the axis with the larger span L in the two pairs of bearings. If the span of two pairs of bearings is the same, any of the axes may be used as the datum axis.

基准轴线一般选两对轴承中跨距 L 较大的那条轴线。如果两对轴承的跨距相同，则可取其中任何一条轴线作基准轴线。

The deviation of the axis parallelism affects the uniformity of the gear load distribution.

轴线平行度偏差影响轮齿载荷分布的均匀性。

The maximum recommended value of $\Delta f_{\Sigma\delta}$ and $\Delta f_{\Sigma\beta}$ is the $f_{\Sigma\delta}$ and $f_{\Sigma\beta}$:

$\Delta f_{\Sigma\delta}$ 和 $\Delta f_{\Sigma\beta}$ 的最大推荐值为 $f_{\Sigma\delta}$ 和 $f_{\Sigma\beta}$：

$$f_{\Sigma\beta}=0.5\left(\frac{L}{b}\right)F_{\beta} \tag{10-12}$$

$$f_{\Sigma\delta}=\left(\frac{L}{b}\right)F_{\beta} \tag{10-13}$$

Where， L —bearing span;

b —gear width;

F_{β} —total deviation of gear helix.

式中，L——轴承跨距；

b——齿宽；

F_{β}——齿轮螺旋线的总偏差。

The acceptance criteria for the parallelism deviation of the gear pair axes are as follows:

齿轮副轴线平行度偏差的合格条件为：

$$\Delta f_{\Sigma\delta}\leqslant f_{\Sigma\delta}\Delta f_{\Sigma\beta}\leqslant f_{\Sigma\beta}$$

10.3.3.3 The Contact spots of gear pair 齿轮副的接触斑点

The contact spots of a gear pair refer to the dispersed contact marks and polished impressions that emerge on the tooth surfaces after the installed gear pair operates under light braking conditions.

齿轮副的接触斑点是指安装好的齿轮副在轻微制动下运转后，齿面上分布的接触擦亮痕迹。

The distribution of contact spots on the expanded map of the wheel is shown in Fig. 10- 13. The magnitude of the contact trace is indicated by the percentage in the tooth height and tooth length directions. In Fig. 10- 13, b_{c1} and b_{c2} are the larger and smaller contact lengths respectively, h_{c1} and h_{c2} are the corresponding length in the tooth height direction. The larger contact and smaller contact heights in the tooth height direction, respectively. If b denotes the tooth width, the contact spot requirements of the gear teeth at different accuracies are listed in Tab. 10-9.

轮齿展开图上的接触斑点分布如图10-13所示。接触痕迹的大小由齿高方向和齿长方向的百分数表示，在图10-13中，b_{c1}、b_{c2}是齿长方向上较大的接触长度和较小的接触长度；h_{c1}、h_{c2}分别表示齿高方向上的较大接触高度和较小接触高度。用b表示齿宽，则齿轮在不同精度时轮齿的接触斑点要求见表10-9。

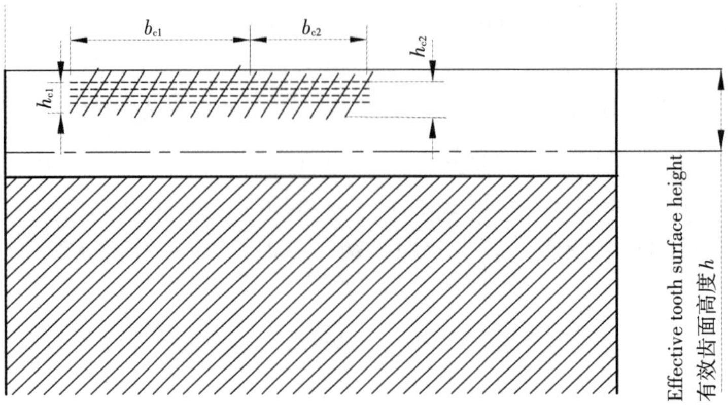

Fig. 10-13 Distribution of contact spots

图10-13 接触斑点分布

Tab. 10-9 Contact spots of the gear after gear assembly

表10-9 齿轮装配后的接触斑点

Precision grade 精度等级	Spur gear 直齿轮				Helical gear 斜齿轮			
	b_{c1}/b	h_{c1}/h	b_{c2}/b	h_{c2}/h	b_{c1}/b	h_{c1}/h	b_{c2}/b	h_{c2}/h
4 and finer ≤4级	50	70	40	50	50	50	40	30
Grade5 and 6 5级和6级	45	50	35	30	45	40	35	20
Grade7 and 8 7级和8级	35	50	35	30	35	40	35	20
Grade 9 ~ 12 9 ~ 12级	25	50	25	30	25	40	25	20

The contact pattern exhibited by a gear pair mounted in a housing serves as an indicator for evaluating the load distribution among its teeth. By measuring the contact pattern between a standard gear and a manufatured gear, one can evaluate the accuracy of the product gear's helix and tooth profile after assembly. The inspection of contact pattern is a relatively simple method, making it suitable for large gears or on-site in scenarios without specialized inspection equipment.

通过安装在箱体的齿轮副接触斑点可评估轮齿间的载荷分布情况，测量齿轮与成品齿轮的接触斑点可评估成品齿轮在装配后的螺旋线和齿廓精度。接触斑点的检查比较简单，经常用在大齿轮或现场没有检查仪的场合。

10.3.4 Determination of Gear Blank Accuracy 齿轮坯精度的确定

A gear blank serves as the foundational workpiece for gear manufacturing before the tooth-cutting process. The precision of this blank exerts a profound influence on the machining, inspection, and assembly accuracy of the resulting gear. Consequently, under specified machining conditions, meticulously controlling the quality of the blank emerges as an efficacious strategy to guarantee and enhance the machining accuracy of the gear teeth.

齿轮坯是指在轮齿加工前供制造齿轮用的工件。齿坯的精度对齿轮的加工、检验和安装精度影响很大。因此，在一定的加工条件下，用控制齿坯质量来保证和提高轮齿的加工精度是一项有效的措施。

The accuracy of a gear blank refers to the error on the datum surfaces that affect the gear processing and transmission quality: damensional deviation, form error, datum surface run-out and surface roughness.

齿坯精度是指在齿坯上影响轮齿加工和齿轮传动质量的基准表面上的误差，包括尺寸偏差、形状误差、基准面的跳动以及表面粗糙度。

10.3.4.1 Blank accuracy of disk gear 带孔齿轮的齿坯公差

Fig. 10-14 shows the common structural form of disk gear, whose datum surfaces are the reference hole (ϕD) for shaft insallation, and the locating end face (S_i) used during tooth cutting, radial datum surface (S_t) and the tooth tip cylinder (ϕd_a).

带孔齿轮的常用结构形式，其基准表面为：齿轮安装在轴上的基准孔（ϕD）、切齿时定位端面（S_i）、径向基准面（S_t）和齿顶圆柱面（ϕd_a），如图10-14所示。

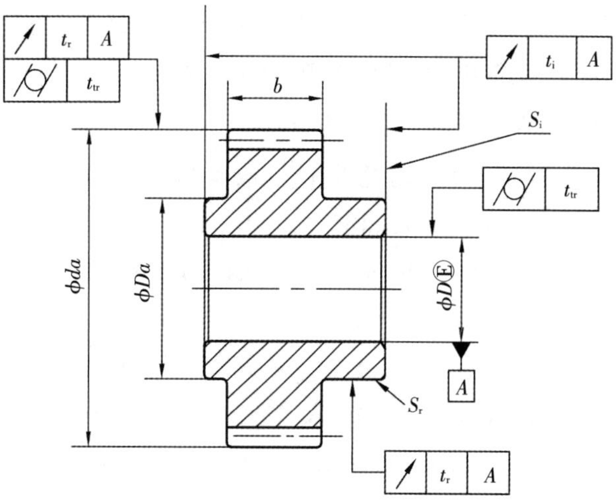

Fig. 10-14 Tolerance of tooth blank with hole gear

图10-14 带孔齿轮的齿坯公差

Tab. 10-10 Gear blank tolerances (from GB/Z 18620.3-2008)

表10-10 齿轮坯公差（摘自GB/Z 18620.3—2008） 单位：μm

	Gear accuracy grade 齿轮精度等级	5	6	7	8	9	10	11	12
Geometric tolerance 尺寸公差	Gear datum hole 齿轮基准孔	IT5	IT6	IT7		IT8		IT9	
	Gear shaft journal 齿轮轴轴颈	It is determined in accordance with the tolerance grade of rolling bearings 按照流动轴承的公差等级确定							
	Addendum circle diameter 齿轮圆直径	IT7	IT8			IT9		IT11	
Dimensition tolerance 几何公差	Characteristic item 特征项目	Cylindricity 圆柱度t_{rt}			Radial run-out 径向圆跳动t_r			Axial run-out 轴向圆跳动t_i	
	Gear datum hole 齿轮基准孔	Smaller of $0.04(L/b)\ F_\beta$ or $0.1F_\beta$ $0.04(L/b)\ F_\beta$或$0.1F_\beta$取两者中之小值			Datum cylinder 基准圆柱面 $0.3F_p$			Datum end surface 基准端面 $0.2(D_d/b)F_\beta$	
	Gear shaft journal 齿轮轴轴颈								
	Addendum circle 齿轮圆								

Notes: ① The three accuracy levels of the gear are different, and the diameter and dimension tolerance of the gear reference hole and the gear shaft neck are determined according to the highest accuracy level.

② When the roof cylinder is not benchmarked, the diameter tolerance of the roof circle is given according to IT11, but not more than $0.1m_n$; The tooth top circle also does not need to specify the cylindrical degree tolerance and the radial circle run-out tolerance.

③ L—bearing hole span; b—tooth width, F_β—total helix deviation, F_p—total deviation of pitch; D_d—base end face diameter.

注：① 齿轮的三项精度等级不同时，齿轮的基准孔、齿轮轴轴颈直径尺寸公差按最高精度等级确定。

② 齿顶圆柱面不作基准时，齿顶圆直径公差按IT11 给定，但不得大于$0.1m_n$；齿顶圆也不需规定圆柱度公差和径向圆跳动公差。

③ L—轴承孔跨距；b—齿宽；F_β—螺旋线总偏差；F_p—螺距累积总偏差；D_d—基准端面直径。

Datum end surface S_i axial circular run-out tolerance t_i to the axis of the datum hole, and the radial datum surface, S_t radial circular run-out tolerance t_t relative to the axis of the datum hole are determined according to the calculation in Tab. 10–10.

基准端面S_i对基准孔轴线的轴向圆跳动公差t_i和径向基准面S_t对基准孔轴线的径向圆跳动公差t_t按表10–10中计算式确定。

If the tip cylinder is used as a processing or measuring datum surface, in addition to the above specified dimensional tolerance, its cylindrical tolerance and the radial circular run-out tolerance relative to the datum hole axis should also be specified. The tolerance values are also determined according to the calculation formulas in Tab. 10–10, At this point, there is no need to specify the radial circular run-out tolerance t_r of the radial datum surface relative to the datum hole axis.

如果齿顶圆柱面作为加工或测量基准面时，除了上述规定的尺寸公差外，还需规定其圆柱度公差和对基准孔轴线的径向圆跳动公差，其公差值也是按表10–10中计算式确定。此时，不必给出径向基准面对基准孔轴线的径向圆跳动公差t_r。

10.3.4.2 Blank accuracy of gear shaft 齿轮轴的齿坯公差

Fig. 10–15 is the common structural form of the gear shaft. The datum surfaces are the two shaft journals（$2\times\phi d$), the axial datum end faces（$2\times S_i$）and the tooth top cylindrical face（ϕd_a for the rolling bearing)。

图10–15为齿轮轴的常用结构形式。基准表面为安装滚动轴承的两个轴颈（$2\times\phi d$)、轴向基准端面（$2\times S_i$）和齿顶圆柱面（ϕd_a)。

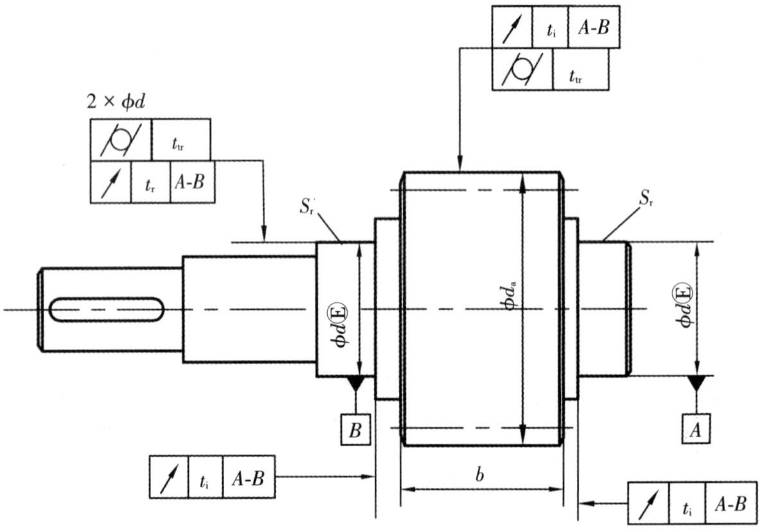

Fig. 10–15 Grey tolerances of the gear shaft

图10–15 齿轮轴的齿坯公差

The dimensional tolerances of the two axle necks（using inclusion requirements）and the diameter of the tooth top circle are selected according to the gear accuracy grade in Tab. 10–10. The cylindricity value（t_{tr}）of the two journal surfaces, the radial run-out tolerance（t_r）of each journal surface relative to their common axis (datum axis), and the axial run-out tolerance（t）of the datum end faces（$2\times S_i$）relative to the common axis of the two journal surfaces are determined according to the calculation formula in Tab. 10–10.

两个轴颈的尺寸公差（采用包容要求）和齿顶圆直径的尺寸公差按齿轮精度等级从表10–10中选取。两轴颈的圆柱度公差t_{tr}、两轴颈分别对它们的公共轴线（基准轴线）的径向圆跳动公差t_r和基准端面（$2\times S_i$）对两轴颈的公共轴线的轴向圆跳动公差t_i，也按表10–10中的计算式确定。

The dimensional tolerance and geometric tolerance of the two shaft necks can also be determined according to the tolerance grade of the rolling bearing.

两个轴颈的尺寸公差和几何公差也可按滚动轴承的公差等级确定。

Tab. 10-11 Recommended limits of surface roughness of tooth surface (From GB/Z 18620.4-2008)

表10-11 齿面表面粗糙度推荐极限值（摘自GB/Z 18620.4—2008）

Accuracy of gear grade 齿轮精度等级	*Ra*			*Rz*		
	Modulus 模数/mm					
	$m \leq 6$	$6 < m \leq 25$	$m > 25$	$m \leq 6$	$6 < m \leq 25$	$m > 25$
3		0.16		—	1.0	
4		0.32		—	2.0	
5	0.5	0.63	0.80	3.2	4.0	5.0
6	0.8	1.00	1.25	5.0	6.3	8.0
7	1.25	1.60	2.0	8.0	11.0	12.5
8	2.0	2.5	3.2	12.5	16	20
9	3.2	4.0	5.0	20	25	32
10	5.0	6.3	8.0	32	40	50

If the tooth top circle is used as the measurement datum, in addition to the above specified size tolerance, cylindricity tolerance and radial circular run-out tolerance relative to the datum axis must be also specified, and the tolerance values are determined using the calculation formula in Tab. 10-11.

如果齿顶圆作为测量基准，除了上述规定的尺寸公差外，还要规定其圆柱度公差和对基准轴线的径向圆跳动公差，其公差值分别按表10-11中计算式确定。

10.3.4.3 Surface roughness profile of gear tooth surface and datum surface 齿轮齿面和基准面的表面粗糙度轮廓

The amplitude parameters for surface roughness of the gear tooth surfaces and datum surfaces are selected from Tab. 10-11 and Tab. 10-12.

齿轮齿面和基准面的表面粗糙度轮廓幅度参数值从表10-11和表10-12中查取。

Tab. 10-12 Recommended upper-limit value of *Ra* for roughness of various datum surfaces of gear

表10-12 齿轮各基准面粗糙度推荐的*Ra*上限值

Each surface *Ra* 表面粗糙度	Gear accuracy grade 齿轮精度等级						
	5	6	7		8	9	
Dentate processing method 齿面加工方法	Gear grinding 磨 齿	Mill or hang teeth 磨或珩齿	Shaving or honing 剃或珩齿	Intensive milling 精插精铣	Plug or roll teeth 插齿或滚齿	Hobbing 滚齿	Gear milling 铣齿
Gear base hole 齿轮基准孔	0.32 ~ 0.63	1.25	1.25 ~ 2.5			5	

Tab. 10-12（Continued）表 10-12（续）

Each surface *Ra* 表面粗糙度	Gear accuracy grade 齿轮精度等级				
	5	6	7	8	9
Gear shaft reference journal 齿轮轴基准轴颈	0.32	0.63	1.25	2.5	
Gear base end surface 齿轮基准端面	2.5 ~ 1.25	2.5 ~ 5		3.2 ~ 5	
Gear top round 齿轮顶圆	1.25 ~ 2.5	3.2 ~ 5			

Example 10-1: Gear precision design examples

A pair of straight cylindrical gears on the drive shaft of the main shaft of a machine tool, $m = 2.75$, $\alpha = 20°$, number of pinion and large gear $z_1 = 26$, $z_2 = 56$, tooth width $b_1 = 28$, aperture $D = 30$ mm, the middle distance between the two bearings L is 90 mm, $n_1 = 1650$ r/min, the gear material is steel, the box material is cast iron, single small batch production, try to design the accuracy of the pinion, and draw the gear parts diagram.

例 10-1：齿轮精度设计实例

某机床主轴箱传动轴上的一对直齿圆柱齿轮，$m = 2.75$，$\alpha = 20°$，小齿轮和大齿 轮齿数 $z_1 = 26$，$z_2 = 56$，齿宽 $b_1 = 28$，孔径 $D = 30$ mm，两轴承中间距离 L 为 90 mm，$n_1 = 1650$ r/min，齿轮材料为钢，箱体材料为铸铁，单件小批量生产，试设计小齿轮的精度，并画出齿轮零件图。

Solution:（1）To determinethe gear accuracy grade.

Since the gear is the drive gear of the machine spindle box, it can be roughly concluded from Tab. 10-4 that the gear accuracy is between grade 3 and grade 8. For further analysis, the gear transmits both movement and power, so its accuracy level can be determined according to the linear speed.

$$v = \frac{\pi d n_1}{1000 \times 60} = \frac{3.14 \times 2.75 \times 26 \times 1650}{1000 \times 60}\ \text{m/s} \approx 6.2\ \text{m/s}$$

Referring to Tab. 10-5, the gear is grade 7 accuracy, Grade 7 according to GB/T 10095.1—2022.

（2）To determine the gear inspection items and allowance values.

① Total cumulative pitch deviation.

According to Tab. 10-1, total cumulative pitch deviation $F_p = 0.038$ mm.

② Single pitch deviation $\pm f_{pt}$.

According to Tab. 10-1, single pitch deviation $\pm f_{pt} = \pm 0.012$ mm.

③ Total profile deviation F_α.

According to Tab. 10-1, total profile deviation $F_\alpha = 0.016$ mm.

④ Total helix deviation F_β.

According to Tab. 10-1, total helix deviation $F_\beta = 0.017$ mm.

（3）To determine tolerance minimum normal backlash $j_{bn\,min}$ and limit deviation of tooth thickness.

Center distance: $a = \frac{m}{2}(z_1 + z_2) = \frac{2.75}{2} \times (26 + 56)m = 112.75$ mm

According to Center distance, minimum normal backlash is seen Eq. (10−2), then

$$j_{bn\min}=\frac{2}{3}\times\left(0.06+0.0005|a|+0.03m\right)$$
$$=\frac{2}{3}\times(0.06+0.0005\times112.75+0.03\times2.75)$$
$$=0.133\text{ mm}$$

According to Eq. (10−3), according to Tab. 10−1, $F_r=0.030$ mm, $F_p=0.038$ mm, $F_\alpha=0.016$ mm, $F_\beta=0.017$ mm, $f_{pt1}=12$ μm, $f_{pt2}=13$ μm, $F_\beta=17$ μm, $L=90$ mm, $b=28$ mm.

$$J_{bn}=\sqrt{0.88\left(f_{pt1}^2+f_{pt2}^2\right)+\left[1.77+0.34(L/b)^2\right]F_\beta^2}$$
$$=\sqrt{0.88\times(12^2+13^2)+\left[1.77+0.34\times(90/28)^2\right]\times17^2}\ \mu\text{m}=42.5\ \mu\text{m}$$

According to Eq. (10−7), from Tab. 10−8, $f_a=27$ μm. Then the upper deviation of tooth thickness :

$$E_{sns}=-\left(\frac{j_{bn\min}+J_{bn}}{2\cos\alpha_n}+|f_a|\tan\alpha_n\right)=-\left(\frac{0.133+0.0425}{2\cos 20}+27\times\tan 20°\right)\text{mm}=-0.107\text{ mm}$$

According to Eq. (10−8) from Tab. 10−7 and Tab. 2−3, the feed tolerance during cutting toothbr = IT9 = 74 μm, then the deviation of tooth thickness

$$T_{sn}=\sqrt{b_r^2+F_r^2}\times2\tan\alpha_n=\sqrt{30^2+74^2}\times2\times\tan20°\ \mu\text{m}=58\ \mu\text{m}$$

According to Eq. (10−5), the lower deviation of tooth thickness

$$E_{sni}=E_{sns}-T_{sn}=(-0.107-0.058)\text{ mm}=-0.165\text{ mm}$$

In the deviation design of gears with medium modules, the limit deviation of the common normal length is usually used to replace the tooth thickness deviation for inspection purposes.

Then according to Eq. (10−7) and (10−8), from Tab. 10−1,

$F_r=30$ μm, then the limit deviations of common normal length deviation

$$E_{bns}=E_{sns}\cos\alpha_n-0.72F_r\sin\alpha_n$$
$$=(-0.107\times\cos20°)-0.72\times0.030\times\sin20°\text{ mm}=-0.108\text{ mm}$$
$$E_{bni}=E_{sni}\cos\alpha_n+0.72F_r\sin\alpha_n$$
$$=(-0.165\times\cos20°)+0.72\times0.030\times\sin20°\text{ mm}$$
$$=0.148\text{ mm}=-0.148\text{ mm}$$

According to Eq. $k=\frac{z\alpha}{180°}+0.5$

Given: $\alpha=20°$, $z_1=26$, the number of striding across teeth

$k=\frac{26\times20°}{180°}$ $k=z/9+0.5=26/9+0.5=3.38$ (rounding the number $k=3$)

The common normal length:

$$W_k=m\left[2.952\times(k-0.5)+0.014z\right]$$
$$=2.75\times\left[2.952\times(3-0.5)+0.014\times26\right]\text{mm}$$
$$=21.297\text{ mm}$$

Then the length and deviation of the public norm are $W_k=21.297_{-0.148}^{-0.108}$ mm .

(4) To determine gear blank accuracy.

① Size tolerance and geometric tolerance of datum hole.

From Tab. 10−10, the size tolerance of datum hole is IT7, and as it is hole-basis, and adopt the envelope requirement, then the size tolerance zone is ϕ30H7, and also from Tab. 10−10, the cylindricity tolerance t is smaller one of $0.04(L/b)F_\beta$ or $0.1F_p$

$$0.04(L/b)F_{\beta}=0.04\times(90/28)\times0.017\ \text{mm}=0.002\ \text{mm},$$
$$0.1F_{p}=0.1\times0.038\ \text{mm}=0.0038\ \text{mm},$$

Thus, geometric tolerance of datum hole choose $t_{ky}=0.002$ mm。

② Size tolerance and geometric tolerance of addendum cylinder.

• Size tolerance of addendum cylinder.

From Tab. 10- 10, the size tolerance of datum hole is IT8, then the size tolerance zone is $\phi 77h8=\phi 77^{0}_{-0.046}$ mm .

• Form tolerance of addendum cylinder.

Cylindricity tolerance of addendum cylinder is smaller one of the same values as the datum hole, so $0.04(L/b)F_{\beta}$ or $0.1F_{p}$, then $t_{ky}=0.002$ mm。

• the radial run-out tolerance of the tooth tip circle relative to the datum hole axis.

From Tab. 10- 13, the radial run-out tolerance of the tooth tip circle relative to the datum hole axis:

$$t_{r}=0.3F_{p}=0.3\times0.038\ \text{mm}=0.011\ \text{mm}$$

③ Run-out tolerance of side datum plane.

• Axial run-out tolerance of the datum end facerelative to the datum hole axis.

From Tab. 10-10, We can get the axial run-out tolerance of the datum end face relative to the datum hole axis.

$t_{i}=0.2(D_{d}/b)F_{\beta}=0.2\times(65/28)\times0.017\ \text{mm}+0.008\ \text{mm}$

• Radial run-out tolerance of datum cylinder.

Since the addendum cylinder surface is utilized as the datum for measurement and machining, there is no need to designate an additional radialdatum surface.

④ Surface roughness of teeth surface and gear blank surface.

The accuracy grade of gear is grade 7, according to Tab. 10-11 and Tab. 10-12.

The roughness of teeth surface is *Ra* 1.25 μm, the surface roughness of datum hole is *Ra* 1.25 μm, the surface roughness of side datum plane is *Ra* 2.5 μm, and the surface roughness of addendum cylinder is *Ra* 3.2 μm.

(5) To determine the accuracy of gear pair.

① Limit deviations of gear pair center distance $\pm f_{\alpha}$. According to Tab. 10- 8, $\pm f_{\alpha}=\pm 27$ μm then $a=112.75\pm0.027$ mm should be indicated on assembly drawing.

② Shaft parallelism tolerance.

• According to Eq. (10-12), the parallelism tolerance of axis on the shaft plane $f_{\Sigma\beta}$, then

$$f_{\Sigma\beta}=0.5(L/b)F_{\beta}=0.028\ \text{mm},\ f_{\Sigma\delta}=2f_{\Sigma\beta}=0.056\ \text{mm}$$

• The parallelism tolerance of axes on the perpendicular plane, according to Eq. (10-13)

$$f_{\Sigma\delta}=2f_{\Sigma\beta}=0.056\ \text{mm}$$

(6) Accuracy requirement of key slot.

The precision requirement of key slot is according to Chapter 9.

(7) Contact spots of gear.

According to Tab. 10-9, the requirements of contact spots of gear are as follows:

Tooth length direction is $b_{e1}/b\geqslant35\%$ and $b_{e2}/b\geqslant35\%$; the extent tooth height dirction $h_{e1}/h\geqslant50\%$ and

$h_{02}/h \geqslant 30\%$.

The limiting deviation of center distance $\pm f_{\alpha}$ and the axial parallelism tolerance $f_{\Sigma\beta}$ and $f_{\Sigma\delta}$, which shall be indicated on the housing drawing.

(8) To draw engineering drawing.

The detail drawing of pinion is shown in Fig. 10-16.

解：(1) 确定齿轮精度等级。

因该齿轮为机床主轴箱传动齿轮，由表10-4可以大致得出，齿轮精度为3~8级，进一步分析，该齿轮既传递运动又传递动力，因此可根据线速度确定其精度等级。

$$v = \frac{\pi\, dn_1}{1000 \times 60} = \frac{3.14 \times 2.75 \times 26 \times 1650}{1000 \times 60}\ \text{m/s} \approx 6.2\ \text{m/s}$$

参考表10-5，该齿轮为7级精度，则齿轮精度表示为7，根据GB/T 10095. 1—2022。

(2) 确定齿轮检验项目及允许值。

① 齿距累积总偏差 F_p 。

由表10-1查得，齿距累积总偏差为 $F_p = 0.038$ mm。

② 单个齿距偏差 $\pm f_{pt}$ 。

由表10-1查得，单个齿距偏差为 $\pm f_{pt} = \pm 0.012$ mm。

③ 齿廓总偏差 F_{α} 。

由表10-1查得，齿廓总偏差为 $F_{\alpha} = 0.016$ mm。

④ 螺旋线总偏差 F_{β} 。

由表10-1查得，螺旋线总偏差为 $F_{\beta} = 0.017$ mm。

(3) 确定最小法向侧隙和齿厚极限偏差。

中心距：$a = \frac{m}{2}(z_1 + z_2) = \frac{2.75}{2} \times (26 + 56)\ \text{m} = 112.75\ \text{mm}$

最小法向侧隙 $j_{bn\min}$ 可由式（10-2）中心距计算确定，即

$$j_{bn\min} = \frac{2}{3} \times \left(0.06 + 0.0005|a| + 0.03m_n\right)$$
$$= \frac{2}{3} \times \left(0.06 + 0.0005 \times 112.75 + 0.03 \times 2.75\right)m = 0.133\ \text{mm}$$

根据式（10-3）和表10-1，查得 $F_r = 0.030$mm, $F_p = 0.038$mm, $F_{\alpha} = 0.016$mm, $F_{\beta} = 0.017$ mm。$f_{pt1} = 12\ \mu\text{m}$, $f_{pt2} = 13\ \mu\text{m}$, $F_{\beta} = 17\ \mu\text{m}$, $L = 90$ mm, $b = 28$ mm,

$$J_{bn} = \sqrt{0.88\left(f_{pt1}^2 + f_{pt2}^2\right) + \left[1.77 + 0.34(L/b)^2\right]F_{\beta}^2}$$
$$= \sqrt{0.88 \times \left(12^2 + 13^2\right) + \left[1.77 + 0.34 \times (90/28)^2\right] \times 17^2}\ \mu\text{m} = 42.5\ \mu\text{m}$$

按式（10-7），由表10-8查得，$f_a = 27\ \mu\text{m}$，则齿厚上偏差为：

$$E_{sns} = -\left(\frac{j_{bn\min} + J_{bn}}{2\cos\alpha_n} + |f_a|\tan\alpha_n\right) = -\left(\frac{0.133 + 0.0425}{2\cos 20^\circ} + 27 \times \tan 20^\circ\right)\text{mm} = -0.107\ \text{mm}$$

按式（10-8），由表10-1查得 $F_r = 30\ \mu\text{m}$，从表10-7和表2-3查得 $b_r = \text{IT9} = 74\ \mu\text{m}$, 因此齿厚公差为：

$$T_{sn} = \sqrt{b_r^2 + F_r^2} \times 2\tan\alpha_n = \sqrt{30^2 + 74^2} \times 2 \times \tan 20^\circ\ \mu\text{m} = 58\ \mu\text{m}$$

按式（10-5），可得齿厚下偏差为：

$$E_{\rm sni}=E_{\rm sns}-T_{\rm sn}=(-0.107-0.058)\ {\rm mm}=-0.165\ {\rm mm}$$

通常对于中等模数齿轮，用公法线长度极限偏差来代替齿厚偏差。

按式（10-7）和（10-8），由表10-1查得 $F_{\rm r}=30\ \mu{\rm m}$，可得公法线长度的上、下偏差为

$$E_{\rm bns}=E_{\rm sns}\cos\alpha_{\rm n}-0.72F_{\rm r}\sin\alpha_{\rm n}$$

$$=(-0.107\times\cos 20^{\circ})-0.72\times0.030\times\sin 20^{\circ}\ {\rm mm}=-0.108\ {\rm mm}$$

$$E_{\rm bni}=E_{\rm sni}\cos\alpha_{\rm n}+0.72F_{\rm r}\sin\alpha_{\rm n}$$

$$=(-0.165\times\cos 20^{\circ})+0.72\times0.030\times\sin 20^{\circ}\ {\rm mm}$$

$$=0.148\ {\rm mm}=-0.148\ {\rm mm}$$

根据公式：跨齿数为 $k=z\alpha/180^{\circ}+0.5$

根据 $\alpha=20^{\circ}$，$z_1=26$，所以

$$k=26/9+0.5=26/9+0.5=3.38\quad（圆整取 k=3）$$

公法线长度为

$$W_{\rm k}=m[2.952\times(k-0.5)+0.014z]$$

$$=2.75\times[2.952\times(3-0.5)+0.014\times26]\ {\rm mm}$$

$$=21.297\ {\rm mm}$$

则公法线长度及其极限偏差为 $W_{\rm k}=21.297^{-0.108}_{-0.148}$ mm。

（4）确定齿轮坯精度和齿轮各表面粗糙度。

① 基准孔的尺寸公差和形状公差。

查表10-10，取基准孔的尺寸公差等级为IT7，并采用包容要求，即

$$\phi30{\rm H7}\ Ⓔ\ \phi30^{+0.021}_{0}\ {\rm mm}\ Ⓔ$$

基准孔的圆柱度公差取 $0.04(L/b)F_{\beta}$ 或 $0.1F_{\rm p}$ 中较小者：

$$0.04(L/b)F_{\beta}=0.04\times(90/28)\times0.017\ {\rm mm}=0.002\ {\rm mm},$$

$$0.1F_{\rm p}=0.1\times0.038\ {\rm mm}=0.0038\ {\rm mm},$$

所以，基准孔的圆柱度公差取 $t_{\rm ky}=0.002$ mm。

② 齿顶圆的尺寸公差和形状公差。

- 齿顶圆的尺寸公差。

查表10-10，取基准孔的尺寸公差等级为IT8，并采用包容要求，孔的公差为 $\phi77{\rm h8}=\phi77^{\ 0}_{-0.046}$ mm

- 齿顶圆的圆柱度公差。

查表10-10，齿顶圆的圆柱度公差取 $0.04(L/b)F_{\beta}$ 或 $0.1F_{\rm p}$ 中较小者：

$$0.04(L/b)F_{\beta}=0.04\times(90/28)\times0.017\ {\rm mm}=0.002\ {\rm mm},$$

$$0.1F_{\rm p}=0.1\times0.038\ {\rm mm}=0.0038\ {\rm mm},$$

取 $t_{\rm ky}=0.002$ mm。

- 齿顶圆对基准孔轴线的径向圆跳动公差。

按表10-10计算得齿顶圆对基准孔轴线的径向圆跳动公差值

$$t_{\rm r}=0.3F_{\rm p}=0.3\times0.038\ {\rm mm}=0.011\ {\rm mm}$$

③ 基准端面的圆跳动公差。

- 基准端面对基准孔轴线的轴向圆跳动公差。

按表10-10确定基准端面对基准孔轴线的轴向圆跳动公差值，即

$$t_{\rm i}=0.2(D_{\rm d}/b)F_{\beta}=0.2\times(65/28)\times0.017\ {\rm mm}=0.008\ {\rm mm}$$

- 径向基准面的圆跳动公差。

由于齿顶圆柱面作为测量和加工基准，因此，不必另选径向基准面。

④ 齿坯表面粗糙度。

齿轮的精度等级为7级，查表10-11，取齿坯表面粗糙度的Ra值为1.25 μm。根据表10-12，基准孔的表面粗糙度为1.25 μm，齿侧基准平面的表面粗糙度为2.5 μm，齿顶圆的表面粗糙度为3.2 μm。

（5）确定齿轮副精度。

① 齿轮副中心距极限偏差$\pm f_\alpha$由表10-8查得：$\pm f_\alpha=\pm 27$ μm，则在图上标注：$a=112.75\pm 0.027$ mm。

② 轴线平行度公差。

• 在轴平面上，轴线平行度公差$f_{\Sigma\beta}$，按式（10-12）确定，即

$$f_{\Sigma\beta}=0.5(L/b)F_\beta=0.028\text{ mm}$$

• 在垂直平面上，轴线平行度公差$f_{\Sigma\delta}$，按式（10-13）确定，即

$$f_{\Sigma\delta}=2f_{\Sigma\beta}=0.056\text{ mm}$$

（6）键槽的精度要求。

键槽的精度要求依据第9章。

（7）轮齿接触斑点。

由表10-9查得轮齿接触斑点要求：在齿长方向上，$b_{e1}/b\geqslant 35\%$和$b_{e2}/b\geqslant 35\%$；在齿高方向上，$h_{e1}/h\geqslant 50\%$和$h_{e2}/h\geqslant 30\%$。

中心距极限偏差$\pm f_\alpha$和轴线平行度公差，$f_{\Sigma\beta}$和，$f_{\Sigma\delta}$在箱体图上注出。

（8）绘制工程图。

图10-16所示为该齿轮的工作图。

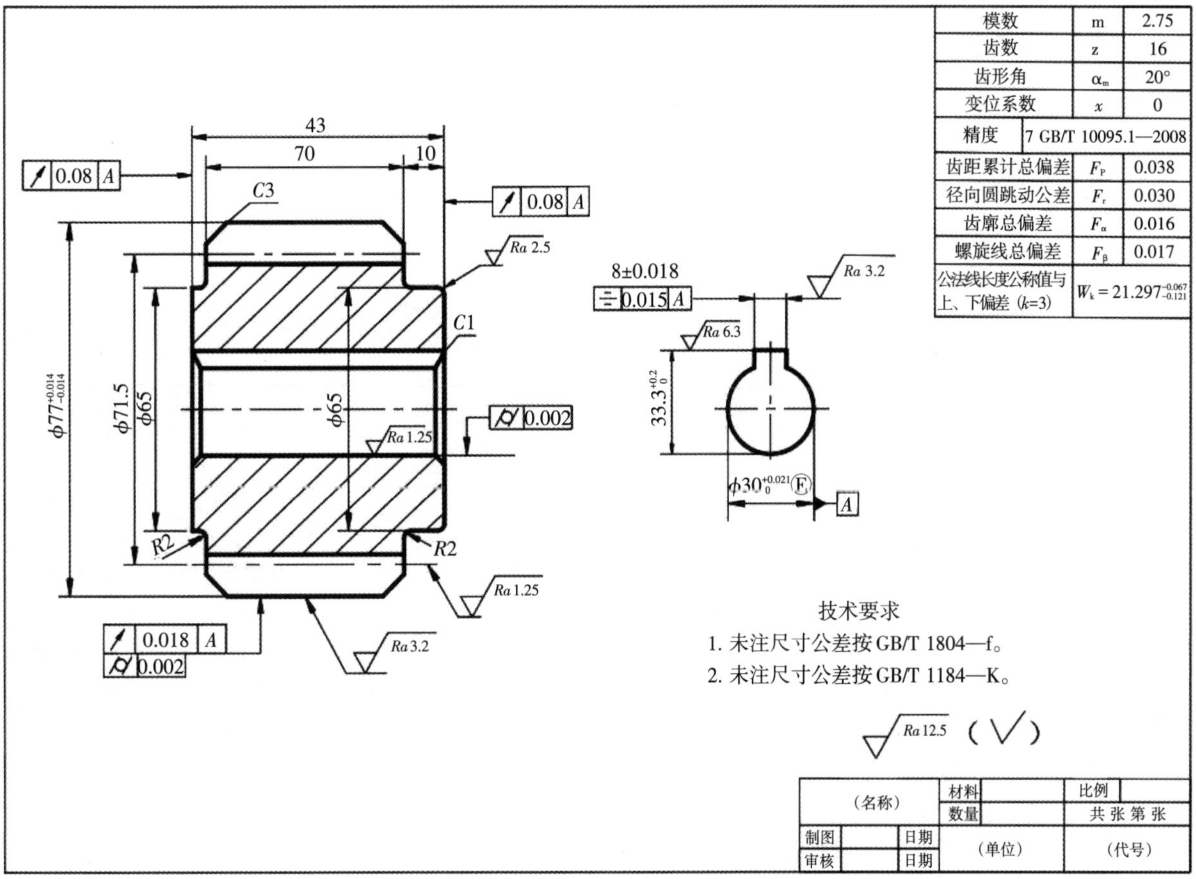

Fig. 10-16 The gear working diagram

图10-16 齿轮工作图

Exercise 10 习题10

10-1 What usage requirements for gear transmission?

齿轮传动有哪些使用要求？

10-2 What accuracy requirements for gear pair?

齿轮副需满足哪些精度要求？

10-3 To sketch the name, symbol and definition of accuracy requirements for cylindrical invoute gear.

请简述圆柱渐开线齿轮精度要求的名称、符号及定义。

10-4 How to control the gear backlash?

如何控制齿轮侧隙？

10-5 A pair of cylindrical spur gears used for a speed reducer, which module $m = 5$ mm, the teeth number is $z_1 = 20$ and $z_2 = 60$ respectively, the bearing span $L = 100$ mm, the face width $b = 50$ mm, the rotate speed of gearn $= 960$ r/min, to design the precision requirements of the gear and to draw the detail drawing of gear.

某减速器中一对圆柱直齿轮，模数$m = 5$，齿数$z_1 = 20$，$z_2 = 60$，轴承跨度$L = 100$，齿面宽$b = 50$，齿轮转速$n = 960$ r/min 。试设计齿轮的精度要求并绘制齿轮的零件详图。

参考文献

[1] 国家标准化管理委员会. 产品几何技术规范(GPS)线性尺寸公差ISO代号体系第1部分:公差、偏差和配合的基础:GB/T 1800.1—2020[S]. 北京:中国标准出版社,2020.

[2] 中国国家标准化管理委员会. 一般公差未注公差的线性和角度尺寸的公差:GB/T 1804—2008[S]. 北京:中国标准出版社,2008.

[3] 中国国家标准化管理委员会. 产品几何技术规范(GPS)几何公差形状、方向、位置和跳动公差标注:GB/T 1182—2018[S]. 北京:中国标准出版社,2018.

[4] 中国国家标准化管理委员会. 产品几何技术规范(GPS)几何公差检测与验证:GB/T 1958—2017[S]. 北京:中国标准出版社,2017.

[5] 中国国家标准化管理委员会. 产品几何技术规范(GPS)基础概念、原则和规则:GB/T 4249—2018[S]. 北京:中国标准出版社,2018.

[6] 中国国家标准化管理委员会.产品几何技术规范(GPS)几何公差最大实体要求(MMR)、最小实体要求(LMR)和可逆要求(RPR):GB/T 16671—2018[S]. 北京:中国标准出版社,2018.

[7] 中国国家标准化管理委员会.产品几何技术规范(GPS)几何公差轮廓度公差标注:GB/T 17852—2018[S]. 北京:中国标准出版社,2018.

[8] ALEX K . Fundamentals of geometric dimensioning and tolerancing[M].2nd.New York: Delmar Cengage Learning, 1997.

[9] The American Society of Mechanical Engineers, Mathematical Definition of Dimensioning and Tolerancing Principles: ASMEY 14.5.1—2019 [S]. New York: The American Society of Mechanical Engineers (ASME), 2019.

[10] 张彦富. 几何量公差与测量技术基础:英文版[M]. 北京:北京航空航天大学出版社,2015.

[11] 张也晗,刘永猛,刘品. 机械精度设计与检测基础[M]. 11版. 哈尔滨:哈尔滨工业大学出版社, 2021.

[12] 金嘉琦,张幼军. 几何精度设计与检测[M]. 2版. 北京:机械工业出版社, 2021.